Optoelectronics and Spintronics in Smart Thin Films

Smart thin films, composed of functional materials deposited in thin layers, have opened new avenues for the development of flexible, lightweight, and high-performance devices. *Optoelectronics and Spintronics in Smart Thin Films* presents a comprehensive overview of this emerging area and details the current and near-future integration of smart thin films in solar cells and memory storage.

- Offers an overview of optoelectronics and spintronics/magnetism.
- Discusses the synthesis of smart nanomaterials.
- Describes deposition techniques and characterization of smart thin films.
- Considers the integration and application of opto-spintronics for technological advancement of solar cells and memory storage devices.

Focused on advancing research on this evolving subject, this book is aimed at advanced students, researchers, and engineers in materials, chemical, mechanical, and electrical engineering, as well as applied physics.

Emerging Materials and Technologies

Series Editor: Boris I. Kharissov

The *Emerging Materials and Technologies* series is devoted to highlighting publications centered on emerging advanced materials and novel technologies. Attention is paid to those newly discovered or applied materials with potential to solve pressing societal problems and improve quality of life, corresponding to environmental protection, medicine, communications, energy, transportation, advanced manufacturing, and related areas.

The series takes into account that, under present strong demands for energy, material, and cost savings, as well as heavy contamination problems and worldwide pandemic conditions, the area of emerging materials and related scalable technologies is a highly interdisciplinary field, with the need for researchers, professionals, and academics across the spectrum of engineering and technological disciplines. The main objective of this book series is to attract more attention to these materials and technologies and invite conversation among the international R&D community.

Biosorbents
Diversity, Bioprocessing, and Applications
Edited by Pramod Kumar Mahish, Dakeshwar Kumar Verma, and Shailesh Kumar Jadhav

Principles and Applications of Nanotherapeutics
Imalka Munaweera and Piumika Yapa

Energy Materials
A Circular Economy Approach
Edited by Surinder Singh, Suresh Sundaramuthy, Alex Ibhadon, Faisal Khan, Sushil Kansal, and S.K. Mehta

Tribological Aspects of Additive Manufacturing
Edited by Rashi Tyagi, Ranvijay Kumar, and Nishant Ranjan

Emerging Materials and Technologies for Bone Repair and Regeneration
Edited by Ashok Kumar, Sneha Singh, and Prerna Singh

Mechanics of Auxetic Materials and Structures
Farzad Ebrahimi

Nanomaterials for Sustainable Hydrogen Production and Storage
Edited by Jude A. Okolie, Emmanuel I. Epelle, Alivia Mukherjee, and Alaa El Din Mahmoud

For more information about this series, please visit:
www.routledge.com/Emerging-Materials-and-Technologies/book-series/CRCEMT

Optoelectronics and Spintronics in Smart Thin Films

James Ayodele Oke and Tien-Chien Jen

CRC Press is an imprint of the
Taylor & Francis Group, an informa business

First edition published 2024
by CRC Press
2385 NW Executive Center Drive, Suite 320, Boca Raton FL 33431

and by CRC Press
4 Park Square, Milton Park, Abingdon, Oxon, OX14 4RN

CRC Press is an imprint of Taylor & Francis Group, LLC

© 2024 James Ayodele Oke and Tien-Chien Jen

Reasonable efforts have been made to publish reliable data and information, but the author and publisher cannot assume responsibility for the validity of all materials or the consequences of their use. The authors and publishers have attempted to trace the copyright holders of all material reproduced in this publication and apologize to copyright holders if permission to publish in this form has not been obtained. If any copyright material has not been acknowledged please write and let us know so we may rectify in any future reprint.

Except as permitted under U.S. Copyright Law, no part of this book may be reprinted, reproduced, transmitted, or utilized in any form by any electronic, mechanical, or other means, now known or hereafter invented, including photocopying, microfilming, and recording, or in any information storage or retrieval system, without written permission from the publishers.

For permission to photocopy or use material electronically from this work, access www.copyright.com or contact the Copyright Clearance Center, Inc. (CCC), 222 Rosewood Drive, Danvers, MA 01923, 978-750-8400. For works that are not available on CCC please contact mpkbookspermissions@tandf.co.uk

Trademark notice: Product or corporate names may be trademarks or registered trademarks and are used only for identification and explanation without intent to infringe.

ISBN: 978-1-032-36220-5 (hbk)
ISBN: 978-1-032-36437-7 (pbk)
ISBN: 978-1-003-33194-0 (ebk)

DOI: 10.1201/9781003331940

Typeset in Times
by codeMantra

Contents

Preface

Welcome to the fascinating world of optoelectronics, spintronics, smart thin films, graphene, carbon nanotubes, chalcogenides, fabrication techniques, characterization techniques, and their applications in solar cells and memory storage. In this preface, we provide a brief overview of these cutting-edge fields that have revolutionized the world of electronics and beyond.

Optoelectronics is the study and application of devices that can both emit and detect light, enabling technologies such as LEDs, lasers diode, photodetectors and solar cells. Spintronics, on the other hand, focuses on utilizing the intrinsic spin of electrons in addition to their charge, leading to novel devices with enhanced functionalities and lower power consumption. Smart thin films, composed of functional materials deposited in thin layers, have opened new avenues for the development of flexible, lightweight, and high-performance devices. Graphene, carbon nanotubes, and chalcogenides, with their exceptional electronic properties, have emerged as promising materials for next-generation electronics. Fabrication techniques play a vital role in realizing these advanced technologies. Characterization techniques allow us to study and understand the behavior and performance of these advanced materials and devices. The integration of smart thin films in solar cells and memory storage has been a game-changer. Opto-spintronics as an emerging technology for solar cells and memory storage.

As you embark on this journey through this book of eight chapters, we hope you will delve into the fascinating world of optoelectronics, spintronics, smart thin films, graphene, carbon nanotubes, chalcogenides, fabrication techniques, characterization techniques, and their applications in solar cells and memory storage. Thanks to the authors of various published articles in which this book was developed. We also hope that this book will be useful for researchers who will be working in this field of research.

James Ayodele Oke
University of Johannesburg
South Africa

Tien-Chien Jen
University of Johannesburg
South Africa

Acknowledgments

The authors would like to appreciate the support and encouragement from family and friends during the development of this book.

Dr. Oke and Prof. Jen would like to acknowledge the financial support from Global Excellence Stature (GES) 4.0 Seed Fund from the University of Johannesburg. Prof. Jen would also like to acknowledge the financial support from National Research Foundation (NRF) of South Africa.

Authors

Dr. James Ayodele Oke received his Ph.D in Physics at the University of South Africa in 2021, which was sponsored by the National Research Foundation (NRF) of South Africa. He is currently a postdoctoral researcher at the Department of Mechanical Engineering Science at the University of Johannesburg, South Africa.

His research interests are in the fields of electronic, magnetic, and optical properties of carbon and chalcogenide materials. He has worked on carbon nanotubes and graphene, and published his work in several peer-reviewed journals and conferences. Dr. Oke has over 14 publications, and some of his research has received special recognition from the American Institute of Physics (AIP) (Scilight) and International Association of Advanced Materials (IAAM).

Prof. Tien-Chien Jen joined the University of Johannesburg in August 2015, and before, he was a faculty member at the University of Wisconsin, Milwaukee. Prof. Jen received his Ph.D. in Mechanical and Aerospace Engineering from UCLA, specializing in thermal aspects of grinding. He has received several competitive grants for his research, including those from the US National Science Foundation, the US Department of Energy, and the EPA. Prof. Jen has brought in $3.0 million of funding for his research and has received various awards for his research, including the NSF GOALI Award. Prof. Jen has established a Joint Research Centre with the Nanjing Tech University of China on "Sustainable Materials and Manufacturing." Prof. Jen is also the Director of the newly established Atomic Layer Deposition Research Centre of the University of Johannesburg. SA National Research Foundation has awarded Prof. Jen a NNEP grant (National Nano Equipment Program) worth USD 1 million to acquire two state-of-the-art atomic layer deposition (ALD) tools for ultra-thin-film coating. These two ALD tools will be the first in South Africa and possibly the first in the African continent. He is currently the research chair (SARChI) of Green Hydrogen in South Africa.

In 2011, Prof. Jen was elected as a Fellow of the American Society of Mechanical Engineers (ASME), which recognized his contributions to the field of thermal science and manufacturing. As stated in the announcement of Prof. Jen Fellow status in the 2011 International Mechanical Engineering and Congress Exposition, "Tien-Chien Jen has made extensive contributions to the field of mechanical engineering, specifically in the area of machining processes. Examples include, but are not limited to, environmentally benign machining, atomic layer deposition, cold gas dynamics spraying, fuel cells and hydrogen technology, batteries, and material processing." In addition, he was inducted to the Academy of Science of South Africa (ASSAf) in 2021. The Academy of Science of South Africa (ASSAf) aspires to be the apex organization for science and scholarship in South Africa, recognized and connected both nationally and internationally. Prof. Jen has written over 360 peer-reviewed articles, including 180 peer-reviewed journal papers, published in Elsevier, ACS, Taylor & Francis, Springer Nature, Wiley, etc. Some of the journals include *Nano-Micro*

Letters, Coordination Chemistry Reviews, Nanotechnology Review, Journal of Molecular Liquids, Journal of Materials Research and Technology, Scientific Reports, International Journal of Heat and Mass Transfer, ASME Journal of Heat Transfer, ASME Journal of Mechanical Design, and *ASME Journal of Manufacturing Science and Engineering*. He has written 16 book chapters and has 5 books published, with the latest book titled *Thin Film Coatings: Properties, Deposition, and Applications*, published by CRC Press/Taylor & Francis in June 2022.

1 Overview of Optoelectronics

1.1 INTRODUCTION TO OPTOELECTRONICS

Optoelectronics combines the principles of electronics and optics to create innovative and efficient devices. It is the study and application of electronic devices and systems that source, detect, and control light, usually considered a sub-field of photonics. It combines optics and electronics to produce devices such as light-emitting diodes (LEDs), laser diodes, photodetectors, and solar cells. These devices convert electrical energy into light or detect light and convert it into electrical signals. They are used in a wide range of applications such as telecommunications, computing, lighting, and renewable energy. The advancement in optoelectronics has led to new possibilities in data communication, medical imaging, and renewable energy solutions.

1.2 LIGHT

Light is a type of electromagnetic radiation that is visible to the human eye [1]. It has both wave-like and particle-like properties and is essential to our perception of the world. In physics, light is often described as a wave that travels through space at a constant speed of 299,792,458 meters per second (m s^{-1}) in a vacuum. This is known as the speed of light and is one of the most fundamental constants in the universe. The wavelength and frequency of light determine its color and energy [2]. Shorter wavelengths and higher frequencies correspond to blue and violet light, while longer wavelengths and lower frequencies correspond to red and infrared (IR) light [3]. In addition to its wave-like properties, light also exhibits particle-like behavior, known as the photoelectric effect [4]. When light strikes a metal surface, electrons are emitted from the surface, a phenomenon known as the photoelectric effect. This behavior led to the development of quantum mechanics, which describes the behavior of matter and energy at the smallest scale [5].

Light plays a crucial role in many areas of physics, including optics, astronomy, and quantum mechanics. In optics, the study of light and its behavior, light is used to form images, transmit information, and measure distances. In astronomy, light from stars and galaxies is used to study the structure and evolution of the universe. In quantum mechanics, light is used to describe the behavior of particles at the quantum scale [6].

The investigation of light throughout history has given rise to various theories and experiments. It began with Sir Isaac Newton, who proposed the corpuscular theory, building upon the ancient Greek concept of matter composed of atoms. Later, Thomas Young conducted interference experiments that effectively demonstrated the wave nature of light. James Clerk Maxwell then established a solid theoretical foundation for the wave nature of light. During the late 1800s and early 1900s, further

DOI: 10.1201/9781003331940-1

experiments were conducted, which revealed that predictions based on a continuous wave model resulted in the well-known "ultraviolet catastrophe." It was at this point that Max Planck and Albert Einstein originated and developed the concept of the photon as the fundamental quantum of propagating electromagnetic energy [7].

Throughout the first half of the 20th century, significant advancements were made in the field of quantization and quantum electrodynamics (QED). Richard Feynman referred to QED as the "best theory we have" for describing the interaction between matter and light. In the latter half of the 20th century, researchers such as Glauber and Yuen contributed to the development of coherent and squeezed optical states. The study of light and electromagnetic fields has a rich history, with the properties of matter being influenced by the quantum mechanical state of its constituent particles and the properties of light being determined by the states of photons. When light interacts with matter, it results in light emission and detection. Similarly, the interaction between matter and the vacuum field leads to spontaneous emission, which is essential for technologies such as LEDs and lasers [7].

Light is a complex and fascinating phenomenon that has been studied by physicists for centuries. Its wave-like and particle-like behavior has led to many important discoveries and advancements in physics and technology. Light continues to be an important area of study and has many practical applications, from creating images and transmitting information to studying the universe [4,6].

1.2.1 Interaction of Light with Matter

The interaction of light with matter is a fundamental concept in physics that describes how electromagnetic radiation, such as light, interacts with materials. This interaction can result in a variety of phenomena, including absorption, reflection, transmission, and scattering [8]. At the most basic level, light interacts with matter through the electric and magnetic fields that make up electromagnetic radiation. When light strikes an object, it can cause the electrons in the material to oscillate, which can result in the absorption of some or all the energy in the light. This absorption can lead to a variety of effects, depending on the nature of the material and the wavelength of the light. When light strikes a smooth surface, such as a mirror or a polished metal, it bounces off at the same angle as it hits the surface, which is known as reflection. This is because the electrons in the material absorb the energy of the light and then re-emit it in a different direction without changing the wavelength or frequency of the light. Another common effect is transmission, which occurs when light passes through a material without being absorbed or reflected. This is often seen in transparent materials such as glass, which allow visible light to pass through while absorbing only a small fraction of the energy. In addition to absorption, reflection, and transmission, light–matter interaction can also result in scattering. Scattering occurs when the light is absorbed and then re-emitted in different directions, often resulting in a change in the wavelength or frequency of the light. This can give rise to a variety of effects, including the blue color of the sky and the colors seen in a rainbow.

Light–matter interaction drives many systems, such as optoelectronic devices such as LEDs and solar cells, biological structures like photosystem II, and potential future quantum devices. The absorption or emission of light typically occurs on the sub-nanometer scale, and the involved processes take place on attosecond to picosecond timescales. Light–matter interaction can be studied at atomic space-time scales by using a scanning tunneling microscope and coupling light into or extracting light from the tunnel junction. Electromagnetic radiation couples with matter through the interaction with charge carriers, leading to excitations such as electronic transitions, collective oscillations, excitons, and spin flips. These excitations can be studied with high spatial and temporal resolution using approaches in which light interacts with the tunnel junction itself or with a quantum system in the junction [9].

The interaction of light with matter plays a critical role in many areas of science and technology, from the design of solar cells and optical fibers to the study of the behavior of atoms and molecules. Understanding these interactions is key to developing new materials and technologies and to advancing our understanding of the physical world.

1.2.1.1 Absorption

Absorption is a fundamental process that occurs when light interacts with matter [9]. It refers to the transfer of energy from a photon of light to an atom, molecule, or solid material, which results in the excitation or ionization of the matter. When light is absorbed by a material, the energy of the photon is converted into potential energy of the excited or ionized matter. This process occurs when the frequency of the photon matches the energy level spacing of the material [10]. The energy of a photon is directly proportional to its frequency, so higher-frequency photons (e.g., blue light) have more energy than lower-frequency photons (e.g., red light) [11]. Therefore, materials absorb different frequencies of light depending on their energy level spacing.

The absorption of light by matter is described by the Beer-Lambert law, which states that the amount of light absorbed by a material is proportional to its concentration and the distance traveled through the material [12]. Mathematically, this law can be expressed as follows [13]:

$$I = I_O e^{(-\alpha c l)} \tag{1.1}$$

where I is the intensity of light after passing through a material, I_O is the initial intensity, α is the absorption coefficient, c is the concentration of the material, and l is the distance traveled through the material.

The absorption coefficient [13], α, is a measure of how strongly a material absorbs light at a given frequency. It depends on the material's energy level spacing, which determines the frequencies of light that it can absorb, as well as its physical properties such as its density and composition.

The absorption of light by matter has many practical applications. For example, it is used in spectroscopy to study the properties of materials, including their chemical composition and molecular structure [14]. In medicine, absorption-based imaging techniques such as X-ray and MRI scans are used to diagnose and treat diseases [15]. In addition, the absorption of light by pollutants in the atmosphere is used to monitor air quality and detect environmental changes [16].

1.2.1.2 Reflection

Reflection is a fundamental process that occurs when light interacts with matter. It refers to the change in direction of a wavefront of light when it encounters a boundary between two materials with different optical properties. This process is a crucial mechanism in many natural and technological processes, ranging from the formation of images in mirrors to the reflection of light by surfaces in optical devices [17].

When light is incident on a surface, part of it is reflected back into the same medium, while the rest is transmitted through the surface or absorbed by it. The amount of light that is reflected depends on the angle of incidence and the optical properties of the materials involved. The angle of incidence is the angle between the incident light ray and the normal to the surface at the point of incidence. The normal is a line perpendicular to the surface at the point of incidence. The reflection of light by a smooth surface follows the law of reflection, which states that the angle of incidence is equal to the angle of reflection [17]. Mathematically, this law can be expressed as follows:

$$\theta_i = \theta_r \tag{1.2}$$

where θ_i is the angle of incidence and θ_r is the angle of reflection.

The law of reflection applies only to smooth surfaces that are much larger than the wavelength of the incident light. If the surface is rough or irregular, the reflection is diffuse, and the angle of reflection is not well defined [17]. In this case, the reflected light is scattered in all directions, leading to a loss of image clarity and resolution [18]. The reflection of light by matter has many practical applications. For example, it is used in the construction of mirrors, which are used in many optical devices, including telescopes, microscopes, and cameras [19]. It is also used in optical fibers, which are used in telecommunications to transmit information over long distances [20].

1.2.1.3 Transmission

Transmission is another important phenomenon that occurs when light interacts with matter. It refers to the ability of a material to allow light to pass through it without changing its direction or scattering it. The amount of light transmitted through a material depends on its properties, such as its thickness, composition, and wavelength [21]. For example, some materials, such as glass, are highly transparent to visible light, while others, such as metals, are highly opaque and do not allow light to pass through them.

The study of transmission is an important part of physics, as it helps us to understand how light interacts with different materials and how it can be used in various applications [22]. For example, in optics, the ability of a lens to transmit light is a crucial factor in determining its quality and performance [22]. The phenomenon of transmission is also used in a variety of applications in everyday life. For example, sunglasses are designed to reduce the amount of visible light transmitted through them, thereby reducing the amount of glare and improving visibility in bright conditions. Similarly, optical fibers are used to transmit information over long distances by guiding light through a transparent core [23]. The ability of a material to transmit light is related to its refractive index, which is a measure of how much the speed of light is reduced when it enters the material. Materials with a high refractive index tend to be more transparent, while materials with a low refractive index tend to be more opaque [24].

1.3 SEMICONDUCTORS

A semiconductor material possesses a continuous crystalline structure, and its energy band configuration plays a crucial role in its properties. When considering an infinite crystal, the energy band structure consists of a continuum of states, which is determined by the crystal's characteristic dimensions being significantly larger than the lattice parameter of the crystal structure. This condition holds when the crystal dimensions are typically greater than a few dozen nanometers. The energy structure of a semiconductor is characterized by a valence band, which corresponds to molecular bonding states, and a conduction band, representing molecular antibonding states. The energy region between the top of the valence band and the bottom of the conduction band is known as the bandgap or forbidden band. This bandgap defines an energy range where electronic states are absent and electrons cannot occupy those energy levels.

A valence band electron remains in a ground state and is localized to a specific atom within the crystal structure. In contrast, a conduction band electron is in an excited state and interacts weakly with the crystal structure. The primary difference between semiconductors and insulators is the size of the bandgap. Semiconductors have a bandgap typically equal to or less than 6 eV, while insulators have a bandgap greater than 6 eV. When the bandgap exceeds this value, the solar spectrum cannot cause inter-band transitions for valence band electrons [25]. Semiconductor materials can be categorized into two main groups: elemental semiconductors (such as silicon, germanium, and diamond) and compound semiconductors (such as SiC, GaAs, InP, InSb, GaN, CdTe, ZnSe, and ZnS) [26]. The electrical conduction properties of a semiconductor can be significantly altered by introducing impurities into its volume. An impurity is considered a donor when it readily releases a free electron into the conduction band.

The impurity's characteristic energy level falls within the bandgap, just below the conduction band. For instance, in compound semiconductors belonging to group IV of the periodic table, such as silicon, the primary donor impurities are those derived from group V of the periodic table, such as phosphorus and arsenic [27]. These impurities replace a silicon atom in the crystal structure, forming four covalent bonds with the surrounding silicon atoms and easily donating their excess electrons to the crystal structure [28]. Consequently, these electrons become free to move with a weak activation energy due to thermal agitation, and this process is called n-type doping [25]. On the other hand, when a group III element, such as boron, is incorporated into the crystal structure of silicon, it forms three covalent bonds, and then acquires an electron from its fourth nearest neighbor silicon atom to complete its outer-shell electronic structure, all subject to a weak thermal activation energy. Such impurities are acceptors, and doping with them is called p-type doping [27]. As a result, a hole carrying a positive elementary charge is left in the silicon crystal structure, corresponding to a vacant energy state in the valence band.

III–V composites typically have donors consisting of group IV (silicon) atoms substituting for group III elements, or group VI elements (S, Se, Te) substituting for group V elements. Acceptors in III–V composites are typically group II (zinc, magnesium) atoms substituting for group III elements. II–VI composites, on the other hand, often have donors from group VII (chlorine, etc.) substituting for group VI elements, while acceptors can belong to either group I (lithium, etc.) or group V (nitrogen, arsenic, phosphorous, etc.). In the latter case, group V elements are substituted for group VI elements in the semiconductor crystal structure, while group I acceptors replace group II elements [29]. The chemical potential or Fermi energy of an intrinsic semiconductor, which is free of n and p impurities, lies in the middle of the material's bandgap. When a moderate n-type doping is introduced, the Fermi level increases from the middle of the bandgap toward the conduction band. The amount of increase in the Fermi level varies with the level of doping [30].

If there is a high level of n-type doping in a semiconductor, the Fermi level may move across the bottom of the conduction band and end up inside it. This is known as the Mott transition and causes the semiconductor to behave like a metal. When this happens, the semiconductor is called a semi-metal and is said to be degenerate [28]. In the case of p-type doping, the semiconductor is considered to degenerate if the Fermi level is positioned below the top of the valence band [27].

Semiconductors operate by creating electron-hole pairs when exposed to light. If a semiconductor material is illuminated by photons with energy equal to or greater than its bandgap, the absorbed photons cause electrons to move from the valence band to the conduction band, where they can travel freely through the crystal structure with the help of an intrinsic or externally applied electric field. The positively charged holes left behind in the valence band can also contribute to electrical conduction by moving from one atomic site to another due to the electric field [27]. The separation of electron-hole pairs generated by light absorption produces a photocurrent, which represents the portion of photogenerated free charge carriers that are collected at the edges of the material by the electrodes of the photodetecting structure. The intensity of this photocurrent, at a given wavelength, increases with the intensity of the incident light [31].

At this level, we can classify photodetectors into two broad categories based on the electric field responsible for the separation of photogenerated electron-hole pairs. The first category is photoconductors, which consist of a single layer of semiconductor with two ohmic contacts. The charge carriers are collected by applying a bias voltage between the contacts, which generates the electric field. The second category is photovoltaic photodetectors, which use the internal electric field of a p-n or Schottky (metal-semiconductor) junction to separate the charges. This includes p-n junction photodetectors (photovoltaic structures composed of a single p-n junction) and p-i-n photodetectors (which include an undoped semiconductor layer between the p and n regions). Additionally, all Schottky junction photodetectors fall under this category, including Schottky barrier photodiodes and metal-semiconductor-metal (MSM) photodiodes [32,33].

1.4 P-N AND P-I-N JUNCTIONS

A p-n junction is formed when a region of p-type doped semiconductor and a region of n-type doping are joined together in a metallurgical linkage. The joining of Fermi levels in equilibrium is primarily achieved through a charge flow between the n and p regions. This results in the formation of a charged capacitor-like region around the junction with no free charge carriers. This region, known as the space charge region (SCR), contains positively ionized donors on the n-side and negatively ionized acceptors on the p-side. The ionized donors and acceptors provide fixed charges, which produce an electric field that curves the energy bands. As a result, an energy barrier is formed between the two regions, with the bottom of the conduction band and the top of the valence band on the n-side being below the corresponding levels on the p-side. This energy barrier is maintained in equilibrium [27,28] (as seen in Figure 1.1).

The width of the SCR decreases as the level of doping in the material increases, while the energy barrier height increases. When an electron-hole pair is produced in the SCR (see Figure 1.1), it is separated by the internal electric field of the junction and does not recombine. These separated charge carriers contribute to the photocurrent, along with those generated within a distance equal to or less than the diffusion length from the junction (see Figure 1.1). Since the band structure of the junction indicates that the photocurrent consists of minority charge carriers, it flows in the opposite direction of the diode bias, which defines the direction of majority charge carrier flow (from the n- to p-region for electrons and vice versa for holes). Additionally, by applying an opposing external electric field, the energy barrier height near the junction and the SCR extent can be increased, enhancing the separation of electron-hole pairs by intensifying the electric field within the junction and thus increasing the photocurrent efficiency [27,28].

It is worth noting that in the case of p-n junction photodetectors, the width of the SCR plays a crucial role when the doping level is moderate [34]. This effect is advantageous as it promotes the photoresponse by ensuring that the generation of electron-hole pairs predominantly occurs within the SCR. One approach to increase the spatial extent of the SCR is by introducing a thin layer of intrinsic semiconductor material, undoped intentionally, between the n and p regions. This structure is referred to as

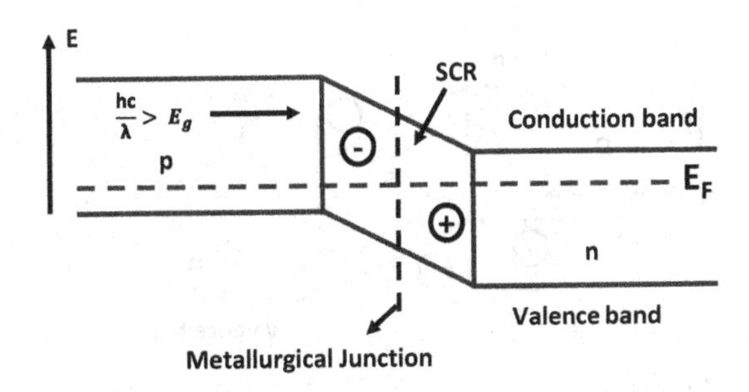

FIGURE 1.1 Structure of the energy bands and the production of photocurrent in a p-n junction.

p-i-n. The p-i-n structure is interesting because it allows for maintaining high doping levels in the n and p regions without significantly reducing the extent of the SCR. The width of the SCR is primarily determined by the thickness of the "i" layer. Moreover, widening the SCR also leads to a reduction in the structure capacitance, making high-speed operational p-i-n structures [34].

The p-i-n structures are widely used in electronic devices, particularly in optoelectronics. The structure consists of a p-region, an intrinsic i-region, and a n-region, which are all doped with different levels of impurities. The intrinsic region is the region between the p and n regions, and it has a low doping concentration, which means that it has a low concentration of free charge carriers [35,36]. One of the main applications of p-i-n structures is in photodiodes, which are used to detect light. In a photodiode, the p-i-n structure is connected to a reverse-bias voltage. When light is incident on the device, it generates electron-hole pairs in the intrinsic region, which are separated by the electric field across the structure. The resulting current can be measured and used to detect the light [35,36].

In a p-i-n structure, the band diagram shows the energy levels of the valence and conduction bands in the p-type, i-type, and n-type regions (see Figure 1.2). In the p-type region, the valence band is close to the Fermi level (EF) due to the higher concentration of holes. The conduction band is separated from the valence band by the bandgap energy (Eg). In the n-type region, the opposite is true, with the valence band far from EF and the conduction band close to EF due to the higher concentration of electrons [26,37]. The i-type region is the intrinsic layer sandwiched between the p-type and n-type layers. In this layer, the Fermi level is located in the middle of the bandgap, and the valence and conduction bands are closely spaced. As a result, carriers have a low probability of recombination and can easily be excited by incident light. When light is incident on the p-i-n structure, photons with energy equal to or greater than the bandgap energy (Eg) can excite electrons from the valence band to the conduction band in the i-type region, creating electron-hole pairs [38].

Another application of p-i-n structures is in solar cells. In a solar cell, the p-i-n structure is used to generate an electric current when exposed to sunlight. The incident light generates electron-hole pairs in the intrinsic region, which are separated by the electric field across the structure [39]. The resulting current can be used to power electronic devices or stored in a battery.

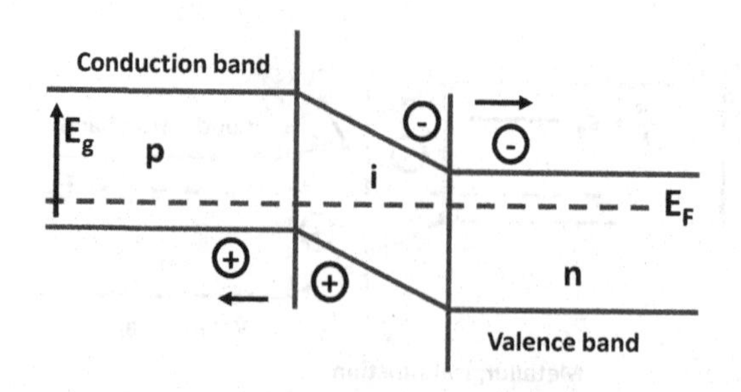

FIGURE 1.2 Structure of the energy bands and the production of photocurrent in a p-i-n junction.

1.5 AVALANCHE EFFECT

The avalanche effect is a phenomenon that occurs in p-i-n structures, which are used in a variety of electronic devices such as photodiodes and solar cells. This effect is the result of the multiplication of charge carriers in the p-region of the structure due to the application of a high voltage. As a result, many electrons and holes are generated, which can lead to breakdown of the structure and damage to the device [27].

The avalanche effect in p-i-n structures occurs when a high voltage is applied across the structure. The high voltage causes the electric field across the structure to increase, which leads to the generation of electron-hole pairs in the p-region through impact ionization. Impact ionization occurs when an electron or hole acquires enough energy to ionize an atom or molecule in the material, resulting in the creation of a new electron-hole pair [25,26]. Once an electron-hole pair is generated in the p-region, it is accelerated by the electric field and can collide with other atoms or molecules, creating additional electron-hole pairs. This process is known as multiplication, and it can lead to an exponential increase in the number of free charge carriers in the p-region [30]. The effect can have a significant impact on the performance of electronic devices. When the number of free charge carriers in the p-region increases, the device can experience breakdown, which can lead to damage or failure [40]. In addition, the avalanche effect can cause noise in the device, which can reduce its sensitivity and accuracy [41].

The avalanche effect, also known as the Townsend mechanism, describes the exponential growth of charged particles in a gas discharge due to ionization processes. In the Townsend mechanism, the ionization of solid or gas molecules occurs through electron impact ionization. The ionization coefficient is a measure of how easily the material can be ionized by an electron or hole [26]. The mathematical expression for the avalanche effect using the Townsend mechanism can be given by the Townsend equation:

$$\alpha = \alpha_0 \exp(\alpha_d) \tag{1.3}$$

Here, α represents the Townsend ionization coefficient, which is the probability of ionizing an atom or molecule per unit distance traveled by an electron. α_0 is the primary ionization coefficient, which is the initial probability of ionization per unit distance. α_d is the product of the electric field strength (E) and the mean free path (λ) of the electrons in the gas.

To mitigate the effects of the avalanche effect, one common technique is to reduce the thickness of the p-region, which reduces the probability of impact ionization. Another technique is to use a lower doping concentration in the p-region, which reduces the ionization coefficient and the probability of impact ionization.

1.6 SCHOTTKY JUNCTION

A Schottky junction is established when a metal and a semiconductor are brought into contact. The formation of a Schottky junction with an n-type semiconductor can be understood through several key phenomena. Figure 1.3 provides a summary of these phenomena.

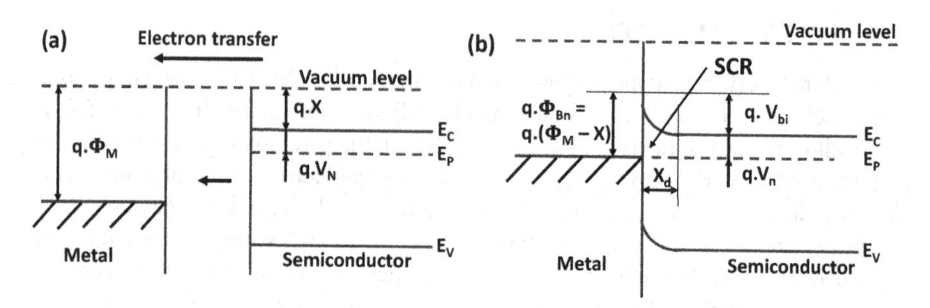

FIGURE 1.3 Schottky junction formation in a n-type semiconductor. (a) electron transfer from a semiconductor to the metal, (b) formation of SCR at the semiconductor

In a state of thermal equilibrium, when the Fermi levels of the metal and the semiconductor reach equilibrium, there is a flow of electronic charge from the semiconductor to the metal (see Figure 1.3a). This occurs when the work function ($q.\Phi_M$) of the metal (with q representing the elementary charge) is higher than the electron affinity (X) of the semiconductor. As a result, a SCR forms at the semiconductor edge adjacent to the junction with a width of X_d (see Figure 1.3b). Only positively ionized donors are present in this region. At the junction between a metal and a semiconductor, a curvature is observed in the energy bands. This curvature gives rise to an energy barrier known as a Schottky barrier between the metal and the semiconductor. The approximate height of this barrier can be determined using the following expression [42]:

$$q \cdot \Phi_{Bn} = q \cdot (\Phi_M - \mathcal{X}) \tag{1.4}$$

At the junction between a metal and a semiconductor in equilibrium, an intrinsic electric field is present near the junction, resembling the electric field observed in a p-n junction. As a result, the phenomenon of photogeneration of charge carriers within and close to the Schottky barrier leads to the emergence of a photocurrent. The electric field within the Schottky junction facilitates the separation of electron-hole pairs, thereby contributing to the generation of the photocurrent [37]. It is important to note that the band diagram of a solar cell can vary depending on the specific type of solar cell, such as Schottky junction solar cells or metal-insulator-semiconductor Schottky junction solar cells. These variations can involve the introduction of impurity energy levels or the use of thin insulating layers to enhance performance and efficiency.

Like the p-n junction, the strength of the internal electric field within the Schottky contact can be adjusted by applying a bias voltage V across the semiconductor and the metal [37] (see Figure 1.4).

When a negative voltage is applied between the metal electrode and the semiconductor in the Schottky contact of an n-type semiconductor, the Schottky junction is reverse-biased. This results in an elevated effective barrier height and an expanded width of the SCR, which is beneficial for photodetection [37]. In the Schottky contact, the flow of majority charge carriers (electrons) is hindered, while only the minority carriers (holes) generated through external excitation (photogeneration) can reach the Schottky contact and generate an electric current.

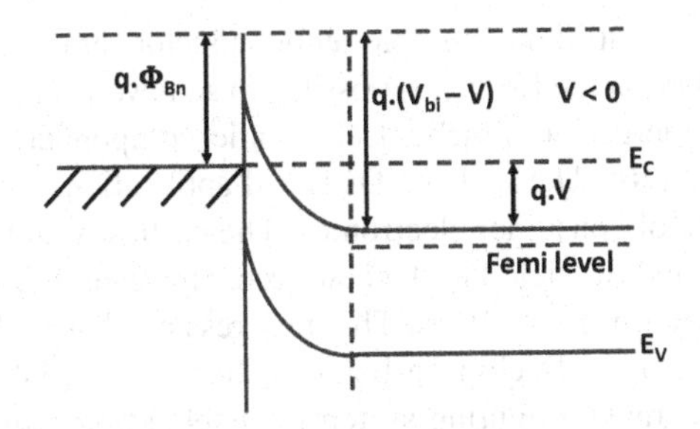

FIGURE 1.4 Schottky junction reverse-bias in a n-type semiconductor.

Like the behavior of a p-n junction, the current in the Schottky junction reverses, flowing toward the Schottky contact from the semiconductor [37]. When illuminating Schottky photodiodes, there are two options: illumination through the front or rear face. In situations where the substrate material is transparent to the detected light (such as sapphire), illumination through the rear face is often preferred. For front-face illumination, a semi-transparent Schottky contact is utilized. This contact consists of an extremely thin metal layer, typically around 100 Å, which is chosen to ensure adequate optical transmission. IR light can be transmitted up to 95% through a 100 Å gold layer, while the transmission percentage in the ultraviolet (UV) range (300–370 nm) is approximately 30% [43]. The maximum gain of p-i-n photodiodes and Schottky photodiodes (excluding avalanche photodiodes) is limited to 1. This means that the gain would be achieved only if all the charge carriers generated by light were collected by the electrodes located at the ends of the device [34].

1.7 OPTOELECTRONIC DEVICES

Optoelectronic devices are electronic components that utilize light (optical) energy to perform a specific function, such as emitting, detecting, or converting light into an electrical signal. Examples include LEDs, laser diodes, photodiodes, phototransistors, and solar cells. These devices have wide applications in fields such as telecommunications, displays, lighting, and renewable energy.

1.7.1 Light-Emitting Diodes

LEDs are solid-state semiconductor devices that consist of a p-n junction with a p-type layer and an n-type layer. The p-type layer contains positively charged holes, while the n-type layer carries mobile electrons. When an electrical current is applied to an LED in the forward bias direction (from the n-region to the p-region), electrons flow from the n-type layer and recombine with holes in the p-type layer. This recombination process releases energy in the form of photons, resulting in the emission of light. The light emitted by an LED corresponds to a narrow band of wavelengths, which is determined by the energy associated with the recombination of electron-hole pairs. This energy is approximately equal to the band gap energy of the semiconductor material used in the LED [44]. By utilizing multiple semiconductors or incorporating a layer of light-emitting phosphor on the LED, it is possible to achieve white light emission [45].

In 1962, LEDs emerged as practical electronic components, initially emitting low intensity IR light [46]. These IR LEDs found application in remote-control circuits for a wide range of consumer electronics. The earliest visible-light LEDs were also low in intensity and restricted to the color red. However, LED performance has improved significantly since the 1980s. This progress has led to the development of high-density, multi-color LED chip-on-board technologies, enabling the transition from red-only LED arrays to lighting systems capable of emitting light in multiple colors. A significant breakthrough occurred in the early 1990s with the development of blue-green LEDs based on aluminum gallium indium nitride (AlGaInN). This advancement allowed for the creation of low-voltage light sources in all three primary colors: red, green, and blue. Consequently, these advancements in LED technology opened new possibilities for the lighting and display industries [44].

In the early stages of LED development, they were primarily used as indicator lamps, serving as replacements for small incandescent bulbs. Additionally, LEDs found their way into seven-segment displays commonly seen in digital clocks [2]. However, subsequent advancements led to the availability of LEDs in various wavelengths, including visible, UV, and IR, with different levels of light output. This expanded range of LEDs enabled the creation of white LEDs suitable for room and outdoor area lighting. Moreover, LEDs have not only revolutionized lighting but have also given rise to new types of displays and sensors. Their high switching rates have proven valuable in advanced communications technology. As a result, LEDs are now employed in a wide range of applications, including aviation lighting, automotive headlamps, advertising, general lighting, traffic signals, camera flashes, lighted wallpaper, horticultural grow lights, and medical devices [47].

While LEDs offer several advantages over incandescent light sources, such as lower power consumption, longer lifetime, improved physical durability, smaller size, and faster switching, they also come with certain limitations. One of the drawbacks of LEDs is their reliance on low voltage and predominantly DC power, as they are generally not suitable for AC power sources. Additionally, LEDs are unable to provide steady illumination when powered by pulsing DC or AC electrical supply sources. Moreover, LEDs have lower maximum operating and storage temperatures compared to incandescent lamps. In contrast, incandescent lamps have the flexibility to operate at a wide range of supply voltages and can utilize both AC and DC currents interchangeably. They can provide steady illumination even when powered by AC or pulsing DC at frequencies as low as 50 Hz. Unlike LEDs, incandescent bulbs can often operate directly from unregulated DC or AC power sources without the need for additional electronic support components [48].

1.7.2 Laser Diode

Laser diodes, also known as semiconductor lasers, are compact and efficient sources of coherent light that are widely used in a variety of applications, including telecommunications, barcode scanners, DVD players, laser pointers, and medical devices.

They are based on the same fundamental principle as other types of lasers, which involves the emission of photons by stimulated emission in a gain medium, but their construction and operation are different [49].

The heart of a laser diode is a p-n junction formed by two types of semiconductor materials, such as gallium arsenide (GaAs) or indium gallium arsenide (InGaAs). When a voltage is applied across the junction, it creates an electric field that separates the positive and negative charges, forming a depletion region with no mobile charge carriers. When the junction is forward-biased, meaning that the positive voltage is applied to the p-side and the negative voltage to the n-side, the depletion region is reduced, and electrons from the n-side and holes from the p-side can recombine in the active region of the diode, where the gain medium is located [43].

The gain medium in a laser diode is usually a thin layer of semiconductor material doped with impurities that create excess electrons and holes, such as GaAs doped with aluminum (AlGaAs) or indium gallium arsenide phosphide (InGaAsP) [37]. When an electron and a hole recombine in the gain medium, they release energy in the form of a photon with a specific wavelength determined by the energy bandgap of the semiconductor material. This photon can stimulate the emission of other photons by the same process of stimulated emission, creating a cascade of photons that amplifies the light and produces coherent radiation [50,51].

To achieve lasing, the gain medium must be surrounded by two mirrors that reflect the light back and forth along the axis of the junction, creating a resonant cavity that enhances the intensity of the light. One of the mirrors is a partially reflective coating on the end of the diode, called the output coupler, that allows some of the light to escape and form the output beam. The other mirror is a high-reflectivity coating on the opposite end, called the back mirror or the cleaved facet, that reflects most of the light back into the cavity. The distance between the two mirrors, called the cavity length, determines the wavelength of the laser emission, and can be controlled by the design of the diode [49,52].

The output power and efficiency of a laser diode depend on several factors, such as the current density, the temperature, the optical losses, and the lifetime of the excited carriers in the gain medium. Higher current densities and lower temperatures generally increase the output power but decrease the lifetime of the gain medium, leading to shorter lifetimes and lower reliability [49,53]. Optical losses, such as absorption and scattering in the material, reduce the intensity of the light and limit the maximum output power. Lifetime degradation mechanisms, such as electromigration and diffusion of the impurities, can also reduce efficiency and output power over time [53].

Despite these limitations, laser diodes are still popular for their small size, low cost, and high performance in many applications. They can be fabricated in large quantities using standard semiconductor processing techniques and can be integrated with other electronic components on a single chip, such as modulators, detectors, and amplifiers. They also offer a wide range of wavelengths, from the near IR to the visible and UV, and can be tailored to specific requirements by adjusting the material composition, the doping concentration, and the cavity design [53].

1.7.3 Modulators

Modulators are devices used in electronics and telecommunications to modify or manipulate signals before transmission. They are essential in many communication systems and are used to alter the characteristics of a signal in order to transmit it over a certain distance or to different devices. The main function of a modulator is to take a low-frequency message signal, which carries the information to be transmitted, and combine it with a high-frequency carrier signal. The resulting signal, which is a combination of the two, is then transmitted over a communication channel [54].

There are different types of modulators depending on the type of signal being modulated, the method used to combine the message and carrier signals, and the modulation technique being used. Some of the most common modulators include the following.

1.7.3.1 Amplitude Modulation

In amplitude modulation (AM), the amplitude of the carrier signal differs in proportion to the signal modulation. This type of modulation is usually enjoyed in radio broadcasting and has the advantage of being easy to implement. However, it is susceptible to noise and distortion.

1.7.3.2 Frequency Modulation

In frequency modulation (FM), the frequency of the carrier signal is varied in proportion to the modulating signal. This type of modulation is commonly used in audio broadcasting and has the advantage of being less susceptible to noise and distortion than AM modulation.

1.7.3.3 Phase Modulation

In phase modulation (PM), the phase of the carrier signal is varied in proportion to the modulating signal. This type of modulation is commonly used in digital communication systems and has the advantage of being highly efficient and resistant to noise and distortion.

1.7.3.4 Pulse Width Modulation

In pulse width modulation (PWM), the duration of the pulses of a fixed carrier signal is varied in proportion to the modulating signal. This type of modulation is commonly used in digital control systems, where it can be used to control the speed of motors and other devices.

1.7.3.5 Quadrature Amplitude Modulation

In quadrature amplitude modulation (QAM), both the amplitude and phase of the carrier signal are varied to encode data onto it. This type of modulation is commonly used in digital communication systems, where it can achieve high data transmission rates.

Modulators play a crucial role in modern communication and electronic systems. In communication, they are used by encoding information onto a carrier signal to enable the efficient and reliable transmission of data over long distances. In electronic systems, modulators can be found in a wide range, including radios, televisions, and cell phones.

1.7.4 Photodetectors

Photodetectors, also known as photosensors, are devices designed to detect and measure light or other forms of electromagnetic radiation. These detectors come in various types, categorized based on the method they employ for detection, such as photoelectric or photochemical effects, as well as different performance metrics, including spectral response. Typically, semiconductor-based photodetectors consist of a p-n junction that facilitates the conversion of incident light photons into an electrical current. When photons are absorbed, electron-hole pairs are generated within the depletion region of the device. Photodiodes and phototransistors are among the most used examples of photodetectors [36].

Several key parameters play a crucial role in defining the behavior of an UV photodetector. These parameters include the response coefficient, which measures the detector's sensitivity to incoming radiation; the gain, which indicates the level of signal amplification; the quantum efficiency (QE), which quantifies the ratio of photons converted into electrical current; the bandwidth, which determines the range of frequencies the detector can effectively detect; the noise equivalent power (NEP), which defines the minimum detectable signal power level; and the detectivity, which combines various factors to assess the detector's overall performance.

1.7.4.1 Response Time

Response time of photodetectors refers to the amount of time it takes for the device to change its output signal from one state to another in response to a change in the light intensity. It is an important parameter for many applications, including optical communication, optical sensing, and imaging systems. The response time can be divided into two components: the rise time, which is the time taken to transition from 10% to 90% of the final output signal, and the fall time, which is the time taken to transition from 90% to 10% of the final output signal. The overall response time is typically taken as the average of these two values [55].

Factors that affect the response time of photodetectors include the type of detector material, the geometry of the device, the bias voltage, and the ambient temperature [36]. For example, photodetectors made from materials with higher electron mobility, such as InGaAs, tend to have faster response times compared to those made from silicon [55].

1.7.4.2 Quantum Efficiency

The concept of QE can be associated with the incident photon to converted electron (IPCE) ratio [56] in a photosensitive device or the TMR effect of a magnetic tunnel junction (MTJ). In the case of a charge-coupled device (CCD) or other photodetectors, QE represents the ratio between the number of collected charge carriers at the device's terminals and the number of photons that strike its photosensitive surface. QE is a dimensionless ratio but is closely connected to responsivity, which is measured in amps per watt. QE is frequently assessed across various wavelengths to evaluate the effectiveness of a device at different photon energy levels, considering that photon energy is inversely related to its wavelength. In the case of semiconductor photodetectors, QE diminishes to zero when the energy of photons falls below the band gap. Photographic films generally exhibit a QE significantly lower than 10% [57], whereas CCDs can achieve a QE exceeding 90% at specific wavelengths.

The QE of photodetectors relies on various factors, including the operating temperature, the material used, and the size of the active area. Photodetectors come in different types, exhibiting different levels of QE. In the visible range, silicon photodiodes generally have a QE of around 50%, whereas high-performance photodetectors like InGaAs photodiodes can achieve QE values of up to 80% in the near-IR range [36]. The determination of QE can be done using the expressions outlined below:

$$QE_{\lambda} = \eta = \frac{N_e}{N_v} \tag{1.5}$$

where N_e is the number of electrons produced and N_v is the number of photons absorbed.

$$\frac{N_v}{t} = \Phi_o \frac{\lambda}{hc} \tag{1.6}$$

Assuming each photon absorbed in the depletion layer produces a viable electron-hole pair and all other photons do not,

$$\frac{N_e}{t} = \Phi_{\xi} \frac{\lambda}{hc} \tag{1.7}$$

where t is the measurement time (in seconds), Φ_o is the incident optical power in watts, and Φ_{ξ} is the optical power absorbed in depletion layer, also in watts.

1.7.4.3 Gain

Gain in a photodetector refers to the amplification of the photocurrent produced in response to an incident light signal. It is a measure of how much the photodetector output increases for a given increase in incident light. The gain can be expressed as the ratio of output photocurrent to input optical power or as the ratio of output voltage to input optical power.

High gain in a photodetector allows for the detection of weak optical signals and improved signal-to-noise ratio (SNR) in the system. There are several types of photodetectors with different gain mechanisms, including photoconductive detectors, photovoltaic detectors, and photodiode amplifiers. The specific type of photodetector used and the gain mechanism employed depend on the application requirements and the desired performance characteristics. It is usually expressed as the ratio of the output photocurrent to the input optical power [43,55]. The gain can be mathematically represented as follows:

$$G = \frac{I}{P} \tag{1.8}$$

where G is the gain, I is the output photocurrent, and P is the input optical power. For a linear photodetector, the gain is constant and independent of the input optical power. However, for some photodetectors, the gain may vary with the input optical power due to nonlinear effects. In such cases, the gain can be represented as a function of input optical power:

$$G = F(P) \tag{1.9}$$

In photodiode amplifiers, the gain can also be represented as the ratio of the output voltage to the input optical power:

$$G = \frac{V}{P} \tag{1.10}$$

where V is the output voltage. In this case, the gain is dependent on the specific amplifier circuit used as well as the input-referred noise and bandwidth of the amplifier.

1.7.4.4 Temporal Response and Bandwidth

The temporal response of a photodetector refers to its ability to accurately measure the amplitude and duration of a light signal over time. The temporal response of a photodetector is an important factor that affects its overall performance, as it can impact the accuracy and precision of the light measurements.

There are several key factors that influence the temporal response of a photodetector, including the type of photodetector, the materials used in its construction, and the operating conditions [58]. For example, some photodetectors are designed for high-speed measurements with fast temporal response times, while others are optimized for low-light conditions and have slower temporal response times. The type of photodetector used also affects its temporal response. For example, photodiodes and phototransistors have fast temporal response times, while photomultiplier tubes (PMTs) are slower but are capable of measuring extremely low light levels [36,55]. The materials used in the construction of a photodetector also play a role in its temporal response. For example, photodetectors made from materials with high carrier mobility will have faster temporal response times as carriers can move quickly through the material. Conversely, materials with low carrier mobility will result in slower temporal response times [55]. The operating conditions of a photodetector also impact its temporal response. For example, increasing the temperature of the photodetector will result in a faster temporal response time, as carriers can move more quickly at higher temperatures [28]. On the other hand, decreasing the temperature of the photodetector will result in a slower temporal response time.

The bandwidth of a photodetector refers to the range of frequencies it can accurately detect and respond to. Bandwidth is a critical parameter in photodetectors, as it determines the speed and sensitivity of the device [36]. In this chapter, we discuss the bandwidth of photodetectors, its importance, and factors that affect it. It is a critical parameter in photodetectors because it determines the device's ability to respond to different frequencies of light. A high-bandwidth photodetector can respond to a wide range of frequencies and provide a more accurate representation of the incoming light signal. This is especially important in applications where the photodetector is used to detect fast-changing signals, such as in communication systems or in high-speed imaging [55,59]. There are several factors that can affect the bandwidth of photodetectors, including material properties, device geometry, and electrical circuit design [59].

1. **Material Properties:** The material properties of the photodetector play a crucial role in determining its bandwidth. Different materials have different electronic and optical properties, and these properties can significantly impact the device's performance. For example, materials with high carrier mobilities and low recombination rates are desirable for high-bandwidth photodetectors.
2. **Device Geometry:** The geometry of the photodetector also plays a role in determining its bandwidth. Devices with larger active areas generally have higher bandwidths than smaller devices, but this also depends on the material properties and electrical circuit design.
3. **Electrical Circuit Design:** The electrical circuit design of the photodetector can also significantly impact its bandwidth. The electrical resistance of the circuit and the capacitance of the device can cause significant changes in the bandwidth of the device.

1.7.4.5 Noise Equivalent Power

In a photodetector, the NEP is a parameter that quantifies the minimum detectable optical power or the sensitivity of the photodetector to optical signals. It represents the power level at which the SNR becomes unity [60]. NEP is defined as the optical power incident on the photodetector that produces an output SNR of 1 ($SNR = 1$), where the signal is the desired optical signal and the noise is the total noise power produced by the photodetector [60]. Mathematically, NEP can be expressed as follows:

$$NEP = \frac{\text{Total noise power}}{\text{Responsivity}} \tag{1.11}$$

The responsivity of a photodetector refers to the ratio of the output current or voltage to the incident optical power. It indicates the efficiency of the photodetector in converting light into an electrical signal. The NEP is typically specified in units of power per square root of a frequency, such as Watts per square root of Hertz ($W/\sqrt{Hz}$). It allows for the comparison of different photodetectors and helps assess their performance in low-light conditions [61].

Lower NEP values indicate higher sensitivity and better performance, as it implies that the photodetector can detect smaller optical power levels. However, it's important to note that NEP is not the only parameter to consider when evaluating a photodetector. Other factors such as speed, linearity, and spectral response are also essential, depending on the specific application requirements [49].

1.7.4.6 Detectivity

Detectivity (D^*) of a photodetector is a measure of its sensitivity to detect low light levels. Detectivity is a measure of the performance of a photodetector in terms of its ability to detect light. It is defined as the minimum amount of light that a photodetector must receive to produce a specified output signal. Detectivity is expressed in units of $\text{cm} \cdot \frac{\sqrt{\text{Hz}}}{\text{W}}$, and is typically used to compare the performance of different photodetector devices [62–64]. The specific detectivity is expressed as:

$$D^* = \frac{\sqrt{A \, \Delta \, f}}{\text{NEP}} \tag{1.12}$$

where A is the area of the photosensitive region of the detector, Δf is the bandwidth, and NEP is the NEP in units [W].

There are several factors that determine the detectivity of a photodetector, including its size, material, and operating temperature. For example, larger photodetectors generally have higher detectivity than smaller ones due to their increased surface area. Similarly, photodetectors made from materials with higher intrinsic carrier concentrations, such as InGaAs and HgCdTe, tend to have higher detectivity [65,66] compared to those made from materials with lower intrinsic carrier concentrations, such as silicon [67]. The operating temperature also plays an important role in the detectivity of a photodetector. Lower temperatures result in lower thermal noise and, therefore, higher detectivity [28,55].

In addition to these intrinsic factors, the detectivity of a photodetector can also be improved using optimized optical and electrical design. For example, anti-reflective coatings can be applied to the surface of the photodetector to minimize loss of light due to reflection, thus increasing the amount of light that is absorbed [68]. The electrical design of the photodetector can also be optimized to minimize noise, such as shot noise and thermal noise, which can degrade the performance of the device [43,69].

The performance of photodetectors can also be improved by using signal processing techniques, such as signal amplification, averaging, and filtering [70]. These techniques can help to increase the SNR [71], which is a key factor in determining the detectivity of the photodetector [72,73]. The detectivity of a photodetector is an important metric for evaluating its performance and comparing the performance of different devices. By understanding the factors that determine the detectivity of a photodetector and by using optimization techniques, it is possible to improve the performance of these devices and increase their sensitivity to light.

1.7.5 Types of Photodetectors

1.7.5.1 Photoconductors

A photoconductor is a type of optical sensor that converts light into an electrical signal by changing its electrical resistance. It is commonly used in applications such as optical scanning, light detection, and scientific instrumentation. Photoconductors are made of materials that change their conductivity when exposed to light, such as selenium, lead sulfide, or indium antimonide [59]. The basic operation of a photoconductor involves exposing the material to light, which increases its conductivity and reduces its resistance [58]. This change in resistance results in a change in the current flowing through the material, which can be used to generate an electrical signal.

The magnitude of the change in resistance is proportional to the intensity of the light, making photoconductors useful for light detection [74].

Photoconductors are often used in optical scanning applications, such as barcode scanning and document scanning [75,76]. In these applications, light from a source such as a laser is directed onto the photoconductor, which is positioned in a moving carriage. As the light travels across the surface of the photoconductor, the change in resistance generates an electrical signal that is proportional to the position of the light [75]. This signal is then used to control the movement of the carriage, allowing for precise and accurate scanning. Another important application of photoconductors is in light detection, where they are used to measure the intensity of light in scientific and industrial applications [28]. Photoconductors are used in light meters, flame detectors, and smoke detectors, among others [37]. In these applications, the photoconductor is exposed to light, and the change in resistance is used to generate an electrical signal that is proportional to the light intensity [28].

The performance of photoconductors is characterized by several key parameters, including light sensitivity, response time, and dark current [27,28]. The light sensitivity is a measure of the change in resistance per unit of light intensity and is a function of the material properties and the active area of the photoconductor. Response time is the time it takes for the photoconductor to reach its steady-state resistance after being exposed to light and is determined by the material properties and the bias voltage applied to the material. Dark current is the current that flows through the photoconductor in the absence of light and is a function of the temperature and the material properties.

1.7.5.2 Photodiodes

Photodiodes are photonic devices that are widely used for converting light into electrical current. They are semiconductor devices that consist of a p-n junction, which enables the conversion of light into electrical current. Photodiodes have a wide range of applications, including optical communication systems, spectrometers, and light detectors. When light is absorbed by a photodiode, it generates an electron-hole pair [36]. This results in the generation of an electrical current, which is proportional to the intensity of the light. This phenomenon is known as the photovoltaic effect [58]. The current generated by a photodiode can be used to detect the presence or absence of light, or to measure the light intensity.

There are several types of photodiodes, each with its own unique properties and applications. The most common types of photodiodes include p-n photodiodes, p-i-n photodiodes, avalanche photodiodes, and Schottky photodiodes [36,58]. p-n photodiodes are the simplest type of photodiode and consist of a p-n junction. They are commonly used for simple light detection applications, such as optical sensors. p-n photodiodes have an intrinsic region between the p and n regions, which increases the sensitivity and reduces the response time of the device. Avalanche photodiodes are designed to operate in reverse-biased mode, where an electric field is high enough to cause an avalanche multiplication of carriers [36,58]. This results in a higher gain and faster response compared to other types of photodiodes. Schottky photodiodes are made of a metal-semiconductor junction instead of a p-n junction, which results in a lower capacitance and faster response time compared to p-n photodiodes [36,58].

Photodiodes have several advantages over other types of light detectors. They are highly sensitive, have a fast response time, and offer a linear response to light intensity. They are also relatively inexpensive to manufacture and are widely available. Photodiodes can be designed to operate in either the photovoltaic mode or the photoconductive mode, depending on the application requirements. In the photovoltaic mode, the photodiode is reverse-biased, which allows for the generation of an electrical current proportional to the light intensity. This mode is commonly used for light detection applications. In the photoconductive mode, the photodiode is forward-biased, which allows for the device to operate as a resistor with a resistance proportional to the light intensity. This mode is commonly used for light measurement applications [55].

Advantages of photodiodes include the following:

- **Fast Response Times:** Photodiodes have fast response times, making them suitable for applications that require quick measurements of light intensity.
- **High Sensitivity:** Photodiodes have high sensitivity to light and can detect light in the near-IR and UV regions, as well as in the visible range.
- **Low Cost:** Photodiodes are relatively inexpensive compared to other types of light detectors.

Disadvantages of photodiodes include the following:

- **Limited Spectral Range:** Photodiodes may not be sensitive to certain wavelengths of light, and their spectral response can vary with temperature.
- **Nonlinear Response:** Photodiodes have a nonlinear response, meaning that their output current is not proportional to the light intensity. This can make it difficult to accurately measure light intensity.

1.7.5.3 Phototransistors

Phototransistors are electronic devices that convert light into electrical signals. They are like photodiodes but have a transistor amplification stage. This allows them to provide a higher output current, making them more suitable for use in analog circuits. Phototransistors are widely used in a variety of applications, such as light detection, switching, amplification, and sensing [77]. There are two main types of phototransistors: n-p-n and p-n-p [78]. n-p-n phototransistors have a p-type base region and n-type emitter and collector regions, while p-n-p phototransistors have an n-type base and p-type emitter and collector regions. The type of phototransistor used will depend on the requirements of the application and the desired output signal [77].

The working principle of a phototransistor is based on the photoelectric effect. When light falls on the base region, it generates electron-hole pairs, which are attracted toward the collector and emitter regions [28]. This creates a small current that is proportional to the intensity of the light. In an n-p-n phototransistor, the base-emitter junction is forward-biased, allowing most of the electrons generated in the base to flow into the emitter. The emitter-collector junction is reverse-biased, so most of the electron's flow into the collector region. The collector current is then amplified by the transistor action, producing a larger output current [28,78]. A phototransistor is a type of bipolar transistor that is sensitive to light. It is used as a light-sensitive switch or amplifier.

The basic equation for phototransistor current is given by [28,78]:

$$I_c = \beta * I_l \tag{1.13}$$

where I_c is the collector current, β is the transistor current gain, and I_l is the light-generated current.

The current gain of the phototransistor can be found using the following equation:

$$\beta = \frac{I_c}{I_l} \tag{1.14}$$

The collector-emitter voltage (V_{ce}) can be calculated as follows:

$$V_{ce} = V_{cc} - (I_c * R_c) \tag{1.15}$$

where V_{cc} is the supply voltage and R_c is the collector resistor.

The voltage across the collector resistor (V_r) can be calculated as follows:

$$V_r = V_{ce} - V_{be} \tag{1.16}$$

where V_{be} is the base-emitter voltage.

Applications of phototransistors are used in a wide range of applications, including:

- **Light Detection:** Phototransistors can be used to detect light levels, for example, in light sensors or photocells.
- **Switching:** Phototransistors can be used to switch electronic circuits, for example, in optical switches.
- **Amplification:** Phototransistors can be used to amplify weak signals, for example, in amplifiers or opto-couplers.
- **Sensing:** Phototransistors can be used to detect changes in light levels, for example, in proximity sensors.

Phototransistors have several advantages over other types of light-sensitive devices, including:

- **High Sensitivity:** Phototransistors are highly sensitive to light, allowing them to detect even low levels of light.
- **Fast Response Time:** Phototransistors have a fast response time, making them suitable for use in high-speed applications.
- **High Output Current:** Phototransistors have a higher output current than photodiodes, making them more suitable for use in analog circuits.

However, phototransistors also have some limitations, including:

- **Limited Bandwidth:** Phototransistors have a limited bandwidth, which may limit their performance in high-frequency applications.

- **Low Sensitivity in the IR Range:** Phototransistors are not as sensitive to light in the IR range as they are to visible light.
- **Dependence on Temperature:** Phototransistors may have a dependence on temperature, which can affect their performance in certain applications.

1.7.6 Solar Cell

Solar cells, also known as photovoltaic cells or PV cells, are electrical devices that convert light energy into electrical energy through the photovoltaic effect [80]. They are essentially p-n junction diodes, although their construction differs slightly from conventional p-n junction diodes [80]. A solar cell consists of a layer of p-type semiconductor placed next to a layer of n-type semiconductor [79]. In the n-type layer, there is an excess of electrons, while the p-type layer has an excess of positively charged holes [79].

When light reaches the p-n junction in a solar cell, the photons easily enter the junction through the thin p-type layer [80]. The incident light carries energy in the form of photons, which provide sufficient energy to the junction, leading to the creation of electron-hole pairs [80]. This process disrupts the thermal equilibrium condition of the junction, resulting in the separation of free electrons in the n-type region and holes in the p-type region [80]. This separation creates a voltage potential across the p-n junction, generating electricity (see Figure 1.5).

The formation of a p-n junction in a solar cell is essential for its operation. The junction acts as a barrier, allowing only the movement of electrons and holes when illuminated by light. This unidirectional movement of charge carriers enables the generation of a current flow in an external circuit connected to the solar cell. In addition to the p-n junction, another important type of junction used in solar cells is the p-i-n junction. The p-i-n junction structure offers certain advantages over the conventional p-n junction in terms of efficiency and performance.

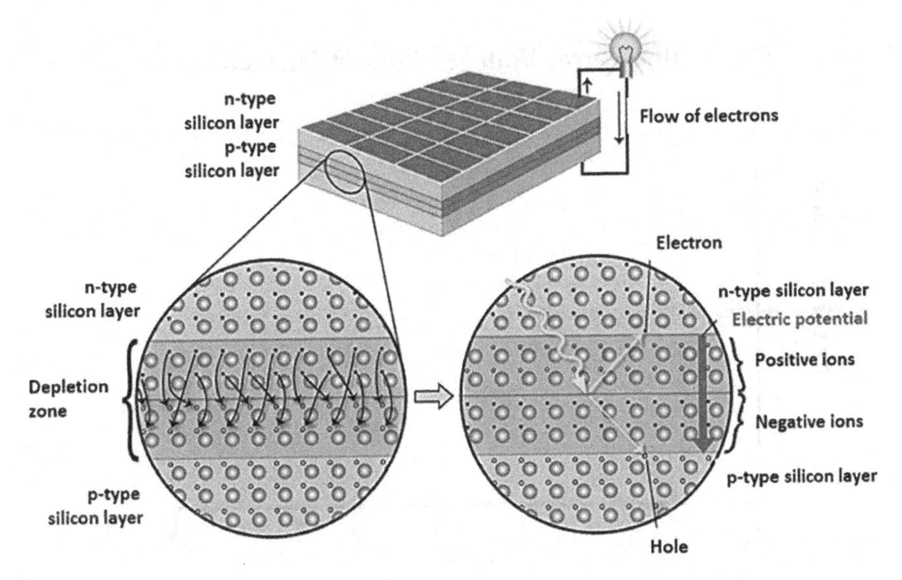

FIGURE 1.5 Schematic of n-type and p-type layers, depletion zone, and junction of a solar cell [79].

In a p-i-n junction solar cell, an intrinsic (i) layer is introduced between the p-type and n-type layers [80]. The intrinsic layer is a semiconductor region with a low concentration of dopants, resulting in fewer impurities and defects [80]. This layer serves as a depletion region that widens the SCR and enhances the collection of charge carriers. The introduction of the intrinsic layer in the p-i-n junction solar cell helps reduce recombination losses and enhances the efficiency of charge carrier collection. The wider depletion region enables more efficient absorption of light and facilitates the generation of electron-hole pairs. The charge carriers generated in the intrinsic layer can then be collected more effectively by the electric field present across the p-i-n junction. Furthermore, the p-i-n junction structure allows for better control over the flow of charge carriers. By adjusting the doping concentrations and thicknesses of the different layers, the performance of the solar cell can be optimized, leading to improved efficiency and performance. The efficiency of a solar cell can be determined using the *I-V* characteristic curve (see Figure 1.6) using the expressions below [81]:

$$P_{max} = I_{max} \times V_{max} \tag{1.17}$$

$$FF = \frac{P_{max}}{I_{SC} \times V_{OC}} \tag{1.18}$$

$$\eta = \frac{P_{max}}{P_{in}} = \frac{FF \times I_{SC} \times V_{OC}}{P_{in}} \tag{1.19}$$

where η is the efficiency, P_{max} is the maximum power, I_{max} is the maximum current, V_{max} is the maximum voltage, I_{SC} is the short circuit current, V_{OC} is the open circuit voltage, FF is the fill factor, and P_{in} is the power of sunlight.

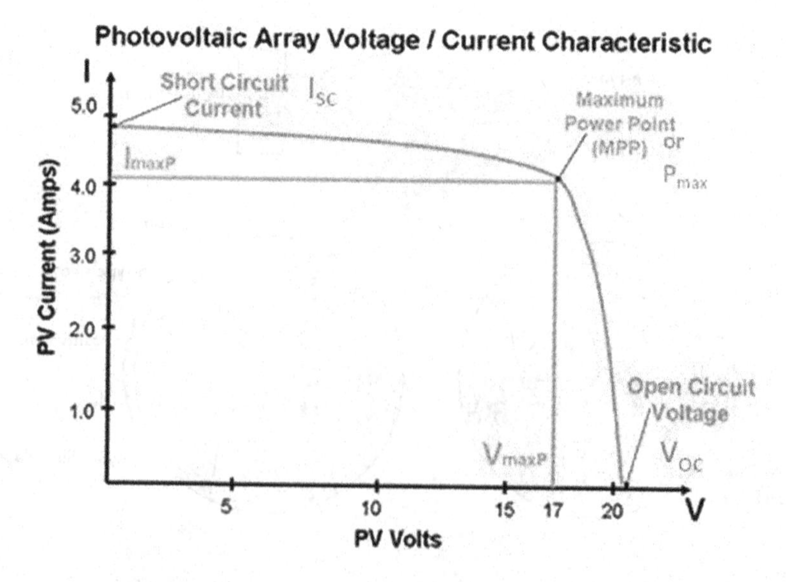

FIGURE 1.6 *I-V* characteristic plot for determination of photovoltaic efficiency [80].

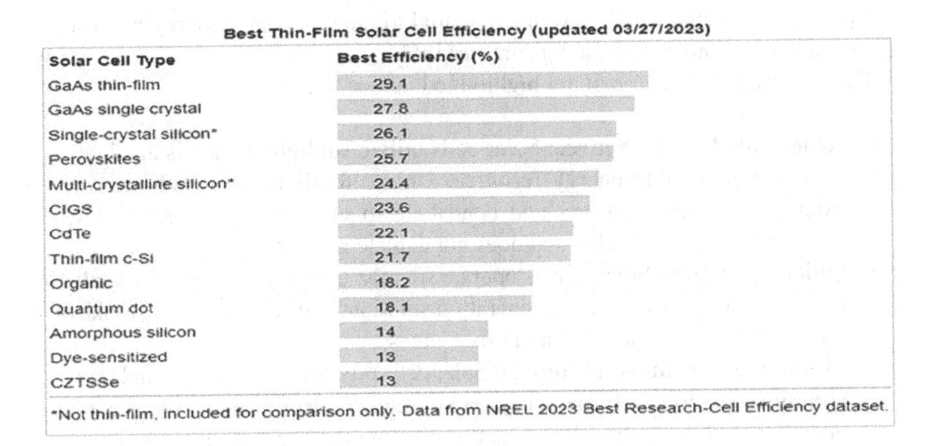

Best Thin-Film Solar Cell Efficiency (updated 03/27/2023)

Solar Cell Type	Best Efficiency (%)
GaAs thin-film	29.1
GaAs single crystal	27.8
Single-crystal silicon*	26.1
Perovskites	25.7
Multi-crystalline silicon*	24.4
CIGS	23.6
CdTe	22.1
Thin-film c-Si	21.7
Organic	18.2
Quantum dot	18.1
Amorphous silicon	14
Dye-sensitized	13
CZTSSe	13

*Not thin-film, included for comparison only. Data from NREL 2023 Best Research-Cell Efficiency dataset.

FIGURE 1.7 Laboratory thin-film solar cell efficiencies [83].

Considering a semiconductor material, the absorption of photons in this material creates electron-hole pairs, and the resulting separation of charges generates a voltage potential that can be utilized as electricity. The design and composition of semiconductor solar cells vary depending on the specific materials used. Silicon is one of the commonly employed semiconductors for solar cells [82]. It offers excellent electrical properties and stability, making it a reliable choice. However, other semiconductor materials (GaAs, CdTe, CIGS, CZTS, etc.) are also being explored to enhance efficiency and reduce the cost of solar cells [82]. Figure 1.7 provides information on the maximum efficiencies of single-junction non-concentrator thin-film cells using various thin-film materials.

It is worth noting that efficiencies achieved in a laboratory setting are generally higher than those of manufactured solar cells, with the latter often exhibiting efficiencies 20%–50% lower [83]. As of 2021, the maximum efficiencies of manufactured solar cells were reported as follows: 24.4% for monocrystalline silicon, 20.4% for polycrystalline silicon, 12.3% for amorphous silicon, 19.2% for CIGS, and 19% for CdTe modules. Among thin-film cell prototypes, the one with the highest efficiency was achieved by First Solar, yielding 20.4% efficiency. This is comparable to the best conventional solar cell prototype efficiency of 25.6% achieved by Panasonic [83].

The efficiency and cost-effectiveness of semiconductor solar cells are key considerations in their development. Researchers focus on improving the efficiency of solar cells by optimizing the optical absorption and transport of photogenerated carriers [84]. Additionally, efforts are made to enhance the cost-effectiveness by exploring different semiconductor materials and manufacturing techniques [82]. The field of semiconductor solar cells is constantly evolving, and ongoing research aims to improve their performance and explore new materials and designs. Thin films, dye-sensitized solar cells, and organic solar cells are areas of active investigation [82]. These alternative solar cell technologies offer potential advantages such as flexibility, lightweight, and easy manufacturing.

The comprehensive understanding of semiconductor solar cells encompasses the history, evolution, and present scenarios of solar cell design, classification, properties, and various semiconductor materials [85]. Researchers explore concepts such as transparent solar cells, which have the potential to integrate solar energy harvesting into windows and other transparent surfaces [85].

The advantages of solar cell are highlighted below:

- **Renewable Energy Source:** Solar cells utilize sunlight, which is an abundant and renewable energy resource. Unlike fossil fuels, which require extensive extraction efforts and contribute to greenhouse gas emissions, solar energy is sustainable and does not deplete natural resources.
- **Quiet Operation:** Solar panels operate silently, unlike wind turbines, making them a noise-free energy generation solution. This characteristic makes them suitable for residential and urban areas.
- **Minimal Environmental Impact:** Solar cells produce clean energy and have a minimal carbon footprint. They do not release harmful emissions during operation, reducing air pollution and mitigating the negative impacts of climate change.
- **Abundant Energy Source:** Sunlight is an immensely abundant energy source. Harnessing just a fraction of the sunlight that reaches the Earth's surface can provide a significant amount of energy to meet human needs.

However, solar cell also has some limitations, including:

- **Initial Cost:** The installation of a solar system requires a significant upfront investment to cover the costs of panels, inverters, batteries, wiring, and installation. The cost of solar panels has been decreasing, but it can still be a barrier for some households.
- **Weather Dependence:** Solar energy generation is dependent on sunlight, and its efficiency varies based on factors such as the amount and quality of direct sunlight received. Cloudy or overcast weather conditions can affect the output of solar panels.
- **Land Use Requirement:** Solar power plants require a significant amount of land to generate electricity on a large scale. This can pose challenges in terms of land availability and potential conflicts with other land uses.
- **Intermittent Power Generation:** Solar energy production is intermittent, as it is available only during daylight hours and is affected by seasonal variations. Energy storage systems or grid integration solutions are necessary to ensure a stable and reliable power supply.

1.8 APPLICATION OF OPTOELECTRONIC

The field of optoelectronics has a wide range of applications, from telecommunications to energy generation, and from sensing to medical devices. The following is a brief overview of some of the applications of optoelectronics.

1.8.1 Telecommunications

Optoelectronics plays a crucial role in telecommunications, as it enables the transmission of high-speed data over long distances without signal degradation. Optical fibers, made of glass or plastic, are used to transmit light signals, which are then converted back into electrical signals at the receiving end. This process enables the transmission of large amounts of data over long distances, making it possible for the internet and other data-intensive applications to exist.

1.8.2 Light-Emitting Diodes

LEDs are small, energy-efficient devices that can emit light in a variety of colors. They are used in a wide range of applications, including lighting, displays, and indicators. In addition, they are used in backlighting for LCD screens as well as in automotive and aviation lighting.

1.8.3 Laser Diode

A laser diode is a type of LED that can emit light at a specific wavelength. They are used in a wide range of applications, including data storage, medical devices, and telecommunications. For example, laser diodes are used in CD and DVD players to read the data stored on the discs.

1.8.4 Photovoltaics

Photovoltaics is the process of converting light into electricity using photovoltaic cells. These cells are made of semiconductors, such as silicon, and are used in solar panels to generate electricity. Photovoltaics is becoming increasingly important as the world looks to reduce its dependence on fossil fuels and find alternative sources of energy.

1.8.5 Optical Sensing

Optical sensors are devices that detect changes in light and convert them into an electrical signal. They are used in a wide range of applications, including medical devices, environmental monitoring, and security systems. For example, optical sensors are used in medical devices to monitor the blood glucose levels of diabetic patients, and in security systems to detect movement.

1.9 CHALLENGES OF OPTOELECTRONICS

Despite its significance in modern technology, there are several challenges that need to be addressed for further advancements in this field.

1.9.1 Power Efficiency

One of the major challenges in optoelectronics is to achieve high power efficiency while converting light into electricity. This requires the development of new materials and devices that can effectively harness light energy and convert it into electricity.

1.9.2 Bandgap Engineering

Another challenge is to control the bandgap in optoelectronic materials. The bandgap is the energy range between the valence and conduction bands in a material and determines its optical and electronic properties. Bandgap engineering enables the design of materials with desired optical and electronic properties, but it is a challenging task due to the complexity of materials and their properties.

1.9.3 Cost-Effectiveness

Optoelectronic devices and systems are still relatively expensive compared to their electronic counterparts. The development of cost-effective materials and manufacturing techniques is crucial for the widespread adoption of optoelectronics in various applications.

1.9.4 Integration with Electronics

Optoelectronics and electronics have different properties and behaviors, making it challenging to integrate these two technologies effectively. The development of integrated optoelectronic systems is crucial for various applications, such as high-speed communication, data processing, and energy harvesting.

1.9.5 Scalability

The scalability of optoelectronic devices is also a challenge. Optoelectronic devices need to be scalable to meet the demands of various applications, ranging from small devices for wearable technology to large systems for energy harvesting and data processing.

1.9.6 Stability

Optoelectronic devices are sensitive to environmental factors such as temperature, humidity, and radiation. Ensuring the stability and reliability of optoelectronic devices under different environmental conditions is crucial for their practical applications.

1.9.7 Robustness and Reliability

Optoelectronic devices are often used in harsh environments, such as in aerospace and industrial applications. Ensuring the robustness and reliability of these devices under these conditions is a major challenge.

1.9.8 Improving Performance

Optoelectronics is constantly evolving, and researchers are working to improve the performance of devices in areas such as speed, bandwidth, and spectral range. For example, improving the speed of data transmission in optical fibers is a major challenge.

1.9.9 Environmental Sustainability

The production of optoelectronic devices can have a significant impact on the environment, and researchers are working to develop environmentally sustainable materials and processes.

1.10 SUMMARY

Optoelectronics is a branch of electronics that deals with the study of the interaction between light and electrical signals. It involves the design, development, and application of devices that can either emit, detect, or control light. Examples of optoelectronic devices include LEDs, laser diodes, modulators, photodetectors, and solar cells. In optoelectronics, the behavior of light is manipulated by electronic devices and vice versa. This enables the conversion of electrical signals to optical signals and vice versa, which is essential in applications such as telecommunications, information processing, and energy conversion. The field of optoelectronics continues to evolve, with researchers and engineers working on new materials and devices that can operate at low cost, higher speeds, lower power consumption, and more efficient energy conversion. With advances in technology, the potential for optoelectronics to revolutionize many industries is huge, and it will be exciting to see what new applications emerge in the future.

REFERENCES

[1] I.L.V CIE, *International Lighting Vocabulary*, Central Bureau ofthe Commission Internationale de l'Eclairage, Kegelgasse, vol. 27, 1987.
[2] J.-P. Uzan, B. Leclercq, B. Mizon, *The Natural Laws of the Universe: Understanding Fundamental Constants*, Springer, 2008.
[3] P.A. Buser, M. Imbert, *Vision*, MIT Press, 1992.
[4] M. Ayene, J. Krick, B. Damitie, A. Ingerman, B. Thacker, A holistic picture of physics student conceptions of energy quantization, the photon concept, and light quanta interference, *International Journal of Science and Mathematics Education*. 17 (2019) 1049–1070.
[5] K. Schaffer, G. Barreto Lemos, Obliterating thingness: an introduction to the 'what' and the 'so what' of quantum physics, *Foundations of Science*. 26 (2021) 7–26.
[6] M.A. Hossain, Everything of the universe is made of light: theory for everything, *Journal of Science and Today's World*. 2 (9) (2013) 1267–1272.
[7] M.A. Parker, *Physics of Optoelectronics*, CRC Press, 2018.
[8] N. Rivera, I. Kaminer, Light-matter interactions with photonic quasiparticles, *Nature Reviews Physics*. 2 (10) (2020) 538–561.
[9] R. Gutzler, M. Garg, C.R. Ast, K. Kuhnke, K. Kern, Light-matter interaction at atomic scales, *Nature Reviews Physics*. 3 (6) (2021) 441–453.
[10] A. Charlesby, *Atomic Radiation and Polymers: International Series of Monographs on Radiation Effects in Materials*, Elsevier, 2016.
[11] V. Peano, C. Brendel, M. Schmidt, F. Marquardt, Topological phases of sound and light, *Physical Review X*. 5 (3) (2015) 31011.
[12] I. Oshina, J. Spigulis, Beer-Lambert law for optical tissue diagnostics: current state of the art and the main limitations, *Journal of Biomedical Optics*. 26 (10) (2021) 100901.
[13] F.L. Galeener, Absorption coefficient, in: R. G. Lerner, G. L. Trigg, *Encyclopedia of Physics*, Wiley-VCH, 1991: pp. 1–175.

[14] J.L. Weishaar, G.R. Aiken, B.A. Bergamaschi, M.S. Fram, R. Fujii, K. Mopper, Evaluation of specific ultraviolet absorbance as an indicator of the chemical composition and reactivity of dissolved organic carbon, *Environmental Science & Technology.* 37 (20) (2003) 4702–4708.
[15] S. Ahn, S.Y. Jung, S.J. Lee, Gold nanoparticle contrast agents in advanced X-ray imaging technologies, *Molecules.* 18 (5) (2013) 5858–5890.
[16] E.G. Snyder et al., The changing paradigm of air pollution monitoring, *Environmental Science & Technology.* 47 (20) (2013) 11369–11377.
[17] F. Osaigbovo, Light and the laws of reflection and refraction as they impact on photography, *Yıldız Journal of Art and Design.* 9 (1) (2022) 49–59.
[18] S. Inoué, *Video Microscopy*, Springer Science & Business Media, 2013.
[19] M. Katz, *Introduction to Geometrical Optics*, World Scientific, 2002.
[20] H. Ohno, H. Naruse, M. Kihara, A. Shimada, Industrial applications of the BOTDR optical fiber strain sensor, *Optical Fiber Technology.* 7 (1) (2001) 45–64.
[21] M. Scalora, M.J. Bloemer, A.S. Pethel, J.P. Dowling, C.M. Bowden, A.S. Manka, Transparent, metallo-dielectric, one-dimensional, photonic band-gap structures, *Journal of Applied Physics.* 83 (5) (1998) 2377–2383.
[22] D. Marcuse, *Light Transmission Optics*, Van Nostrand Reinhold, 1982.
[23] D.J. Richardson, J.M. Fini, L.E. Nelson, Space-division multiplexing in optical fibres, *Nature Photonics.* 7 (5) (2013) 354–362.
[24] S. Johnsen, E.A. Widder, The physical basis of transparency in biological tissue: ultrastructure and the minimization of light scattering, *Journal of Theoretical Biology.* 199 (2) (1999) 181–198.
[25] C. Kittel, P. McEuen, *Introduction to Solid State Physics*, John Wiley & Sons, 2018.
[26] D.A. Neamen, *Semiconductor Physics and Devices Basic Principles*, Tata McGraw Hill Publishing, 1992.
[27] S.O. Kasap, *Electronic Materials and Devices*, McGraw-Hill, New York, 2006.
[28] B.G. Streetman, S. Banerjee, *Solid State Electronic Devices*, vol. 4, Prentice Hall, 2000.
[29] J. Singh, *Electronic and Optoelectronic Properties of Semiconductor Structures*, Cambridge University Press, 2007.
[30] S.M. Sze, *Semiconductor Devices*: *Physics and Technology*, John Wiley & Sons, 2008.
[31] N.W. Ashcroft, N.D. Mermin, *Solid State Physics*, Holt Saunders International Editions, New York, 1976.
[32] M. Johnson, *Photodetection and Measurement*: *Maximizing Performance in Optical Systems*, McGraw-Hill Education, 2003.
[33] D. Birtalan, W. Nunley, *Optoelectronics*: *Infrared-Visable-Ultraviolet Devices and Applications*, CRC Press, 2018.
[34] D. Decoster, J. Harari, *Optoelectronic Sensors*, John Wiley & Sons, 2013.
[35] J. Piprek, *Semiconductor Optoelectronic Devices*: *Introduction to Physics and Simulation*, Elsevier, 2013.
[36] B.E.A. Saleh, M.C. Teich, *Fundamentals of Photonics*, John Wiley & Sons, 2019.
[37] S.M. Sze, Y. Li, K.K. Ng, *Physics of Semiconductor Devices*, John Wiley & Sons, 2021.
[38] B. Anderson, R. Anderson, *Fundamentals of Semiconductor Devices*, McGraw-Hill, Inc., 2004.
[39] A. Reinders, P. Verlinden, W. Van Sark, and A. Freundlich, *Photovoltaic Solar Energy*: *From Fundamentals to Applications*, John Wiley & Sons, 2017.
[40] S.M. Sze, M.-K. Lee, *Semiconductor Devices*: *Physics and Technology*: *Physics and Technology*, Wiley Global Education, 2012.
[41] A. Van der Ziel, *Noise in Solid State Devices and Circuits*, Wiley-Interscience, 1986.
[42] S.M. Sze, K.K. Ng, Chapter 3: Metal-semiconductors contacts, in: *Physics of Semiconductor Devices*, 3rd ed., John Wiley & Sons, 2007: pp. 134–196.
[43] P. Bhattacharya, *Semiconductor Optoelectronic Devices*, Prentice-Hall, Inc., 1997.

[44] N.G. Yeh, C.-H. Wu, T.C. Cheng, Light-emitting diodes-their potential in biomedical applications, *Renewable and Sustainable Energy Reviews.* 14 (8) (2010) 2161–2166.

[45] P. Sharma, M. Khan, A. Choubey, LED revolution: deep UV LED, *International Journal of Engineering and Technology.* 6 (2019) 6486.

[46] T.M. Okon, J.R. Biard, *The First Practical LED*, Edison Tech Center, vol. 9, 2015.

[47] E.A. Pelaez, E.R. Villegas, LED power reduction trade-offs for ambulatory pulse oximetry, in: *2007 29th Annual International Conference of the IEEE Engineering in Medicine and Biology Society*, 2007: pp. 2296–2299.

[48] J. Dakin, R.G.W. Brown, *Handbook of Optoelectronics (Two-Volume Set)*, CRC Press, 2006.

[49] L.A. Coldren, S.W. Corzine, M.L. Mashanovitch, *Diode Lasers and Photonic Integrated Circuits*, John Wiley & Sons, 2012.

[50] G.P. Agrawal, N.K. Dutta, *Semiconductor Lasers*, Springer Science & Business Media, 2013.

[51] M.O. Scully, M.S. Zubairy, *Quantum Optics*, American Association of Physics Teachers, 1999.

[52] W.T. Silfvast, *Laser Fundamentals*, Cambridge University Press, 2004.

[53] P.W. Epperlein, *Semiconductor Laser Engineering, Reliability and Diagnostics: A Practical Approach to High Power and Single Mode Devices*, John Wiley & Sons, 2013.

[54] J.G. Webster, *The Measurement, Instrumentation and Sensors Handbook*, vol. 14. CRC Press, 1998.

[55] M. Razeghi, *Fundamentals of Solid State Engineering*, Springer, 2006.

[56] S.E. Shaheen, C.J. Brabec, N.S. Sariciftci, F. Padinger, T. Fromherz, J.C. Hummelen, 2.5% efficient organic plastic solar cells, *Applied Physics Letters.* 78 (6) (2001) 841–843.

[57] F. Träger, *Springer Handbook of Lasers and Optics*, vol. 2, Springer, 2012.

[58] S.O. Kasap, *Optoelectronics and Photonics: Principles and Practices*, 2nd ed., Pearson, 2012.

[59] A. Ghatak, K. Thyagarajan, *An Introduction to Fiber Optics*, Cambridge University Press, 1998.

[60] J. Amorim, G. Baravian, M. Touzeau, J. Jolly, Two-photon laser induced fluorescence and amplified spontaneous emission atom concentration measurements in O2 and H2 discharges, *Journal of Applied Physics.* 76 (3) (1994) 1487–1493.

[61] R.A. Lewis, A review of terahertz detectors, *Journal of Physics D: Applied Physics.* 52 (43) (2019) 433001.

[62] R.C. Jones, Proposal of the detectivity D** for detectors limited by radiation noise, *Journal of the Optical Society of America.* 50 (11) (1960) 1058–1059.

[63] R.C. Jones, Quantum efficiency of photoconductors, *Proceedings of IRIS.* 2 (1957) 9.

[64] V. Mackowiak, J. Peupelmann, Y. Ma, A. Gorges, *NEP-Noise Equivalent Power*, vol. 56, Thorlabs, Inc, 2015.

[65] P. Wang et al., Sensing infrared photons at room temperature: from bulk materials to atomic layers, *Small.* 15 (46) (2019) 1904396.

[66] A. Rogalski, Infrared detectors: status and trends, *Progress in Quantum Electronics.* 27 (2–3) (2003) 59–210.

[67] J. Piotrowski, W. Galus, M. Grudzien, Near room-temperature IR photo-detectors, *Infrared Physics.* 31 (1) (1991) 1–48.

[68] Y.H. Ghymn, K. Jung, M. Shin, H. Ko, A luminescent down-shifting and moth-eyed anti-reflective film for highly efficient photovoltaic devices, *Nanoscale.* 7 (44) (2015) 18642–18650.

[69] S. Ghosh, A. Bhattacharyya, G. Sen, B. Mukhopadhyay, Optimization of different structural parameters of GeSn/SiGeSn quantum well infrared photodetectors (QWIPs) for low dark current and high responsivity, *Journal of Computational Electronics.* 20 (2021) 1224–1233.

[70] S. Assefa, F. Xia, Y.A. Vlasov, Reinventing germanium avalanche photodetector for nanophotonic on-chip optical interconnects, *Nature.* 464 (7285) (2010) 80–84.
[71] G.M. Hieftje, Signal-to-noise enhancement through instrumental techniques. II. Signal averaging, boxcar integration, and correlation techniques, *Analytical Chemistry.* 44 (7) (1972) 69A–78a.
[72] B. Müller, U. Renz, Development of a fast fiber-optic two-color pyrometer for the temperature measurement of surfaces with varying emissivities, *Review of Scientific Instruments.* 72 (8) (2001) 3366–3374.
[73] S. Huang et al., Black silicon photodetector with excellent comprehensive properties by rapid thermal annealing and hydrogenated surface passivation, *Advanced Optical Materials.* 8 (7) (2020) 1901808.
[74] J. Hecht, *Understanding Fiber Optics*, Jeff Hecht, 2015.
[75] W.J. Smith, *Modern Optical Engineering: The Design of Optical Systems*, McGraw-Hill Education, 2008.
[76] W.T. Welford, R. Winston, *Optics of Nonimaging Concentrators. Light and Solar Energy*, Academic Press, 1978.
[77] A. Ahmad, *Handbook of Optomechanical Engineering*, CRC Press, 2017.
[78] B. Razavi, *Fundamentals of Microelectronics*, John Wiley & Sons, 2021.
[79] How a Solar Cell Works, American Chemical Society, https://www.acs.org/education/resources/highschool/chemmatters/past-issues/archive-2013-2014/how-a-solar-cell-works.html, (accessed 27 April 2023).
[80] Electrical4U, Solar Cell: Construction, Working Principle, and Types of Solar Cell, https://www.electrical4u.com/solar-cell/, (accessed 27 April 2023).
[81] B. Qi, J. Wang, Fill factor in organic solar cells, *Physical Chemistry Chemical Physics.* 15 (2013) 8972–8982.
[82] M.P. Paranthaman, W.-N Winnie, N.B. Raghu, eds., *Semiconductor Materials for Solar Photovoltaic Cells*, vol. 218, Springer International Publishing, 2016.
[83] Thin-Film Solar Cells, https://en.wikipedia.org/wiki/Thin-film_solar_cell, (accessed 27 April 2023).
[84] E. Rahman, N. Alireza, Semiconductor thermionics for next generation solar cells: photon enhanced or pure thermionic?, *Nature Communications.* 12 (1) (2021) 4622.
[85] S. Patil, J. Rushi, P. Nisarg, D. Archan, D. Ishan, B. Kshitij, Flexible solar cells, in: M. I. Ahamed, R. Boddula, M. Rezakazemi, *Fundamentals of Solar Cell Design*, John Wiley & Sons, 2021: pp. 505–536.

2 Overview of Spintronics

2.1 INTRODUCTION TO SPINTRONICS/MAGNETISM

Spintronics, also known as spin electronics, is a field of study that focuses on the active control and manipulation of the intrinsic spin of electrical charge and its associated magnetic moment in solid-state devices. Unlike traditional semiconductor electronics that focus on the charge of the carriers (electrons or holes), spintronics emphasizes the upspin or downspin of the carriers. Spintronics utilizes the electron's spin degree of freedom for information processing, rather than its charge degree of freedom, which is the basis for conventional electronics. In an ideal scenario, spin current would solely flow in the spintronics circuit without any charge current, resulting in no heat being generated or wasted. Recently, there has been significant interest in the concept of spintronics [1]. The manipulation and control of spin-polarized charge carriers, or electron spin, in semiconductors and metals through electrical methods are key aspects of spintronics. Spintronics aims to utilize the spin degree of freedom of electrons for information processing, storage, and transfer in electronic devices. By harnessing the spin of electrons, spintronics offers the potential to create practical semiconductor and metal spintronic devices that can revolutionize information processing and storage technologies. Unlike traditional electronics that rely solely on the charge of electrons, spintronics takes advantage of both the charge and spin properties of electrons to encode and process information. This property has the potential to offer solid-state devices and other spin-dependent devices a diverse range of functionalities. Electron spin is characterized by a magnetic field that has two possible positions, referred to as "up" and "down" (as shown in Figure 2.1). This introduces two additional binary states to the traditional high and low-logic values

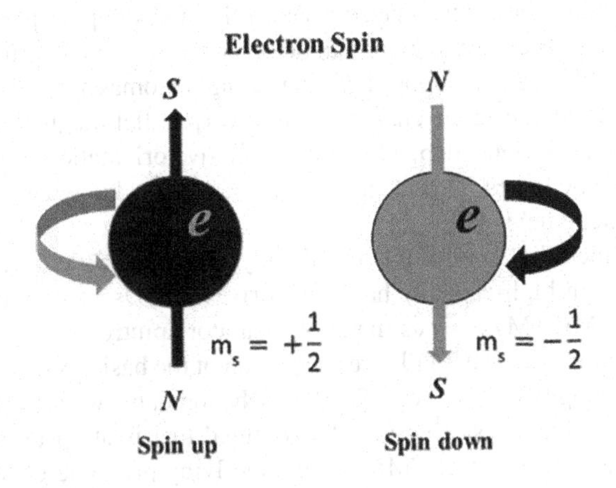

FIGURE 2.1 Spin-up and spin-down electrons.

DOI: 10.1201/9781003331940-2

represented by simple currents. By including the spin state, a bit can have four possible states, named "up-high," "up-low," "down-high," and "down-low." These four states represent quantum bits or qubits [2].

Condensed matter physics investigates the behavior of electrons, which possess two fundamental characteristics: charge and spin. While the subject primarily focuses on the electron's charge, magnetism arises from its spin property. Unbalanced electron spins lead to local magnetic moments in individual atoms. At low temperatures, when neighboring atoms have an exchange coupling, a macroscopic magnetic order can emerge. If the coupling is positive, the magnetic moments align in parallel (ferromagnetism), whereas if it's negative, they align antiparallel (antiferromagnetism). The strength of the coupling affects the critical temperature at which this magnetic order disappears. In ferromagnetic materials (FMs), the Curie temperature (TC) characterizes this critical temperature, while in antiferromagnetic materials, it is known as the Néel temperature (TN). Prior to the discovery of giant magnetoresistance (GMR), the charge and spin attributes of electrons were commonly believed to be independent of each other, and little attention was given to their correlation [3].

Magnetoresistance (MR) refers to the alteration of electric conductivity caused by the presence of a magnetic field, and it is a term commonly used in this context. The characteristics of various MR effects differ based on the material. For instance, metallic, semiconducting, and insulating materials exhibit distinct MR effects. FMs with metallic conductance demonstrate the anisotropic magnetoresistance (AMR) effect, where the conductance depends on the relative angle between the magnetization and electric current. Typically, resistance is lower when the current flows in a direction perpendicular to the magnetization direction rather than parallel. Spin-orbit interactions are considered to be the origin of the AMR effect. Although the MR ratio resulting from AMR is relatively small, such as a few percent for $Ni_{80}Fe_{20}$ alloy (permalloy) at room temperature, this phenomenon is highly beneficial in technical applications, such as in sensors [3].

Spin transport and relaxation in semiconductors and metals are crucial aspects of solid-state physics research and are essential to the advancement of electronic storage technology. Spintronics has played a vital role in developing prototype devices used in the industry, such as the giant magnetoresistive sandwich structure memory-storage cell. This device is composed of alternating ferromagnetic and nonmagnetic metal layers, and the resistance changes from low (parallel magnetizations) to high (antiparallel magnetizations) depending on the relative orientation of the magnetizations in the magnetic layers. This change in resistance, also known as MR, can detect changes in magnetic fields [2].

The field of metallic spintronics has already yielded practical devices, such as GMR read heads in high-capacity hard disk drives (HDDs), and magnetic random access memory (MRAM) devices made of insulator spintronics, specifically magnetic tunnel junctions (MTJs). MTJs are a variation of the basic spin valve that exhibit giant tunneling magnetoresistance (TMR), a phenomenon where electrons tunnel through a thin insulator, resulting in a TMR signal that is almost 100 times larger than that of a GMR spin valve. TMR is the underlying principle of MRAM, which is a type of nonvolatile memory that stores data using magnetic moments rather than electrical charges. The spin of electrons in FMs can be externally modified by applying a magnetic field, and the new spin values are retained even after the field is removed, allowing the encoded information to be securely stored without requiring power and without the risk of demagnetization [2].

However, in the current research, the main focus is on semiconductor spintronics. Unlike charge distribution in conventional electronics, creating an inhomogeneous spin distribution does not require an energy penalty, but spin is not conserved, unlike charge. Hence, the field of semiconductor spintronics focuses on tackling essential challenges related to the controlled manipulation of electron spin at specific positions, the transportation of spins between different locations within a conventional semiconductor setup, achieving all-electrical spin control through spin-orbit interactions, exploring diluted magnetic semiconductors, and utilizing fixed or mobile spin qubits for quantum computing. Spintronics also holds promise for applications such as high-speed magnetic filters, sensors, quantum transistors, and spin qubits for quantum computers [4,5]. In addition, spintronic devices have the potential to revolutionize information technology in the 21st century by enabling quantum computing and quantum communication through electronic solid-state devices. However, before practical spintronic devices can be realized, there is a need for further fundamental research to understand various aspects of spintronics, including spin coherence, spin entanglement, spin dynamics, spin relaxation, spin transport, and so on.

2.2 REQUIRED CHARACTERISTICS OF SPINTRONICS

Fundamental aspects of spintronics encompass the generation of carrier-spin polarization, spin coherence, spin entanglement, control over spin and charge dynamics, spin injection, and spin-polarized transport in electronic materials.

2.2.1 Spin Polarization

Spin polarization refers to the alignment of the intrinsic angular momentum or spin of elementary particles in a specific direction. In the case of FMs, this property is linked to the magnetic moment of conduction electrons, leading to the generation of spin-polarized currents. The degree of spin polarization can be quantified as $PX = X_S/X$, where X represents a particular quantity, and X_S and $X_{-\lambda} + X_{-\lambda}$ denote the spin-resolved λ components of X. To avoid any confusion, it is essential to specify both the spin-resolved components and the relevant physical quantity X. Conventionally, the ↑ or plus sign (numerical value +1) is used to denote spin-up, while the ↓ or minus sign (numerical value −1) represents spin-down, relative to the chosen axis of quantization. In the context of ferromagnetic metals, the ↑ (↓) notation is used to describe carriers with magnetic moments parallel (antiparallel) to the magnetization, or carriers with majority or minority spin. Other particles, such as holes, nuclei, and excitations, can also exhibit spin polarization, and the same principles apply [6]. In the context of semiconductors, the terms "majority" and "minority" typically refer to the relative populations of the carriers. Meanwhile, the quantum numbers m or 1 and k or 2 correspond to the mj quantum numbers with respect to the z-axis, which is aligned with the direction of light propagation or the applied magnetic field [7].

2.2.2 Spin Relaxation

Once spin-polarized carriers are generated, it is crucial to determine their spin orientation memory, or how long they retain their spin state. For electronic applications, it is especially crucial because if the spins relax too quickly, the spin-polarized current will not travel far enough to be practically useful in a device [1]. Spin-spin relaxation refers to the phenomenon where the transverse component of the magnetization vector, M_{xy}, exponentially decays toward its equilibrium value in processes such as nuclear magnetic resonance (NMR) and magnetic resonance imaging. The spin-spin relaxation time, T_2, serves as a time constant that characterizes the decay of the signal. On the other hand, the spin-lattice relaxation time, T_1, represents the duration required for the magnetic resonance signal to irreversibly decay from its initial value (around 37% or $1/e$) after the longitudinal magnetization has been tipped toward the magnetic transverse plane. The decay of the magnetic resonance signal can be described by the relationship below [1,2]:

$$M_{xy}(t) = M_{xy}(0)\, e^{-t/T2} \tag{2.1}$$

The above equation represents T_2 relaxation, which typically occurs more rapidly than T_1 recovery. Different materials exhibit different T_2 values, reflecting variations in their relaxation times. During T_2 relaxation, the excited nuclear spins, which are partially oriented in the transverse plane, interact with each other due to local magnetic field inhomogeneities at micro- and nanoscales. As a result, the accumulated phases of these spins deviate from their expected values. While the slow- or non-varying component of this deviation is reversible, some net signal will inevitably be lost due to short-lived interactions, such as collisions and diffusion, that occur in heterogeneous spaces. It is important to note that T_2 decay does not occur solely due to the tilting of the magnetization vector away from the transverse plane [1].

To the electronic band structure, spin relaxation is highly sensitive. The spins of conduction electrons decay because of the momentum scattering and spin-orbit interaction. Impurity scattering causes spin relaxation at low temperatures ($T \leq 20$ K) and is not dependent on temperature. At higher temperatures, spin coherence of electrons is lost due to collisions with phonons. Phonons can induce a spin flip because electronic Bloch states are not spin Eigen states in the presence of a spin-orbit coupling. The spin relaxation rate $1/T_1$ increases with temperature, with the growth becoming linear above the Debye temperature. This mechanism is the most crucial spin relaxation mechanism in metals and semiconductors with inversion symmetry [8].

Various spin relaxation processes play significant roles in solids, including Elliot-Yafet, D'yakonov-Perel', Bir-Aronov-Pikus, and hyperfine interactions. These mechanisms restore the equilibrium of the nonequilibrium distribution of spins generated by interfaces or spin injections, which can hinder spintronics. Researchers and developers often aim to suppress these effects, although relaxation can be beneficial for the rapid operation of some devices [2].

2.2.3 Spin Injection

To achieve spin injection in nonmagnetic materials (NMMs), a nonequilibrium spin population, known as spin accumulation, is typically created at the interface with a magnetic electrode. The rate of spin injection is affected by the spin relaxation and dephasing mechanisms present in the NM, which work to restore equilibrium in the accumulated spin population. For spintronic devices, it is crucial that nonequilibrium electronic spins in semiconductors and metals have relatively long lifetimes of around 1 ns. However, in confined semiconductor heterostructures, spin lifetimes can increase to hundreds of nanoseconds, allowing for the transport of coherent spin packets over distances of hundreds of micrometers. Suppression of spin relaxation and dephasing mechanisms is often an important area of research and development, but in some cases, relaxation can aid in fast device operation [2].

2.2.3.1 Ohmic Injection

In FMs, a spin-polarized electric current is produced due to the significant difference in the electrical conductivity of majority-spin (spin-up) and minority-spin (spin-down) electrons. Although the formation of an ohmic contact between an FM and a semiconductor is the simplest method for spin injection, it can result in spin-flip scattering and the loss of spin polarization due to heavily doped semiconductor surfaces in typical metal-semiconductor ohmic contacts.

A recent study by Schmidt et al. [9] has highlighted a crucial issue with ohmic spin injection across ideal FM-nonferromagnet (NFM) interfaces, building on previous work in the field [7,10–12]. Efficient spin injection depends on the ratio of the conductivities of the FM and NFM electrodes, σF and σN, respectively. When $\sigma F \leq \sigma N$, as in the case of a typical metal, efficient spin injection can occur, but when the NFM electrode is a semiconductor, $\sigma F \gg \sigma N$, and the spin injection efficiency will be very low. However, a ferromagnet with nearly 100% spin-polarized conduction electrons could enable efficient spin injection in diffusive transport. Many materials have such half-metal-ferromagnetic properties [13,14]. Johnson et al. [15,16] have proposed an approach that may overcome this obstacle to spin injection by utilizing the spin splitting of electrons in a semiconductor 2D quantum well structure due to the spin-orbit effect arising from an asymmetry in the confining potential. The result can be an inducement of a nonequilibrium spin polarization if the 2D electron gas is carrying a current [17]. However, there have been suggestions of an alternative, local-Hall-effect explanation for the data and other questions regarding this approach, as observed in the ohmic contact experiments where the small percentage change in device resistance with changes in ferromagnet orientation has been noted [18–21].

2.2.3.2 Tunnel Injection

Alvarado and Renaud [22] used a ferromagnetic tip scanning tunneling microscope (STM) to demonstrate that vacuum tunneling can effectively inject spins into a semiconductor. A recent study has explored the influence of surface structure on spin-dependent STM tunneling [23]. Additionally, the development of ferromagnetic-insulator-ferromagnetic tunnel junctions with high MR has revealed that tunnel barriers can maintain the spin polarization during tunneling, indicating that tunneling may be a more efficient method of achieving spin injection than diffusive transport. Rashba's theoretical work in 2000 [24] provides insight into the potential effectiveness of tunnel injection. If the impedance of the barrier at an interface is high enough, the (spin-dependent) density of electronic states of the two electrodes involved in the tunneling process determines the transport across the interface. The current through the barrier is then small enough for the electrodes to remain in equilibrium, and the relative (spin-dependent) conductivities of the electrodes have little impact on the spin-dependent transport across the interface. Consequently, both metal-insulator-semiconductor tunnel diodes and metal-semiconductor Schottky barrier diodes that employ a ferromagnetic electrode can be expected to be an effective means of injecting spins into a semiconductor system [2].

2.2.3.3 Ballistic Electron Injection

To achieve spin injection, an alternative method to tunnel injection is to utilize the ballistic regime for spin injection across ferromagnet-semiconductor interfaces. The probability of spin-dependent interfacial ballistic electron transmission is determined by the difference between the two-spin conduction subbands of the ferromagnet and the conduction band of the semiconductor. The interface's transmission and reflection probabilities are commonly assumed to be determined by the conservation of transverse momentum of the incident electron [25,26]. It is essential to have a low probability of an injected spin-polarized electron being elastically scattered back into the ferromagnetic injector. A three-dimensional ballistic point contact between a ferromagnet and a semiconductor should suffice for efficient spin injection. However, if the design requires the spin-dependent capture of an injected carrier by another ferromagnetic electrode, then the transport through the semiconductor region must be entirely ballistic. Recent studies have shown that point contacts between ferromagnetic and non-FMs can result in high spin-polarized currents (up to 40%) being injected into the non-FM [27,28].

2.2.3.4 Hot Electron Injection

A spin technique that is different from tunnel injection is the use of polarized "hot" electron injection. This technique involves injecting electrons into a ferromagnetic layer at energies that are much higher than EF, resulting in highly polarized electron currents [29–31]. The inelastic mean free paths of majority-spin and minority-spin electrons differ significantly, which allows for a ballistic electron current that is over 90% polarized to pass through a 3 nm Co layer. This highly polarized current can then pass through a metal-semiconductor interface and enter the semiconductor material with a high degree of polarization (90%) [31], with the transmission probability being determined by energy and momentum constraints imposed by the band structure difference between the metal and semiconductor at the interface. However, the efficiency of this technique is relatively low. Additionally, if there is substantial spin-flip scattering at the interface, the highly polarized current may be lost. The injection energy can be adjusted by the tunnel injection bias relative to the bottom of the semiconductor conduction band [2].

2.2.4 Spin Transport

The question of whether the quasi-independent electron model is sufficient to explain experimental results in semiconductor systems has been of great interest in spin transport theory. When spin-polarized carriers are present, modified charge transport and intrinsic spin transport arise, which are absent in the unpolarized case. Both of these aspects provide information about the degree of spin polarization, which can be utilized in spintronics. In any material with an imbalance of spin populations at the Fermi level, spin-polarized transport will occur naturally. This imbalance is often observed in FMs, where the density of states for spin-up and spin-down electrons is nearly identical, but the states are shifted in energy relative to each other [2].

The quasi-independent electron model has been of particular interest in the study of spin transport theory in semiconductor systems. It has been investigated whether this model can fully account for experimental results, or whether many-body or correlated electron processes play a role. The presence of spin-polarized carriers gives rise to modified charge transport and intrinsic spin transport, which are absent in the case of unpolarized carriers. The degree of spin polarization can be used in spintronics and naturally occurs in materials with an imbalance of spin populations at the Fermi level, such as FMs. The density of states available to spin-up and spin-down electrons is usually nearly identical, but the states are shifted in energy with respect to each other, causing an unequal filling of the bands and a net magnetic moment. This also leads to unequal numbers, characters, and mobilities of spin-up and spin-down carriers, which can produce a net spin polarization in transport measurements. However, the sign and magnitude of the polarization depend on the specific measurement being made. Ferromagnets can be used to inject spin-polarized carriers into semiconductors, superconductors, or normal metals, or to tunnel through an insulating barrier. In each case, the specific spin-polarized carriers and electronic energy states associated with each material must be identified. The search for 100% spin-polarized conducting materials continues, but materials such as Fe, Co, Ni, and their alloys, which have a polarization of 40%–50% [32], are suitable for developing technologically useful devices. Here, polarization P is defined in terms of the number of carriers with spin-up or spin-down. When dealing with solid-state devices, it is useful to imagine the electron current as 100% spin-polarized due to its effects. In this scenario, the carriers are restricted to states that have spins parallel to the spin direction of the states available at the Fermi level. If the magnetization of the materials is changed, the spin direction of these states will also change [2].

2.2.5 Spin Detection

One possible method for detecting spin populations in semiconductors is to exploit the spin-dependent transport properties of semiconductor ferromagnet interfaces. However, using ohmic contacts for spin collection can be problematic, and more effective spin collection and detection may require a ballistic or tunneling contact from the semiconductor to a ferromagnet. To achieve efficient spin-dependent extraction of the injected spin-polarized current, the tunnel barrier must be sufficiently thin so that spin-dependent tunneling transport into the ferromagnetic electrode is more likely than spin relaxation within the semiconductor [24]. Another spin detection technique is to measure the chemical potential of the nonequilibrium spin populations using a potentiometric measurement with a ferromagnetic electrode [16,18]. For a complete spin transistor device, according to Tang et al. [33], electrical field tunable spin precession will only be detectable if there is ballistic transport throughout the device structure. Additionally, they suggest that a very narrow, single- or few-electron channel device structure will be necessary, in agreement with the initial proposal by Datta et al. [34].

2.2.6 Spin Transfer

The transport of spin-polarized current from a relatively thick ferromagnetic layer through a nonmagnetic layer to a thin-film "free" nanomagnet can generate strong, uniform spin-wave precessional modes in the nanomagnet via spin-dependent scattering of the polarized current [35,36]. Moreover, in the absence of a strong external magnetic field, this process can lead to the reversal of the orientation of the magnetic moment of the free nanomagnet, where the final orientation depends on the direction of the current flow [37]. Known as "spin transfer," this phenomenon presents opportunities for the development of new nanoscale devices for memory and other spin electronics applications [38]. In addition to direct current-addressable magnetic memory, one potential application is the utilization of spin transfer to excite a uniform spin wave in a nanomagnet, which can serve as a precessing spin filter to introduce a coherent spin pulse into a semiconductor structure [2].

2.2.7 Spin Coherence

The technique of using optical pulses to create a superposition of the basis spin states, defined by an applied magnetic field, allows for the observation of electronic spin precession (coherence) in semiconductors, heterostructures, and quantum dots. Through experimentation, it has been found that certain doping concentrations result in significantly enhanced spin lifetimes in semiconductors, with some exceeding 100 ns. Additionally, this phenomenon persists in heterostructures and quantum dots at room temperature, suggesting potential practical applications for coherent quantum magneto-electronics [2].

2.2.8 Spin Accumulation

The injection of spin-polarized current from a ferromagnetic film to a nonmagnetic film at a rate faster than the spin polarization causes the nonequilibrium magnetization to build up in a region of thickness Ls. The nonequilibrium magnetization results in inequivalent chemical potentials for the upspin and downspin subbands of the normal metal. However, the chemical potential of the ferromagnet is held in equilibrium by the intrinsic ferromagnetic nonmagnetic metal interface, which creates an internal electric field associated with the nonequilibrium spin accumulation. Due to the electron carrying both spin and charge, the gradient of spin density creates an electric field that can generate current flow or voltage differences. If the magnetic moments of the two ferromagnetic layers are parallel, a positive current is generated in the detector arm, while antiparallel magnetic moments result in a negative current. The device operates as a nonvolatile computer memory element, storing information through the orientation of the second layer. By modulating the direction of magnetization in the second layer, the current flow through the detector undergoes bipolar modulation [2].

2.2.9 Spin Generation

There are various methods to generate spin-polarized electrons in NM materials, namely, mechanical rotation, electrical generation, spin-orbit effects, electric field application, spin-band splitting, electromagnetic wave application, influence of thermal gradient, and geometrical phese. These methods are described in [39]. The most commonly used method is spin injection from an FM material, which can be conventional FM metals (e.g., Fe, Co, Ni, and Gd), half-metallic ferromagnets (HMF), or dilute magnetic semiconductors (DMS) connected to a NM metal or semiconductor via an ohmic contact or a tunnel barrier. In addition, a stray field at the edge of an FM can be utilized to induce a population difference in spin-polarized electrons in an NM material. Electromagnetic waves, such as circularly polarized light and microwaves, can also excite spin-polarized electrons in semiconductor, based on an optical selection rule. The reverse effect generates circularly polarized light emission through a spin-polarized electron current. Furthermore, spin generation by electromagnetic waves can be extended to include spin pumping and high-frequency spin induction. Furthermore, the spin Seebeck and Nernst effects have been discovered to generate a spin-polarized carrier flow in response to a thermal gradient, offering potential for energy harvesting [40].

In spintronics, the efficiency of spin-current generation is the most important performance metric for device applications [39,41]. The efficiency (η) of generation is typically defined as the spin current produced per unit energy input. For example, in devices where spin current is produced from a charge current via the spin Hall effect, the efficiency of charge-to-spin current conversion is defined as the ratio of the electron-spin current density generated (js) to the electron-charge-current density introduced (jc): $\eta = js/jc$ [42]. The magnitude of js is commonly inferred from a measured voltage and depends on the theoretical model used to interpret it. In some devices with in-plane current flow, such as those used for spin-orbit torque, spin-torque ferromagnetic resonance, and spin Hall measurements, measuring js is challenging, and models like parallel conduction are used to estimate efficiency (η), leading to overestimation. Methods that generate spin current without interfaces have higher efficiency than those that involve interfaces. This makes them more suitable for use in devices [40].

For instance, the efficiency of the interface between a FM and a NM for spin generation via spin-orbit effects and electromagnetic wave applications is generally limited to about 20%, according to references [43,44]. Nonetheless, it has been theoretically predicted that in a MTJ featuring coherent tunneling across a MgO barrier, the efficiency (η) could potentially achieve near-perfect levels of approximately 100%. It is important to note that the article by Tashiro et al. [44] calculated efficiency as a ratio between the absorbed and introduced microwave power, providing an indicative efficiency. The limited efficiency is mainly due to interfacial spin scattering caused by defects and contamination. The use of a highly spin-polarized FM, such as a half-metallic Heusler alloy, has been shown to increase the efficiency to nearly 30% [45]. Furthermore, the effective spin polarization in FM can be increased by scattering asymmetry, for instance, 94% was reported at a Co/Ni interface [46]. Theoretical predictions suggest that an efficiency increase up to 100% can be achieved by using coherent tunneling with a MgO barrier [47,48].

According to reference [49], it has been predicted that under specific conditions, NM could achieve 100% efficiency in generating spin currents. For instance, experiments have shown that a topological insulator can generate a spin current with an efficiency (η) of up to approximately 60% [50], which currently represents the highest reported value. However, this value is still a topic of debate [51]. Furthermore, the theoretical prediction is that a NM material with a significant spin-orbit coupling can generate a mechanically induced spin current with an efficiency of up to 100% [40], which depends on the efficiency of the electrical motor that drives the rotation of an object [52]. Therefore, a detailed discussion of the spin-current generation efficiency of each method is essential, even if some systems may be challenging to implement experimentally, as they may hold the key to the development of future spintronic applications.

2.3 MAGNETISM

Various materials exhibit different types of magnetism, which can be categorized as diamagnetism, paramagnetism, ferromagnetism, antiferromagnetism, and ferrimagnetism. Additionally, there are other forms of magnetism, such as superparamagnetism, metamagnetism, and parasitic magnetism. The main categories are further explained below:

2.3.1 Diamagnetism

The diamagnetic characteristic is associated with atoms that have completely filled electron shells. In diamagnetic materials, the constituent atoms or molecules have all of their electrons paired up in such a way that their magnetic dipole moments cancel out each other. Consequently, there are no dipoles to be aligned by an external magnetic field. In fact, a diamagnetic substance's electrons' orbital motion changes in response to an applied magnetic field, opposing it. This property can be explained as an instance of Lenz's law, which states that a magnetic field applied to a conductor produces an electromotive force that opposes the applied field. Various diamagnetic materials exist, with the strongest being metallic bismuth and organic molecules like benzene. Other examples include metals such as cadmium, copper, silver, tin, and zinc.

2.3.2 Paramagnetism

Paramagnetic substances possess intrinsic permanent magnetic moments, which tend to align themselves parallel to an external magnetic field, thus increasing the lines of force in the direction of the applied field. This results in an overall magnetic moment that adds to the magnetic field. The paramagnetic property is not influenced by the applied field but is temperature-dependent. For ideal paramagnetic materials, the relative permeability $J.lr > I$ and the susceptibility χ are small positive values that vary inversely with absolute temperature. Paramagnetic materials usually contain transition metals or rare earth materials with unpaired electrons. They are governed by Curie's law, which states that the susceptibility χ is proportional to $1/T$, with T being the absolute temperature.

2.3.3 Ferromagnetism

A substance that exhibits ferromagnetism has a magnetic moment even when there is no external magnetic field present. This occurs due to the strong interaction between the magnetic moments of the constituent atoms or electrons within the substance. As a result of this interaction, these magnetic moments align parallel to each other, creating many individual magnetic dipole moments that can interact over long ranges, resulting in areas of magnetism known as domains. In FMs, these domains contain dipoles that are all aligned, and they tend to align with an applied magnetic field. When the domains within an FM are reoriented from a magnetized state back to a demagnetized state, the energy expended causes a lag in response, which is known as hysteresis. However, above the Curie temperature (T_c), this loss becomes complete. Metallic iron has a T_c of approximately 770°C, and ferromagnetism is mostly exhibited by metals and alloys. Examples of FMs include iron, cobalt, nickel, and chromium dioxide.

2.3.3.1 Magnetic Moment

Magnetic moment, or magnetic dipole moment, is a property that measures an object's tendency to align with a magnetic field. It is defined as the strength and orientation of the magnet or object that generates a magnetic field. The magnetic moment can be generated in two ways: through the motion of electric charge and through spin angular momentum. For instance, when an electric current flows through a wire, it creates a magnetic field (H). The current density (J) is equivalent to the curl of the magnetic field ($\nabla \times H$) in the presence of a stable electric field [53]. In addition, introducing a current (I) into a finite loop of area (A) with charge (q) and angular velocity (ω) produces a magnetic moment (m). Electrons in a circular path with charge ($q = -e$), velocity ($v = \omega \times r$), and mass (me), as illustrated in Figure 2.2, also produce an orbital magnetic moment (m_o) and orbital angular momentum (l) [53].

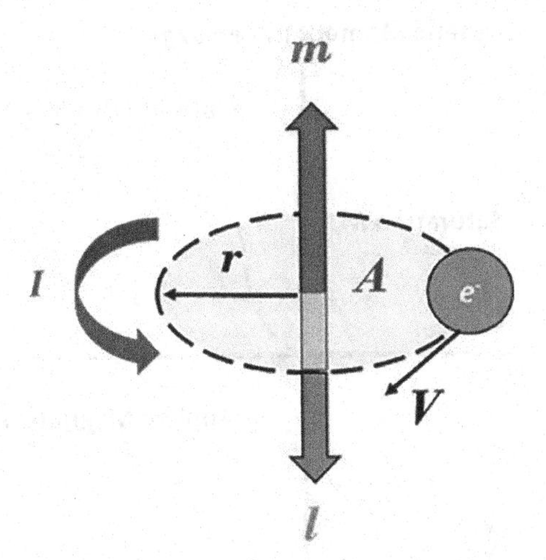

FIGURE 2.2 Production of angular momentum (l) and magnetic moment (m) through current flowing a loop.

2.3.3.2 Magnetic Domains

The concept of a magnetic domain refers to a region within a magnetic material where the magnetization is uniform and aligned in the same direction. This creates a single domain with magnetic moments pointing in the same direction. However, when the material is cooled below the T_c it becomes divided into smaller regions, known as multi-domains. These multi-domains are separated by domain walls, where the magnetization in one domain may be parallel or in a different direction to another domain [53]. The behavior of the material is determined by the structure of the magnetic domain, and the magnetization in each domain can be categorized into different angles, such as single-, 180°, and 90° wall domains [54]. Figure 2.3 displays these categories.

2.3.3.3 Magnetic-Hysteresis Loop

The magnetic-hysteresis (M-H) loop is a graphical representation that displays the magnetization and demagnetization characteristics of a material, as shown in Figure 2.4. The acceleration of domain walls, when a magnetic field is applied, is in opposition to the movement of crystal defects. The purpose of this acceleration is to direct the magnetic

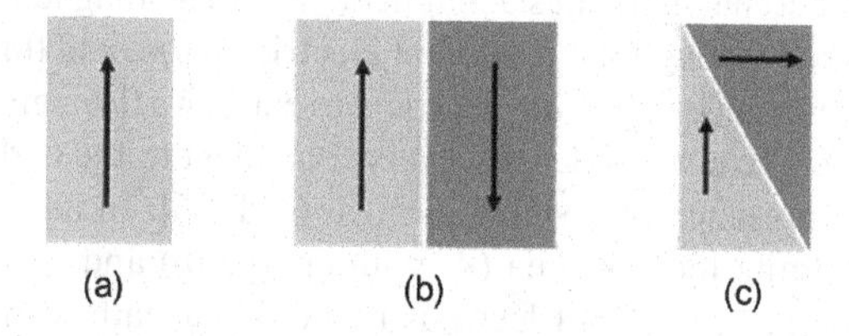

FIGURE 2.3 (a) Single-wall domain. (b) 180° wall domain. (c) 90° wall domain.

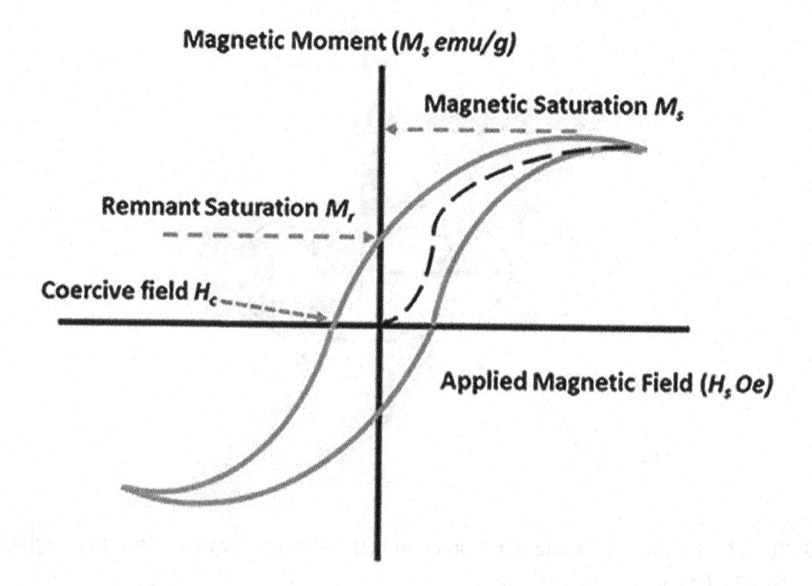

FIGURE 2.4 M-H loop of a magnetic material.

moment toward the crystal axis, which is closely aligned with the field direction, thus forming a single domain. An increase in the magnetic field strength can then cause the single domain to rotate from its easy direction to a parallel direction. Once the magnetization in the material reaches its magnetic saturation point (M_s), the material will retain its magnetization even after the applied field is removed, known as remanent magnetization (M_r). To reduce the retained magnetization to zero, the process must be reversed, requiring the application of a coercive field (H_c). This mechanism is similar to the electrical hysteresis loop [53].

2.3.4 Antiferromagnetism

Antiferromagnetism is a type of ordered magnetism, like ferromagnetism and ferrimagnetism, in which the magnetic moments of atoms or molecules, which are associated with the electron spin, align in a regular pattern with neighboring spins pointing in opposite directions. Lev Landau first introduced this phenomenon in 1933 [55]. Antiferromagnetic order can usually exist at low temperatures but disappears at and above the Néel temperature, which is named after Louis Néel, who first identified this type of magnetic ordering [56]. Typically, the material is paramagnetic above the Néel temperature.

2.3.5 Ferrimagnetism

Ferrimagnets, like ferromagnets, can retain their magnetization even in the absence of an external magnetic field. However, they share a similarity with antiferromagnets in that neighboring pairs of electron spins tend to point in opposite directions. This does not create a contradiction because, in the optimal geometric arrangement, the sublattice of electrons pointing in one direction contributes more magnetic moment than the sublattice pointing in the opposite direction. Most ferrites are ferrimagnetic, and the first magnetic substance ever discovered, magnetite, was initially thought to be a ferromagnet. Louis Néel was the first to discover ferrimagnetism and subsequently disproved magnetite's classification as a ferromagnet.

2.4 SPINTRONIC DEVICE

There have been various proposals and examinations of spin devices for a variety of purposes, including nonvolatile MRAM [4], MTJ-based programmable spintronic logic devices [5,6], rotational speed control systems, and positioning control devices in robotics [8]. The discovery of important phenomena such as GMR [2] and TMR [9,10] has made these advancements possible. While a spin field effect transistor (spin-FET) may not necessarily require less energy to switch compared to a traditional metal-oxide semiconductor field effect transistor (MOSFET) [11], it is thought to be better suited for certain applications due to its oscillatory transfer characteristics, such as single-stage frequency multiplier. Semiconductors utilizing spin-based computing can merge information storage and processing into a single material, which allows for logic in memory computing and the use of spintronics in bio-inspired applications [12].

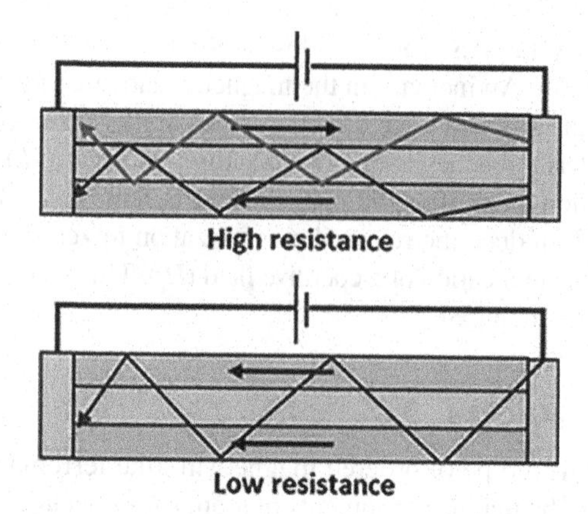

FIGURE 2.5 Schematic of magnetization: high resistance (anti-aligned) and low resistance (aligned).

2.4.1 Giant Magnetoresistive

GMR is a quantum mechanical effect that occurs in multilayer structures made up of alternating conductive layers that are ferromagnetic or nonmagnetic. The discovery of GMR earned Albert Fert and Peter Grünberg the Nobel Prize in Physics in 2007. GMR is characterized by a significant change in electrical resistance when the magnetization of adjacent ferromagnetic layers is either parallel or antiparallel. The resistance is lower when the layers are parallel and higher when they are antiparallel, and an external magnetic field can control the magnetization direction. The effect is due to the scattering of electrons and their spin orientation. In a two-layer system, known as a "spin valve," the magnetic moment of one ferromagnetic layer is difficult to reverse, while the other is easy to reverse. The easily reversible layer acts as the valve control and can be manipulated by an external magnetic field. This device can measure and monitor fields and has many potential applications. Although the orientation shown in Figure 2.5 provides more useful resistance, the physical mechanism behind spin-polarized transport is complex. In anti-aligned films, spin exclusion is observed, leading to high interface scattering and the "channeling" of current into narrowed pathways, as shown in Figure 2.5. Both of these resistance-generating mechanisms are eliminated when the films become aligned, resulting in a decrease in device resistance [2].

2.4.2 Magnetic Tunnel Junctions

Functional devices such as GMR read heads in large-capacity HDDs and MRAM have already been achieved in metallic spintronics, while insulator spintronics rely on MTJs. The basic spin valve has evolved into a related MTJ structure that exhibits giant TMR, where electrons tunnel through a thin insulator. This leads to a TMR signal that is nearly 100 times larger than that from a GMR spin valve. TMR is also the foundation of MRAM, a nonvolatile memory that stores data using magnetic moments rather than electrical charges [2].

2.4.2.1 Fabrication of MTJs

Developing spintronic devices requires the creation of MTJs with high TMR ratios, which can be achieved through various nanotechnology-based deposition methods such as molecular-beam epitaxy, magnetron sputtering, electron-beam evaporation, and CVD. The MTJs are prepared using photolithography, and their main components consist of FM and insulator layers. The FM layers can be produced using sputter deposition techniques such as magnetron sputtering and ion-beam deposition, and the main fabrication issue is ensuring proper magnetic alignment and thickness with deposition rates in the angstrom-per-second range. While the best way to fabricate insulating layers is still under research, Al_2O_3 tunnel barriers have been successfully produced by depositing a metallic aluminum layer ranging from 5 to 15 Å thickness, and alternative methods such as ion-beam oxidation, glow discharge, plasma, atomic-oxygen exposure, and ultraviolet-stimulated oxygen exposure are also being explored. Since Julliere's initial report on TMR in 1975, numerous studies have been conducted on the property, particularly with regards to Al_2O_3 insulating layers. The deposition process must ensure the control of the magnetic properties of the magnetic layers, which requires meeting specific requirements such as maintaining inherent magnetic anisotropy during the deposition process, which can be accomplished through the application of a magnetic field. Key factors for controlling magnetic anisotropy include material thickness and uniformity, coercivity, and magneto-restriction [57].

2.4.2.2 Tunnel Magnetoresistance

The TMR is a phenomenon that occurs in a component called a MTJ, which is composed of two FMs separated by a thin insulating layer. When the insulator is thin enough, electrons can pass through it from one ferromagnet to the other, which is not allowed in classical physics. Thus, TMR is a quantum mechanical effect.

The TMR was first observed by Michel Julliere in 1975 in Fe/Ge-O/Co-junctions at a low temperature, but the effect was not significant [58]. Terunobu Miyazaki in 1991 observed a 2.7% change in resistance at room temperature, and later in 1994, Miyazaki found an 18% change in junctions of iron separated by an amorphous aluminum oxide insulator [59]. 11.8% change in junctions with electrodes of CoFe and Co was also observed by Jagadeesh Moodera [60]. The highest TMR observed with aluminum oxide insulators at that time was around 70% at room temperature.

Since 2000, researchers have been developing tunnel barriers made of crystalline magnesium oxide (MgO). Several researchers [47,48] in 2001 made the theoretical prediction that using iron as the ferromagnet and MgO as the insulator, the TMR could reach several thousand percent. The same year, Bowen et al. [61] reported experiments showing a significant TMR in an MTJ made of Fe/MgO/FeCo(001). Parkin [62] and Yuasa and his team [63] were successful in creating Fe/MgO/Fe junctions in 2004 that exhibited over 200% TMR at room temperature. Ikeda et al. [64] in 2008 observed TMR effects of up to 604% at room temperature and more than 1100% at 4.2 K in junctions of CoFeB/MgO/CoFeB.

2.4.2.3 Spin Transfer Torque

The memory element in modern MRAM technology is made up of a spin-torque switchable nanomagnet, which acts as the "free layer" (FL) and is implemented in a

MTJ. The magnetic moment's orientation of the nanomagnet in relation to the "reference layer" (RL) of the MTJ represents the memory cell's bit-state. The MTJ has a TMR value, which is calculated by expressing the resistance contrast between the FL-RL's parallel (P) or antiparallel aligned (AP) states as a ratio $(R_{AP} - R_P)/R_P \times 100\%$. This resistance signal can be used to read the MTJ's bit-state. In order to record a bit, the MTJ is subjected to a bias voltage exceeding a particular threshold in the intended direction. This results in a substantial flow of tunnel spin-current through the FL, which then changes its orientation to match the spin polarization direction [65].

2.4.2.4 Spin Current

Due to the rapid advancements in the creation of nanoscale MTJs, spin has become a topic of great interest. Spin is a quantum mechanical property that provides an additional degree of freedom for electrons to interact with magnetic fields. Stern and Gerlach [66] in 1922 provided direct experimental evidence for the existence and quantization of electron spin. Michel Julliere [58] later reported the first evidence of spin-dependent tunneling in 1975. Berger [67] in 1978 proposed that spin-polarized currents act on the local magnetization of ferromagnets, leading to GMR. One of the significant properties of spin is its weak interaction with other spins and the environment, which results in a long relaxation time, making it a crucial parameter in the fields of spin transport and quantum computing. Incorporating spins into existing electronics requires addressing issues such as efficient spin injection, transport, control, manipulation, and detection of spin-polarized current. The use of magnetism in spintronics provides an attractive pathway for designing semiconductor spintronic devices, as the spin can be generated and manipulated by an electric field through spin-orbit coupling. Although the spin-orbit interaction on mobile electrons was theoretically proven several decades ago, its practical application is still in its early stages [2].

2.4.3 Spin Field Effect Transistor (FET)

The spin-FET device is a three-terminal structure in which the flow of spin-polarized current between the drain and source terminals is controlled by the gate terminal [34]. Figure 2.6 illustrates the structure of the spin-FET, in which the source and drain terminals are made of FMs, such as iron or cobalt. The ferromagnetic source acts as a spin polarizer, while the ferromagnetic drain acts as a spin filter, only allowing electrons with a single polarization to pass and rejecting all others with different polarizations. The semiconducting channel is sandwiched between the ferromagnetic source and drain. A metallic gate terminal is positioned over the semiconducting channel, separated by an oxide layer, as shown in Figure 2.6. In a one-dimensional channel, the gate voltage produces an effective magnetic field perpendicular to the channel's direction of current flow due to the Rashba effect. Carriers are injected with their polarizations parallel to the channel, and their spins precess around the Rashba field. The gate voltage can control the angle of spin precession by modulating the strength of the Rashba field and the precessional frequency. When the angle of spin precession is an even multiple of π, the transistor conducts current, whereas when it is an odd multiple of π, the current is blocked [68].

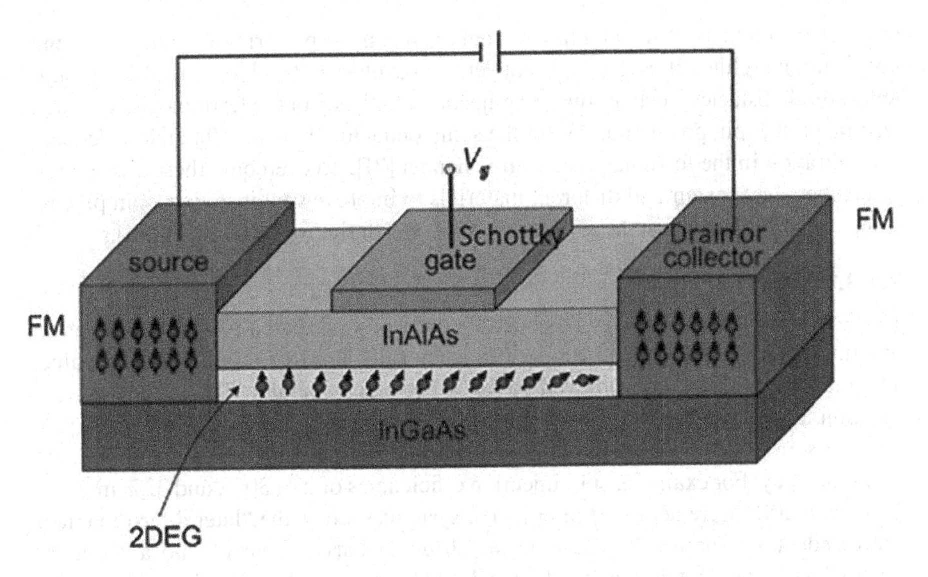

FIGURE 2.6 Datta-Das structure of spin-FET based on InGaAs/InP heterostructure [69].

In the presence of Dresselhaus interaction, an effective magnetic field arises along the channel. However, this field does not affect spin precession, as the spins are already polarized along the channel. Spin-FETs have been developed based on gate modulation of Dresselhaus interaction [70], but they require a different structure from the more common Datta-Das transistor, which relies on the Rashba effect [34]. A spin-FET is a normally ON transistor, and unlike a conventional MOSFET, its source and drain terminals are made up of FMs rather than semiconductors [34]. The source terminal serves as a spin polarizer, polarizing the electrons in a single orientation [71]. The spin-polarized carriers then move through the semiconductor channel with the assumption that their spin polarization does not change, though this is not the case in reality. The carriers are later collected at the drain terminal, which is set to accept the spin-polarized carriers from the source terminal. The drain terminal only accepts electrons with spin polarization in the single direction set by the source terminal and rejects those whose spin polarization changes. Applying a suitable voltage to the gate terminal of the spin-FET [72] induces a magnetic field in the channel through the Rashba effect [73]. This field causes a change in the spin polarization of electrons passing through the channel, resulting in their rejection by the drain terminal. Consequently, the device turns OFF as the electrons cannot flow [74].

The spin-FET is considered a hybrid device as it relies on the movement of electrons through the channel from the source to the drain and does not fully utilize the spin property of electrons for information flow. In 1990, the concept of spin-FET was introduced by Datta-Das [34] and has since been an area of interest for many researchers [75], who have proposed different models and made efforts to enhance its performance and architecture [76]. Several research groups have tested various channel materials, such as InAs, InAlAs, InP, and InGaAs, in spin-FET to improve their performance [69]. Although spintronics and spin-FET have been extensively researched, there is still potential for enhancing their performance and exploring novel circuit architectures that have not yet been explored [77]. The spin-FET encounters several obstacles, such as low spin injection efficiency of the ferromagnetic source terminal [78], spin precession within the semiconductor channel [79], and inadequate spin filtration in the ferromagnetic drain channel [71]. To overcome these challenges, researchers have examined different materials to improve spin injection, spin precession, and spin filtration in the source, channel, and drain, respectively [80,81].

2.4.3.1 Spin Injection

100% spin injection efficiency has been theoretically predicted [47,48]. However, in practical systems, the spin injection efficiency falls short of this theoretical value. The maximum efficiency achieved so far is between 85% and 90% in the "heterojunction ferromagnet/organic semiconductor" system [82]. Looking back at the past few years, practical systems have made progress toward achieving higher spin injection efficiency. For example, spin injection efficiencies of 2%, 30%, and 32% in 2001, 2002, and 2005 were achieved in practical systems such as the "lateral ferromagnet/semiconductor structure," "magnetic metal/tunnel barrier contact into a semiconductor," and "semiconductor employing MgO," respectively [83–85]. Additionally, the "ferromagnet/organic semiconductor heterojunction" system has achieved a high spin injection efficiency of 85%–90% [82]. Based on the progress made in practical systems over the years, it is possible that achieving the ideal value of 100% spin injection efficiency may be attainable within the next 5–8 years. The impact of spin injection efficiency on circuit performance is significant. A lower spin injection efficiency leads to a lower conductance on/off ratio, as seen in the example of a 70% spin injection efficiency system [86], which would have an on/off ratio of conductance of about 3:1. Such a low ratio would make the spin-FET unsuitable for digital logic.

2.4.3.2 Spin Filtering

Theoretical predictions suggest that the spin filtering efficiency (SFE) can be as high as 100% in a "hybrid spin filter" system [87] and 96% in a CrO_2 hydrogenated silicene system [88]. However, in practical systems, the highest SFE value achieved so far is 90% in an Al/EuS/Au junction system back in 1988 [89]. While there was early success in spintronic device development, the goal of reducing power usage in electronic devices has made achieving high SFE values challenging. Over the past couple of decades, there has been steady progress in achieving high SFE values. For example, in 2005–2006, SFE values of 25%, 19%, and 22% were achieved in systems of ferromagnetic $BiMnO_3$ tunnel barrier [90], $La_{2/3}Sr_{1/3}MnO_3/NiFe_2O_4/Au$ [91], and ferrimagnetic-$NiFe_2O_4$ tunnel barrier [92], respectively. Moreover, a $PtCoFe_2O_4$ Al_2O_3/Co MTJ system achieved an SFE value of 25% in 2007 [93]. In 2010, there was a 100% increase from previous SFE values, leading to an SFE value of 44% [94]. Although reaching an ideal SFE value of 100% is still a long way off, the steady progress in this area is promising, and it may be possible to approach the ideal SFE value by the end of the decade. The effect of SFE on circuit performance is significant, as a lower SFE value means that fewer spin-polarized electrons accumulate at the drain side, decreasing the device's current drive. This can result in the device being unable to drive the second stage [68].

2.4.3.3 Rashba Effect

The flow of electrons in the channel is governed by the Rashba effect, but the spin polarization of electrons may change when spin-orbit interactions occur outside of the Rashba effect, leading to electron rejection by the drain terminal. This will result in a decrease in current drive, and if the interactions are as strong as the Rashba effect, it may switch the spin-FET from ON to OFF state, ultimately disrupting the predicted circuit functioning [68].

2.4.3.4 Dresselhaus Effect

When the Dresselhaus interaction is present, an effective magnetic field is created along the channel. This field does not affect spin precession since the spins are polarized along the channel, assuming the Rashba and Dresselhaus fields are balanced. However, if these fields are unbalanced, the Dresselhaus field can be utilized to switch the spin-FET from ON to OFF or vice versa. Some spin-FETs operate by modulating the Dresselhaus interaction with the gate [70], but using the Dresselhaus effect to modulate the spin-FET requires a distinct structure from the commonly used Datta-Das transistor, which relies on the Rashba effect. The device's performance is crucial as it serves as the fundamental building block of a circuit, and any decline in its performance will inevitably impact the circuit's overall performance [68].

2.5 APPLICATIONS OF SPINTRONICS

2.5.1 HDD Read Head

In 1956, IBM introduced the HDD, which had the capability to store 4.4 MB of data on 50 discs with a density of kbit in^{-2}. IBM then introduced the GMR spin valve HDD read head in 1997, while Seagate introduced the TMR head in 2005. The MgO-based TMR technology is still used in modern HDDs, which have an areal density of over 600 Gbit in^{-2} [95]. However, as the areal density of storage keeps increasing, the resistance-area (RA) product of the MTJ tunnel barrier needs to decrease. This is because the size of the MTJ reader reduces, which requires the RA product to decrease to maintain the sensor impedance at a few tens of Ohms to match the pre-amp impedance. However, for HDD applications above 2 Tbit in^{-2}, using MgO-based MTJ becomes increasingly challenging since the barrier thickness becomes ultra-low (below 1 nm). This not only affects its reliability but also reduces its TMR [96]. Toshiba has suggested utilizing spin valves that include a nano-oxide spacer layer (NOL) composed of an AlOx oxide layer with metallic Cu pinholes. These pinholes function to constrain the current path perpendicularly to the GMR stack in a localized manner. To create this NOL layer, an $Al_{1-x}Cu_x$ alloy spacer layer [with $x=(2–3)\%$] is oxidized [97].

The junction comprising $Co_{0.5}Fe_{0.5}$ (2.5)/NOL/$Co_{0.5}Fe_{0.5}$ (2.5) (with thickness in nm) has shown RA of (0.5–1.5) Ω·μm and MR of (7–10)% at RT. These values do not meet the requirements for the 2 Tbit in^{-2} HDD, and the reliability of such devices is questionable due to the high current density passing through the pinholes. Therefore, there is a pressing need for further enhancements in these junctions. Recently, there is a significant amount of research being conducted on spin valves that are based on

Heusler alloys with Ag spacer layer [98], which could constitute an alternative. The growth of Heusler alloys with Ag spacer layer-based spin valves typically requires high-temperature deposition or annealing, which is not compatible with the head fabrication process. In addition, the total thickness of the reader in HDD must be minimized to below 20 nm, as it determines the reader shield-to-shield spacing and linear downtrack resolution. This creates an additional constraint in the design of the reader MR stack.

To achieve a higher areal density in HDDs, there is a trilemma that must be resolved between grain size, writability, and thermal stability [99]. One way to enhance writability is to facilitate the magnetization switching during the writing process by providing extra energy to the local magnetization through microwave or heat transfer. A technique called heat-assisted magnetic recording (HAMR) has been suggested for this purpose [100,101]. A plasmonic antenna is used to transfer heat to the media using a laser beam. This creates a thermal gradient of up to 10 K nm^{-1}, which heats up the media locally and reduces the thermal stability of the data bit to be written. This makes it easier to switch the magnetization. There have been successful demonstrations of this technique [102]. TDK and Seagate both demonstrated new HAMR drives in 2012, with TDK achieving an areal density of 1.5 Tbit in^{-2} and a bit-error rate of 10^{-2}. However, a significant challenge was the reliability of the head, particularly the plasmonic antenna, which reaches high temperatures. Despite this, both companies were able to demonstrate head lifetimes exceeding 1,000 hours [103]. Zhu et al. proposed microwave-assisted magnetic recording (MAMR) as another approach for energy-assisted recording [104]. The MAMR technique utilizes microwaves generated by a spin transfer oscillator, which is patterned in the write gap of the write head. This helps to decrease the switching field by a factor of 10 [105].

2.5.2 Magnetic Sensors

Magnetic sensors are utilized for detecting various parameters such as position, angle, rotation, and magnetic fields. There are three main types of magnetic sensors based on Hall, AMR, and GMR effects. Silicon is patterned into a crossbar to fabricate Hall sensors, and replacing Si with compound semiconductors such as InAs and GaAs enhances their sensitivity. However, these sensors have certain limitations, such as the large temperature dependence of their output in a finite magnetic field, their operational temperature range being between 230 and 390 K, and their detectable field range being between 10^2 and 10^3 kOe [40].

Magnetic sensors with high sensitivity are essential for magnetoencephalography, a technique capable of detecting small magnetic fields with a spatial resolution of approximately 5 mm and a temporal resolution of approximately 1 ms [106]. The electric dipole moment generated by activated synapse voltages creates a magnetic field that can be detected as a voltage or brain wave or as a magnetic field induced by current resulting from the synapse voltage, known as magnetoencephalography. Brain wave measurements cannot precisely locate the activated synapse in the brain due to the high electrical resistance of the scalp and crania and the low resistance of cerebrospinal fluid. However, a magnetic field can penetrate these elements, allowing for accurate determination of the activated synapse with magnetoencephalography. The orientation of the electric dipole moments in the cerebral cortex can vary depending on its shape, making it difficult for magnetic sensors to accurately detect them. A magnetic sensor used in magnetoencephalography can detect the dipole that generates a magnetic field perpendicular to the scalp with high sensitivity but may struggle to detect the dipole that generates a field parallel to the scalp. To achieve $fT/Hz^{1/2}$ resolution, both superconducting quantum interference devices (SQUIDs) and optically pumped atomic magnetometers are currently being utilized. Cryogenic temperatures are required for the operation of SQUID, while the atomic magnetometer is not suitable for miniaturization as it needs a minimum size of around 1 cm^3 to maintain its sensitivity [40]. As a result, it is crucial to develop new magnetic sensors that offer higher spatial and time resolution. Such sensors can be arranged in an array form with different tilted angles to overcome the insensitivity issue. Recent studies have shown that a MTJ, which consists of Si/SiO_2//Ta (5)/Ru (10)/$Ni_{0.8}Fe_{0.2}$ (70)/Ru (0.9)/$Co_{0.4}Fe_{0.4}B_{0.2}$ (3)/MgO (1.6)$Co_{0.4}Fe_{0.4}B_{0.2}$ (3)/Ru (0.9)/$Co_{0.75}Fe_{0.25}$ (5)/$Ir_{0.22}Mn_{0.78}$ (10)/Ta (5)/Ru (30) (thickness in nm), has a noise density of 14 $pT/Hz^{1/2}$ at a single frequency of 10 Hz. Developing such sensors with higher resolution can be beneficial for improving the insensitivity issue [40]. More sensitivity enhancement is required to expand the range of applications.

2.5.3 Magnetic Random Access Memories

MRAM is a type of memory that does not require power to keep data stored and is designed to be read without causing damage to the data. It works by using magnetic anisotropy energy to preserve information and the MR principle to retrieve it [107]. A standard MRAM cell comprises a magnetoresistive element and a transistor, which is similar to a DRAM (dynamic RAM) cell that has a transistor and a capacitor. In a DRAM, the memory state is defined by the charge held in the capacitor, while in MRAM, the resistance of the magnetoresistive element determines the state of 1 or 0. However, since the voltages of two states have a low absolute difference, a transistor is essential for each MRAM cell. Additionally, the transistor supplies the current needed for the write operation. A memory device must meet three main criteria: (i) the device should be capable of storing information, and if it can do so for an extended period of time without a power source, it is considered a nonvolatile memory device; (ii) there must be mechanisms to read information from the device; and (iii) there must be mechanisms to write information onto the device. Researchers have explored and designed various MRAM schemes to fulfill these requirements. In MRAM, the following functions are performed: (i) The resistance difference between two states of a magnetoresistive device is detected for the read operation. (ii) Magnetic retention properties, resulting from the magnetic anisotropy of the storage layer, are responsible for information storage. (iii) Changing the orientation of the magnetization of the storage layer, which can be achieved by inducing a magnetic field, is performed for the write operation [108].

To store information in MRAM, the spin valve structure was the first storage element that was utilized. This structure is mainly composed of two ferromagnetic layers separated by a nonmagnetic conductive layer. The ferromagnetic layers are referred to as the free/soft layer and the hard/pinned layer, respectively. An AFM layer is used in proximity to or in contact with the pinned layer to ensure that the magnetization direction of the layer remains fixed during the memory device's operation. The fundamental concept behind MRAM storage is the energy barrier (EB) that must be surpassed to flip the magnetization of a single-domain magnet from one orientation to another. If the EB is high enough to overcome the external stray fields and thermally assisted magnetization reversal, the magnetization will remain fixed in a particular orientation. Although the method of material design and information recording is different, this storage principle is comparable to that utilized in magnetic recording [108]. A MRAM stores "0" and "1" states by varying the magnetization direction of the FL, while the magnetization direction of the reference layer or pinning layers (PL) remains fixed. The reference layer is made of materials that have a high EB to ensure its direction is never changed. On the other hand, the FL is designed with materials that have a magnetic anisotropy just high enough to store the magnetization for several years (usually 10 years in magnetic recording). To store information in MRAM, the necessary EB is usually directly proportional to the product of the magnetic anisotropy constant Ku and the volume of the FL, denoted as *V*. To ensure effective storage for a period exceeding 10 years, the required energy level must exceed the thermal energy kBT by a significant margin (60 times in this case). The thermal stability factor D (5EB/kBT) is used to express the relationship between the EB and the thermal energy kBT, with the possibility of the EB differing from KuV in some cases. Although high anisotropy is preferable for information storage, it is necessary to maintain a balance, as excessively high anisotropy would limit the ability to orient the magnetization direction for writing 0 and 1 states. There are two types of MTJs that differ based on their storage mechanisms. The first type is the in-plane MTJ, where the magnetization of the ferromagnetic layers is situated in the film plane. The second type is the perpendicular MTJ, where the magnetization is oriented perpendicularly to the film plane [108].

2.6 BENEFITS AND CHALLENGES

Reasons for the development of spintronics include the following factors: the limitations of Moore's law, reduced power consumption, lower electric current requirements, enhanced device speed, increased storage capacity, smaller device sizes, decreased heat dissipation, nonvolatile spintronic memory, and faster spin manipulation for improved reading and writing speeds. However, spintronics encounters various challenges that need to be addressed, such as spin generation and injection, long-distance spin transport, and manipulation and detection of spin orientation. To overcome these hurdles, researchers have proposed new concepts and materials for spintronics, including half metals, spin-gapless semiconductors, and bipolar magnetic semiconductors. Additionally, topological insulators can be considered as a special category of spintronic materials, utilizing their surface states for pure spin generation and transportation [2]. The design of these spintronic materials heavily relies on the utilization of first-principles calculations, which play a crucial role in the development process.

2.7 FUTURE WORKS IN SPINTRONICS

As the field of spintronics continues to grow, there are several potential future works that could significantly advance the technology.

2.7.1 Developing New Materials

Spintronics requires materials that can conduct electrons with a high spin polarization, which means that the electrons' spin states remain aligned as they move through the material. Researchers are exploring new materials that have better spin transport properties, such as topological insulators and two-dimensional materials like graphene.

2.7.2 Creating New Devices

There is significant potential for creating new spintronic devices, such as spin valves, spin transistors, and spin filters. These devices could be used for data storage, logic operations, and communication applications. Researchers are working on developing more efficient and versatile devices that can be integrated into existing electronic systems.

2.7.3 Improving Spin Manipulation Techniques

In spintronics, it is essential to manipulate the spin of electrons in a controlled manner. Researchers are exploring new techniques for controlling and manipulating the spin of electrons, such as spin-orbit torque, spin Hall effect, and spin pumping.

2.7.4 Enhancing Spin Detection Methods

Detecting the spin of electrons is crucial for spintronics. Researchers are working on developing more sensitive and efficient spin detection techniques, such as spin-resolved photoemission spectroscopy and spin-dependent tunneling spectroscopy.

2.7.5 Developing Spin-Based Quantum Computing

Spintronics has significant potential for quantum computing applications, which rely on the manipulation and control of individual quantum states. Researchers are exploring the use of spin qubits for quantum computing, which could provide a more robust and scalable approach than existing qubit technologies.

2.7.6 Integrating Spintronics with Other Technologies

Spintronics could be integrated with other technologies, such as photonics and plasmonics, to create new types of hybrid devices. These hybrid devices could offer enhanced performance and functionality compared to standalone spintronic or photonic devices.

2.8 SUMMARY

Spintronics, also referred to as spin electronics, is an expanding field of study that focuses on utilizing the intrinsic spin of electrons as a means to convey and process information. Unlike conventional electronics, which rely solely on the electrical charge of electrons, spintronics employs the spin of electrons to encode data. This innovative approach has the potential to bring about significant advancements in various technological domains, including data storage, computing, and sensing. One of the primary advantages of spintronics lies in its potential for reduced power consumption compared to traditional electronics, making it more energy efficient. Furthermore, spintronics holds the promise of enabling faster data transfer and processing speeds, opening new possibilities for enhanced performance. Additionally, spintronics research may lead to the development of novel sensor technologies. Among the most promising applications of spintronics is its utilization in MRAM, which has the capacity to supplant conventional computer memory systems. MRAM possesses nonvolatile properties, allowing it to retain information even in the absence of power while also exhibiting faster access times in comparison to other nonvolatile memory alternatives. Moreover, spintronics is currently being explored for its potential in quantum computing, where it could play a crucial role in controlling the spin of qubits. This area of research holds immense promise for revolutionizing computing capabilities and expanding the horizons of information processing.

REFERENCES

[1] G.A. Prinz, Spin-polarized transport, *Physics Today*. 48 (1995) 58–63.
[2] S.C. Ray, *Magnetism and Spintronics in Carbon and Carbon Nanostructured Materials*, Elsevier, 2020.
[3] T. Shinjo, *Nanomagnetism and Spintronics*, Elsevier, 2013.
[4] G. Burkard, D. Loss, D.P. DiVincenzo, Coupled quantum dots as quantum gates, *Physical Review B*. 59 (1999) 2070.
[5] D. Loss, D.P. DiVincenzo, Quantum computation with quantum dots, *Physical Review A*. 57 (1998) 120.
[6] P.M. Tedrow, R. Meservey, Spin polarization of electrons tunneling from films of Fe, Co, Ni, and Gd, *Physical Review B*. 7 (1973) 318.
[7] B.T. Jonker, S.C. Erwin, A. Petrou, A.G. Petukhov, Electrical spin injection and transport in semiconductor spintronic devices, *MRS Bulletin*. 28 (2003) 740–748.
[8] R.J. Elliott, Theory of the effect of spin-orbit coupling on magnetic resonance in some semiconductors, *Physical Review*. 96 (1954) 266.
[9] G. Schmidt, D. Ferrand, L.W. Molenkamp, A.T. Filip, B.J. Van Wees, Fundamental obstacle for electrical spin injection from a ferromagnetic metal into a diffusive semiconductor, *Physical Review B*. 62 (2000) R4790.
[10] M. Johnson, R.H. Silsbee, Thermodynamic analysis of interfacial transport and of the thermomagnetoelectric system, *Physical Review B*. 35 (1987) 4959.
[11] P.C. Van Son, H. Van Kempen, P. Wyder, Boundary resistance of the ferromagnetic-nonferromagnetic metal interface, *Physical Review Letters*. 58 (1987) 2271.
[12] S. Hershfield, H.L. Zhao, Charge and spin transport through a metallic ferromagnetic-paramagnetic-ferromagnetic junction, *Physical Review B*. 56 (1997) 3296.
[13] R.A. De Groot, F.M. Mueller, P.G. Van Engen, K.H.J. Buschow, New class of materials: half-metallic ferromagnets, *Physical Review Letters*. 50 (1983) 2024.

[14] K.P. Kämper, W. Schmitt, G. Güntherodt, R.J. Gambino, R. Ruf, Cr O 2-a new half-metallic ferromagnet?, *Physical Review Letters.* 59 (1987) 2788.
[15] M. Johnson, Theory of spin-dependent transport in ferromagnet-semiconductor heterostructures, *Physical Review B.* 58 (1998) 9635.
[16] M. Johnson, Spin injection and detection in a ferromagnetic metal/2DEG structure, *Physica E: Low-Dimensional Systems and Nanostructures.* 10 (2001) 472–477.
[17] L.E. Vorob'ev, E.L. Ivchenko, G.E. Pikus, I.I. Farbshteĭn, V.A. Shalygin, A. V Shturbin, Optical activity in tellurium induced by a current, *Soviet Journal of Experimental and Theoretical Physics Letters.* 29 (1979) 441.
[18] P.R. Hammar, M. Johnson, Potentiometric measurements of the spin-split subbands in a two-dimensional electron gas, *Physical Review B.* 61 (2000) 7207.
[19] P.R. Hammar, B.R. Bennett, M.J. Yang, M. Johnson, Observation of spin injection at a ferromagnet-semiconductor interface, *Physical Review Letters.* 84 (2000) 5024.
[20] F.G. Monzon, H.X. Tang, M.L. Roukes, Magnetoelectronic phenomena at a ferromagnet-semiconductor interface, *Physical Review Letters.* 84 (2000) 5022.
[21] B.J. Van Wees, Comment on "Observation of spin injection at a ferromagnet-semiconductor interface," *Physical Review Letters.* 84 (2000) 5023.
[22] S.F. Alvarado, P. Renaud, Observation of spin-polarized-electron tunneling from a ferromagnet into GaAs, *Physical Review Letters.* 68 (1992) 1387.
[23] V.P. LaBella, D.W. Bullock, Z. Ding, C. Emery, A. Venkatesan, W.F. Oliver, G.J. Salamo, P.M. Thibado, M. Mortazavi, Spatially resolved spin-injection probability for gallium arsenide, *Science.* 292 (2001) 1518–1521.
[24] E.I. Rashba, Theory of electrical spin injection: tunnel contacts as a solution of the conductivity mismatch problem, *Physical Review B.* 62 (2000) R16267.
[25] G. Kirczenow, Ideal spin filters: a theoretical study of electron transmission through ordered and disordered interfaces between ferromagnetic metals and semiconductors, *Physical Review B.* 63 (2001) 54422.
[26] D. Grundler, Ballistic spin-filter transistor, *Physical Review B.* 63 (2001) 161307.
[27] S.K. Upadhyay, A. Palanisami, R.N. Louie, R.A. Buhrman, Probing ferromagnets with Andreev reflection, *Physical Review Letters.* 81 (1998) 3247.
[28] S.K. Upadhyay, R.N. Louie, R.A. Buhrman, Spin filtering by ultrathin ferromagnetic films, *Applied Physics Letters.* 74 (1999) 3881–3883.
[29] D.J. Monsma, J.C. Lodder, T.J.A. Popma, B. Dieny, Perpendicular hot electron spin-valve effect in a new magnetic field sensor: the spin-valve transistor, *Physical Review Letters.* 74 (1995) 5260.
[30] R. Jansen, O.M.J. Van't Erve, S.D. Kim, R. Vlutters, P.S. Anil Kumar, J.C. Lodder, The spin-valve transistor: fabrication, characterization, and physics, *Journal of Applied Physics.* 89 (2001) 7431–7436.
[31] W.H. Rippard, R.A. Buhrman, Spin-dependent hot electron transport in Co/Cu thin films, *Physical Review Letters.* 84 (2000) 971.
[32] R.J. Soulen Jr, J.M. Byers, M.S. Osofsky, B. Nadgorny, T. Ambrose, S.F. Cheng, P.R. Broussard, C.T. Tanaka, J. Nowak, J.S. Moodera, Measuring the spin polarization of a metal with a superconducting point contact, *Science.* 282 (1998) 85–88.
[33] H.X. Tang, F.G. Monzon, R. Lifshitz, M.C. Cross, M.L. Roukes, Ballistic spin transport in a two-dimensional electron gas, *Physical Review B.* 61 (2000) 4437.
[34] S. Datta, B. Das, Electronic analog of the electro-optic modulator, *Applied Physics Letters.* 56 (1990) 665–667.
[35] L. Berger, Emission of spin waves by a magnetic multilayer traversed by a current, *Physical Review B.* 54 (1996) 9353.
[36] J.C. Slonczewski, Excitation of spin waves by an electric current, *Journal of Magnetism and Magnetic Materials.* 195 (1999) L261–L268.

[37] J.A. Katine, F.J. Albert, R.A. Buhrman, E.B. Myers, D.C. Ralph, Current-driven magnetization reversal and spin-wave excitations in Co/Cu/Co pillars, *Physical Review Letters.* 84 (2000) 3149.
[38] W. Weber, S. Riesen, H.C. Siegmann, Magnetization precession by hot spin injection, *Science.* 291 (2001) 1015–1018.
[39] A. Hirohata, K. Takanashi, Future perspectives for spintronic devices, *Journal of Physics D: Applied Physics.* 47 (2014) 193001.
[40] A. Hirohata, K. Yamada, Y. Nakatani, I.-L. Prejbeanu, B. Diény, P. Pirro, B. Hillebrands, Review on spintronics: principles and device applications, *Journal of Magnetism and Magnetic Materials.* 509 (2020) 166711.
[41] A. Hirohata, *Book Review: Spin Current*, Frontiers in Physics, 2018.
[42] E. Lesne, Y. Fu, S. Oyarzun, J.C. Rojas-Sánchez, D.C. Vaz, H. Naganuma, G. Sicoli, J.-P. Attané, M. Jamet, E. Jacquet, Highly efficient and tunable spin-to-charge conversion through Rashba coupling at oxide interfaces, *Nature Materials.* 15 (2016) 1261–1266.
[43] M.-H. Nguyen, M. Zhao, D.C. Ralph, R.A. Buhrman, Enhanced spin Hall torque efficiency in Pt100− x Al x and Pt100− x Hf x alloys arising from the intrinsic spin Hall effect, *Applied Physics Letters.* 108 (2016) 242407.
[44] T. Tashiro, S. Matsuura, A. Nomura, S. Watanabe, K. Kang, H. Sirringhaus, K. Ando, Spin-current emission governed by nonlinear spin dynamics, *Scientific Reports.* 5 (2015) 1–9.
[45] T. Kimura, N. Hashimoto, S. Yamada, M. Miyao, K. Hamaya, Room-temperature generation of giant pure spin currents using epitaxial Co2FeSi spin injectors, *NPG Asia Materials.* 4 (2012) e9.
[46] J. Bass, CPP magnetoresistance of magnetic multilayers: a critical review, *Journal of Magnetism and Magnetic Materials.* 408 (2016) 244–320.
[47] W.H. Butler, X.-G. Zhang, T.C. Schulthess, J.M. MacLaren, Spin-dependent tunneling conductance of Fel MgOl Fe sandwiches, *Physical Review B.* 63 (2001) 54416.
[48] J. Mathon, A. Umerski, Theory of tunneling magnetoresistance of an epitaxial Fe/MgO/Fe (001) junction, *Physical Review B.* 63 (2001) 220403.
[49] A. Kirihara, K. Kondo, M. Ishida, K. Ihara, Y. Iwasaki, H. Someya, A. Matsuba, K. Uchida, E. Saitoh, N. Yamamoto, Flexible heat-flow sensing sheets based on the longitudinal spin Seebeck effect using one-dimensional spin-current conducting films, *Scientific Reports.* 6 (2016) 1–7.
[50] K. Kondou, R. Yoshimi, A. Tsukazaki, Y. Fukuma, J. Matsuno, K.S. Takahashi, M. Kawasaki, Y. Tokura, Y. Otani, Fermi-level-dependent charge-to-spin current conversion by Dirac surface states of topological insulators, *Nature Physics.* 12 (2016) 1027–1031.
[51] D. Loss, P.M. Goldbart, Persistent currents from Berry's phase in mesoscopic systems, *Physical Review B.* 45 (1992) 13544.
[52] K. Yasuda, A. Tsukazaki, R. Yoshimi, K. Kondou, K.S. Takahashi, Y. Otani, M. Kawasaki, Y. Tokura, Current-nonlinear Hall effect and spin-orbit torque magnetization switching in a magnetic topological insulator, *Physical Review Letters.* 119 (2017) 137204.
[53] A. Ahmed, Synthesis and characterization of magnetic carbon nanotubes, Doctoral Dissertation, McMaster University, 2017, pp. 1–181.
[54] J. A. Oke, Functionalized multiwall carbon nanotubes for electronic and magnetic applications, Doctoral Dissertation, University of Johannesburg, 2020.
[55] L.D. Landau, A possible explanation of the field dependence of the susceptibility at low temperatures, *Physikalische Zeitschrift der Sowjetunion.* 4 (1933) 675.
[56] L. Néel, Propriétés magnétiques des ferrites; ferrimagnétisme et antiferromagnétisme, *Annales de Physique*, 12 (1948) 137–198.

[57] E. Titus, R. Krishna, J. Grácio, M. Singh, A.L. Ferreira, R.G. Dias, Carbon nanotube based magnetic tunnel junctions (MTJs) for spintronics application, in: *Electronic Properties of Carbon Nanotubes*, IntechOpen, 2011.

[58] M. Julliere, Tunneling between ferromagnetic films, *Physics Letters A*. 54 (1975) 225–226.

[59] T. Miyazaki, N. Tezuka, Giant magnetic tunneling effect in Fe/Al2O3/Fe junction, *Journal of Magnetism and Magnetic Materials*. 139 (1995) L231–L234.

[60] J.S. Moodera, L.R. Kinder, T.M. Wong, R. Meservey, Large magnetoresistance at room temperature in ferromagnetic thin film tunnel junctions, *Physical Review Letters*. 74 (1995) 3273.

[61] M. Bowen, V. Cros, F. Petroff, A. Fert, C. Martınez Boubeta, J.L. Costa-Krämer, J.V. Anguita, A. Cebollada, F. Briones, J.M. De Teresa, Large magnetoresistance in Fe/MgO/FeCo (001) epitaxial tunnel junctions on GaAs (001), *Applied Physics Letters*. 79 (2001) 1655–1657.

[62] S.S.P. Parkin, C. Kaiser, A. Panchula, P.M. Rice, B. Hughes, M. Samant, S.-H. Yang, Giant tunnelling magnetoresistance at room temperature with MgO (100) tunnel barriers, *Nature Materials*. 3 (2004) 862–867.

[63] S. Yuasa, T. Nagahama, A. Fukushima, Y. Suzuki, K. Ando, Giant room-temperature magnetoresistance in single-crystal Fe/MgO/Fe magnetic tunnel junctions, *Nature Materials*. 3 (2004) 868–871.

[64] S. Ikeda, J. Hayakawa, Y. Ashizawa, Y.M. Lee, K. Miura, H. Hasegawa, M. Tsunoda, F. Matsukura, H. Ohno, Tunnel magnetoresistance of 604% at 300 K by suppression of Ta diffusion in Co Fe B/ Mg O/ Co Fe B pseudo-spin-valves annealed at high temperature, *Applied Physics Letters*. 93 (2008) 82508.

[65] J.Z. Sun, Spin-transfer torque switched magnetic tunnel junction for memory technologies, *Journal of Magnetism and Magnetic Materials*. 559 (2022) 169479.

[66] S. Pakvasa, The Stern-Gerlach experiment and the electron spin, ArXiv Preprint ArXiv:180509412 (2018).

[67] L. Berger, Low-field magnetoresistance and domain drag in ferromagnets, *Journal of Applied Physics*. 49 (1978) 2156–2161.

[68] G.F.A. Malik, M.A. Kharadi, F.A. Khanday, N. Parveen, Spin field effect transistors and their applications: a survey, *Microelectronics Journal*. 106 (2020) 104924.

[69] R.M.S. De Andrade, *Spin-FET Based on InGaAs/InP Heterostructure*, Master Dissertation, Lunds Universitet, 2011.

[70] S. Pramanik, C.G. Stefanita, S. Patibandla, S. Bandyopadhyay, K. Garre, N. Harth, M. Cahay, Observation of extremely long spin relaxation times in an organic nanowire spin valve, *Nature Nanotechnology*. 2 (2007) 216–219. https://doi.org/10.1038/nnano.2007.64.

[71] Y. Sato, S. Gozu, T. Kita, S. Yamada, Study for realization of spin-polarized field effect transistor in In0. 75Ga0. 25As/In0. 75Al0. 25As heterostructure, *Physica E: Low-Dimensional Systems and Nanostructures*. 12 (2002) 399–402.

[72] S. Meena, S. Choudhary, Enhancing TMR and spin-filtration by using out-of-plane graphene insulating barrier in MTJs, *Physical Chemistry Chemical Physics*. 19 (2017) 17765–17772.

[73] J. Nitta, T. Bergsten, Electrical manipulation of spin precession in an InGaAs-based 2DEG due to the Rashba spin-orbit interaction, *IEEE Transactions on Electron Devices*. 54 (2007) 955–960.

[74] L.-T. Chang, I.A. Fischer, J. Tang, C.-Y. Wang, G. Yu, Y. Fan, K. Murata, T. Nie, M. Oehme, J. Schulze, Electrical detection of spin transport in Si two-dimensional electron gas systems, *Nanotechnology*. 27 (2016) 365701.

[75] J.H. Kim, J. Bae, B.-C. Min, H. Kim, J. Chang, H.C. Koo, All-electric spin transistor using perpendicular spins, *Journal of Magnetism and Magnetic Materials*. 403 (2016) 77–80.
[76] S. Shirotori, H. Yoda, Y. Ohsawa, N. Shimomura, T. Inokuchi, Y. Kato, Y. Kamiguchi, K. Koi, K. Ikegami, H. Sugiyama, Voltage-control spintronics memory with a self-aligned heavy-metal electrode, *IEEE Transactions on Magnetics*. 53 (2017) 1–4.
[77] S. Bandyopadhyay, B. Das, A.E. Miller, Supercomputing with spin-polarized single electrons in a quantum coupled architecture, *Nanotechnology*. 5 (1994) 113.
[78] S. Takahashi, S. Maekawa, Spin injection and detection in magnetic nanostructures, *Physical Review B*. 67 (2003) 52409.
[79] H.C. Koo, J.H. Kwon, J. Eom, J. Chang, S.H. Han, M. Johnson, Control of spin precession in a spin-injected field effect transistor, *Science*. 325 (2009) 1515–1518.
[80] S. Sugahara, M. Tanaka, A spin metal-oxide-semiconductor field-effect transistor using half-metallic-ferromagnet contacts for the source and drain, *Applied Physics Letters*. 84 (2004) 2307–2309.
[81] M.A. Kharadi, G.F.A. Malik, K.A. Shah, F.A. Khanday, Sub-10-nm silicene nanoribbon field effect transistor, *IEEE Transactions on Electron Devices*. 66 (2019) 4976–4981.
[82] B. Dlubak, M.-B. Martin, C. Deranlot, B. Servet, S. Xavier, R. Mattana, M. Sprinkle, C. Berger, W.A. De Heer, F. Petroff, Highly efficient spin transport in epitaxial graphene on SiC, *Nature Physics*. 8 (2012) 557–561.
[83] S.A. Crooker, M. Furis, X. Lou, C. Adelmann, D.L. Smith, C.J. Palmstrøm, P.A. Crowell, Imaging spin transport in lateral ferromagnet/semiconductor structures, *Science*. 309 (2005) 2191–2195.
[84] X. Jiang, R. Wang, R.M. Shelby, R.M. Macfarlane, S.R. Bank, J.S. Harris, S.S.P. Parkin, Highly spin-polarized room-temperature tunnel injector for semiconductor spintronics using MgO(100), *Physical Review Letters*. 94 (2005) 1–4. https://doi.org/10.1103/PhysRevLett.94.056601.
[85] A.T. Hanbicki, B.T. Jonker, G. Itskos, G. Kioseoglou, A. Petrou, Efficient electrical spin injection from a magnetic metal/tunnel barrier contact into a semiconductor, *Applied Physics Letters*. 80 (2002) 1240–1242.
[86] G. Salis, R. Wang, X. Jiang, R.M. Shelby, S.S.P. Parkin, S.R. Bank, J.S. Harris, Temperature independence of the spin-injection efficiency of a MgO-based tunnel spin injector, *Applied Physics Letters*. 87 (2005) 262503.
[87] P. LeClair, J.K. Ha, H.J.M. Swagten, J.T. Kohlhepp, C.H. Van de Vin, W.J.M. De Jonge, Large magnetoresistance using hybrid spin filter devices, *Applied Physics Letters*. 80 (2002) 625–627.
[88] M.A. Kharadi, G.F.A. Malik, F.A. Khanday, K.A. Shah, Hydrogenated silicene based magnetic junction with improved tunneling magnetoresistance and spin-filtering efficiency, *Physics Letters A*. 384 (2020) 126826.
[89] J.S. Moodera, X. Hao, G.A. Gibson, R. Meservey, Electron-spin polarization in tunnel junctions in zero applied field with ferromagnetic EuS barriers, *Physical Review Letters*. 61 (1988) 637.
[90] M. Gajek, M. Bibes, A. Barthélémy, K. Bouzehouane, S. Fusil, M. Varela, J. Fontcuberta, A. Fert, Spin filtering through ferromagnetic Bi Mn O 3 tunnel barriers, *Physical Review B*. 72 (2005) 20406.
[91] U. Lüders, M. Bibes, K. Bouzehouane, E. Jacquet, J.-P. Contour, S. Fusil, J.-F. Bobo, J. Fontcuberta, A. Barthélémy, A. Fert, Spin filtering through ferrimagnetic NiFe 2 O 4 tunnel barriers, *Applied Physics Letters*. 88 (2006) 82505.
[92] S. Matzen, J.-B. Moussy, P. Wei, C. Gatel, J.C. Cezar, M.A. Arrio, P. Sainctavit, J.S. Moodera, Structure, magnetic ordering, and spin filtering efficiency of NiFe2O4 (111) ultrathin films, *Applied Physics Letters*. 104 (2014) 182404.

[93] A.V Ramos, M.-J. Guittet, J.-B. Moussy, R. Mattana, C. Deranlot, F. Petroff, C. Gatel, Room temperature spin filtering in epitaxial cobalt-ferrite tunnel barriers, *Applied Physics Letters*. 91 (2007) 122107.
[94] Y.K. Takahashi, S. Kasai, T. Furubayashi, S. Mitani, K. Inomata, K. Hono, High spin-filter efficiency in a Co ferrite fabricated by a thermal oxidation, *Applied Physics Letters*. 96 (2010) 72512.
[95] S. Mao, Y. Chen, F. Liu, X. Chen, B. Xu, P. Lu, M. Patwari, H. Xi, C. Chang, B. Miller, Commercial TMR heads for hard disk drives: characterization and extendibility at 300 gbit 2, *IEEE Transactions on Magnetics*. 42 (2006) 97–102.
[96] M. Takagishi, K. Yamada, H. Iwasaki, H.N. Fuke, S. Hashimoto, Magnetoresistance ratio and resistance area design of CPP-MR film for 2-5$\hbox {Tb/in}^{2} $ read sensors, *IEEE Transactions on Magnetics*. 46 (2010) 2086–2089.
[97] H.N. Fuke, S. Hashimoto, M. Takagishi, H. Iwasaki, S. Kawasaki, K. Miyake, M. Sahashi, Magnetoresistance of FeCo nanocontacts with current-perpendicular-to-plane spin-valve structure, *IEEE Transactions on Magnetics*. 43 (2007) 2848–2850.
[98] A. Hirohata, W. Frost, M. Samiepour, J. Kim, Perpendicular magnetic anisotropy in Heusler alloy films and their magnetoresistive junctions, *Materials*. 11 (2018) 105.
[99] H.J. Richter, The transition from longitudinal to perpendicular recording, *Journal of Physics D: Applied Physics*. 40 (2007) R149.
[100] H. Katayama, S. Sawamura, Y. Ogimoto, J. Nakajima, K. Kojima, K. Ohta, New magnetic recording method using laser assisted read/write technologies, *Journal of the Magnetics Society of Japan*. 23 (S1) (1999) 233–236.
[101] R.E. Rottmayer, S. Batra, D. Buechel, W.A. Challener, J. Hohlfeld, Y. Kubota, L. Li, B. Lu, C. Mihalcea, K. Mountfield, Heat-assisted magnetic recording, *IEEE Transactions on Magnetics*. 42 (2006) 2417–2421.
[102] M.A. Seigler, W.A. Challener, E. Gage, N. Gokemeijer, G. Ju, B. Lu, K. Pelhos, C. Peng, R.E. Rottmayer, X. Yang, Integrated heat assisted magnetic recording head: design and recording demonstration, *IEEE Transactions on Magnetics*. 44 (2007) 119–124.
[103] G. Ju, Y. Peng, E.K.C. Chang, Y. Ding, A.Q. Wu, X. Zhu, Y. Kubota, T.J. Klemmer, H. Amini, L. Gao, High density heat-assisted magnetic recording media and advanced characterization-progress and challenges, *IEEE Transactions on Magnetics*. 51 (2015) 1–9.
[104] Y. Tang, J.-G. Zhu, Narrow track confinement by AC field generation layer in microwave assisted magnetic recording, *IEEE Transactions on Magnetics*. 44 (2008) 3376–3379.
[105] T. Seki, K. Utsumiya, Y. Nozaki, H. Imamura, K. Takanashi, Spin wave-assisted reduction in switching field of highly coercive iron-platinum magnets, *Nature Communications*. 4 (2013) 1–6.
[106] S. Baillet, Magnetoencephalography for brain electrophysiology and imaging, *Nature Neuroscience*. 20 (2017) 327–339.
[107] J. Åkerman, Toward a universal memory, *Science*. 308 (2005) 508–510.
[108] R. Sbiaa, H. Meng, S.N. Piramanayagam, Materials with perpendicular magnetic anisotropy for magnetic random access memory, *Physica Status Solidi (RRL)-Rapid Research Letters*. 5 (2011) 413–419.

3 Overview of Smart Nanomaterials and Thin Films

3.1 INTRODUCTION TO SMART NANOMATERIALS

Precise categorization of smart materials can be challenging due to the absence of a universally clear definition. However, they are commonly understood as materials that exhibit controlled reactions to external physical or chemical stimuli, enabling them to perform specific tasks. Nonetheless, this broad definition fails to distinguish smart materials from conventional ones, considering that most materials do respond in some way to such stimuli. It may be more appropriate, then, to focus on the concept of "smart behavior," which refers to the material's ability to sense environmental stimuli and respond in a useful, reliable, and reproducible manner [1]. Examples of such behavior can be observed in nanomaterials that display photoresponsiveness, electro-responsiveness, or magneto-responsiveness.

Materials exhibiting photoresponsiveness can be characterized as those capable of reacting to external light stimuli. On the other hand, electro-responsive materials are those that undergo changes in size or shape when subjected to an applied electric field. These emerging material classes have garnered significant attention due to their immense potential across a wide range of applications. These applications encompass diverse fields such as sensor technology, drug delivery systems, actuation mechanisms, robotics, optical systems, and energy-harvesting methods [2].

The utilization of magnetic smart nanomaterials has proven highly advantageous in various industrial and commercial sectors, including photonic and electronic devices, magnetic storage, and biomedical theranostics. These materials, referred to as magneto-responsive materials, can respond to applied magnetic fields as stimuli. Among them, magnetic nanoparticles (MNPs) have gained significant attention as an emerging platform due to their magnetic responsiveness and their potential for diverse applications [2].

This chapter considers some smart nanomaterials, namely graphene, carbon nanotubes (CNTs), and chalcogens, that can be doped to transform their properties into a fast response to light, electrical, and magnetic fields as stimuli.

DOI: 10.1201/9781003331940-3

3.2 GRAPHENE, CARBON NANOTUBES, AND CHALCOGENIDES

3.2.1 Overview of Graphene

3.2.1.1 History of Graphene

The origins of graphene can be traced back to 1859, when Benjamin Collins Brodie conducted experiments involving graphite and strong acids. During his research, Brodie obtained a substance that he referred to as "carbonic acid." At that time, he believed he had discovered a new form of carbon called "graphon" with a molecular weight of 33. However, contemporary understanding suggests that what Brodie had observed was a suspension of minute crystals of graphene oxide (GO). In other words, he had come across a graphene sheet densely coated with hydroxyl and epoxide groups [3].

Throughout the following century, several scientific papers emerged that discussed the layered structure of graphite oxide [4]. The exploration of graphene theory began in 1947 when P.R. Wallace investigated it as a fundamental concept to comprehend the electronic properties of three-dimensional graphite [5]. A pivotal moment in the history of graphene occurred when it was demonstrated that the substance previously referred to as "carbonic acid" comprised of individual atomic planes suspended in a solution [4]. In 1948, G. Ruess and F. Vogt utilized transmission electron microscopy (TEM) and, by drying a droplet of a graphene oxide suspension on a TEM grid, successfully observed folded flakes with thicknesses as low as a few nanometers [6]. These investigations were further pursued by Ulrich Hofmann's research group. In 1962, Hofmann and Hanns-Peter Boehm conducted studies to isolate the thinnest possible fragments of reduced graphite oxide, ultimately identifying some of these fragments as monolayers [7].

In the 1970s, a significant breakthrough occurred with the development of epitaxial graphene, wherein single layers of graphite were grown on the surface of other materials [8]. Epitaxial graphene shares the same single-atom-thick hexagonal lattice of sp^2-bonded carbon atoms as free-standing graphene. However, it is important to note that substantial charge transfers take place from the underlying substrate to the epitaxial graphene. Additionally, in certain cases, the d-orbitals of the substrate atoms mix with the π orbitals of graphene, leading to a significant modification in the electronic structure of epitaxial graphene.

The observation of single layers of graphite embedded within bulk materials using TEM and within soot obtained through chemical exfoliation marked a significant development. Initial attempts to produce thin graphite films through mechanical exfoliation began in 1990 [9], but it wasn't until 2004 that monolayers of graphene were successfully generated. The unambiguous identification of graphene monolayers in TEM, accomplished by counting folding lines, occurred four decades after the 1962 study [4,10,11]. In 2004, Andre Geim and Konstantin Novoselov's research group achieved the first unambiguous production and identification of graphene. However, they acknowledge the earlier experimental discovery of graphene in 1962 by Hanns-Peter Boehm and his colleagues. The theoretical exploration of graphene was initiated by P.R. Wallace in 1947 [4,7]. The term "graphene" was introduced by Boehm et al. in 1986, derived from a combination of the word "graphite" and a suffix referring to polycyclic aromatic hydrocarbons [12].

3.2.1.2 Graphene

Carbon exhibits various forms, including amorphous carbon, diamond, and graphite. Among the well-known allotropes of carbon are graphite and diamond. Graphene materials can be derived from graphite through a process involving the oxidation of graphite using a strong oxidant. This oxidation enables the exfoliation of graphite in water, resulting in the formation of a layer or a few layers of graphene referred to as graphene oxide (GO) [13]. Producing a single layer of graphene from multilayers of pristine graphite poses significant challenges, making graphene relatively expensive and difficult to obtain. As mentioned earlier, GO provides a cost-effective alternative for graphene-based materials since it can be produced using comparatively inexpensive chemical techniques [13]. GO can be described as a two-dimensional material with a carbon atomic arrangement similar to that of pristine graphene. Both graphene and GO have garnered significant attention due to their immense potential for applications across various fields [14].

Following the discovery of graphene and GO, numerous graphene-/GO-based materials have been proposed due to their wide range of potential applications. These applications encompass touch screens, batteries, electronic devices, energy storage devices, non-volatile memory devices, biosensors, and biomedical applications [13]. The qualities and electrical properties of GO, including its resistance switching behavior, can be enhanced through modifications of the attached chemical groups. Notably, GO exhibits unique resistance swapping behaviors by removing or adding oxygen-containing groups, which differ from the conventional filament-dominated resistance swapping observed in other materials [15,16]. The exceptional physical-chemical properties and ultrathin thickness (approximately 1 nm) of GO contribute to its attractiveness. A layer of GO can be described as a graphene sheet with epoxide, hydroxyl, and/or carboxyl groups attached to both sides of its surface [17].

There are several methods available to convert graphite oxide into GO, including common techniques such as sonication, stirring, or a combination of both. Although graphite oxide shares similar chemical properties with GO, they differ structurally. The primary distinction lies in the interplanar distance between the atomic layers, which is influenced by water intercalation during the oxidization process. This increased distance, caused by oxidation, disrupts the sp^2 bonding network, resulting in graphite oxide and GO being referred to as insulators [13]. Consequently, GO exhibits poor conductivity. However, most properties of pristine graphene can be restored by reducing GO using various techniques such as heat, light, or chemical reduction. The reduced form of GO obtained through these processes is commonly known as reduced graphene oxide (rGO). It is also referred to as chemically modified graphene, chemically converted graphene, or functionalized graphene [13].

3.2.1.3 Synthesis of GO

The synthesis of GO is done through the exfoliation of graphite crystals. Graphite comprises carbon atoms in three dimensions with multiple layers of graphene sheets. The graphite material can be oxidized with a relevant oxidizing agent to form graphite oxide [18]. The use of oxygenated group materials for oxidizing graphite leads to the attachment of oxygenated related bonds at the edge of the graphite structure [13]. The oxidation process causes an expansion in the layer separation, thereby making graphite oxide hydrophilic [13].

The hydrophilic property of graphite oxide enables easy exfoliation in water using sonication technique [19]. The exfoliation gives single layer or few layers of graphene with oxygen functional groups at the edge and basal planes (GO). The first method for synthesizing graphite oxide was proposed by Brodie [20]. The procedure in Brodie's method involves mixing a 1:3 proportion of graphite and potassium chlorate ($KClO_3$) and reacting with fuming nitric acid (HNO_3) for about 4 days at 60°C. Brodie's method was improved in 1938 by Staundenmaier, where the fuming nitric acid was replaced by sulfuric acid [21]. The procedure was performed using multiple aliquots of $KClO_3$. The reaction was considered hazardous since it resulted in explosions. Staudenmaier later invented a new technique that was less hazardous and easier to implement. However, it became obsolete as new techniques became available [22]. Hummer and Offeman introduced a technique that involves oxidizing graphite by using potassium permanganate ($KMnO_4$) in the presence of sulfuric acid [22]. In the Hummers method, 100 g of graphite powder, 2.3 L of H_2SO_4, 50 g of sodium nitrate ($NaNO_3$), and 300 g of $KMnO_4$ were used for the synthesis of GO. Hummer's method is a modified Staundenmaier method where $KMnO_4$ is used as a substitute to $KClO_3$ to eradicate the explosive tendencies of the chemical reaction [21]. Furthermore, $KClO_3$ was used to substitute the fuming HNO_3 to eradicate the foggy acidic fumes that were generated after the reaction. The entire reaction procedure lasted for a couple of hours, and a superior grade of GO was produced.

Hummer's method is, however, limited due to issues of toxicity. Toxic gases such as NO_2 and N_2O_4 are produced after the reaction, and therefore, this method is not conducive to health safety. Furthermore, the reaction led to the production of more oxygen content compared with Brodie's method [21,23], which makes the method less favorable for GO synthesis. Scientific advancements have led to modification of the Hummers method, which is less toxic [24]. In situations where large amounts of graphene are required, especially for industrial purposes, GO can prove to be an adequate alternative. Various methods for GO reduction have been employed, such as chemical and heating methods. Some methods include the following [13]: (1) treatment of GO with hydrazine hydrate while maintaining the reduced solution for ≈24 hours at 100°C; (2) exposition of GO residue to hydrogen plasma for a couple of seconds; (3) annealing of GO in distilled water at desired temperature and heating duration; (4) direct furnace heating of GO residue to extreme temperatures; and (5) use of linear sweep voltametry to enable GO reduction.

3.2.1.4 Structure of GO

The unique characteristics of GO can be better comprehended by examining its structure and how it interacts with other substances when functionalized. Understanding the precise chemical structure of GO has proven challenging due to its non-stoichiometric nature, which means the ratio of atoms is not based on simple integer values [18]. To shed light on the complexities of GO's structure, various chemical models have been proposed and documented in the literature. One such model, known as

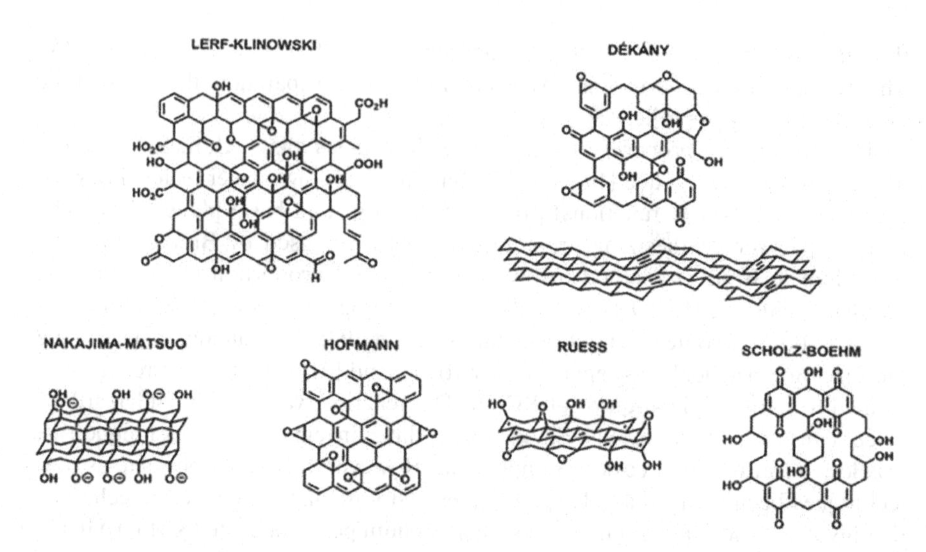

FIGURE 3.1 Several chemical structures of GO [25].

the Hofmann structural model, illustrates the presence of epoxy functional groups in GO. These groups consist of oxygen atoms bonded to two adjacent carbon atoms through single bonds, forming a network across the basal planes of graphite. The overall chemical formula assigned to this structure is C_2O [25]. A visual representation of the Hofmann model can be seen in Figure 3.1. The Ruess model presents an alternative perspective on the structure of GO, deviating from the previously proposed Hoffmann model. According to the Ruess model, the basal-plane structure of GO incorporates hydroxyl functional groups, which account for the presence of hydrogen atoms within the material. This incorporation of hydroxyl groups results in a change from the original sp^2 hybridization to a sp^3 hybridization system, accompanied by lattice repetition. In the Ruess model, the suggested structural modification involves substituting the first and fourth positions of the cyclohexane ring with epoxides at the first and third positions while maintaining a hydroxyl group at the fourth position [25].

Scholz and Boehm put forth a model suggesting the replacement of epoxide and ether functional groups in GO with hydroxyl groups positioned at the fourth position of 1,2-oxidized cyclohexane rings. This modification transforms the structure into a corrugated carbon layer, characterized by interconnected ribbons of quinoidal structure and opened cyclohexane rings [26]. Another model proposed by Lerf and Klinowski offers enhanced reliability by discarding the lattice-related model and introducing a non-stoichiometric amorphous system [27]. In their study, Lerf and Klinowski employed solid-state nuclear magnetic resonance techniques to characterize the structure of GO. The experimental procedures utilized by Dreyer et al. and Gao provided a brief overview of their work [25,26]. The results from Dreyer et al. [3] revealed significant interplatelet hydrogen bonding facilitated by the presence of epoxide and alcohol functional groups, contributing to the tightly packed structure of GO. Further investigations conducted by Lerf et al. [27] suggested that the double bond within GO could be either aromatic or conjugated in nature.

Szabo-Dekany [3] proposed a two-domain model for GO, which consists of trans-linked cyclohexyl species interspersed with tertiary alcohols and 1,3-ethers in one domain. The other domain comprises a corrugated arrangement of keto/quinoidal species, lacking carboxylic acid groups. The model also suggests that additional oxidation leads to the destruction of alkenes present in quinones. This process involves the formation of 1,2-ethers and potential aromatic pockets during the initial oxidation step in GO synthesis [25]. Generally, GO is considered amorphous, and its structure heavily depends on the synthesis technique and the extent of oxidation. The degree of oxidation encompasses not only oxygen but also other functional groups [28].

3.2.1.5 Properties of GO

3.2.1.5.1 Electrical Properties

Graphene is a remarkable material that possesses excellent electrical conductivity of 6,500 S m^{-1} [29] and high electron mobility of 25 $m^2 V^{-1} s^{-1}$ [30]. It is composed of two-dimensional layers of sp^2 carbon, each only one atom thick. When incorporated into polymers, graphene has been found to significantly enhance their electrical conductivity, even at low filler concentrations. For example, with just 1% volume fraction in polystyrene, the conductivity can reach 0.1 S m^{-1} [31]. However, during the fabrication of GO, the sp^2 bonding orbitals of graphene are disrupted, resulting in the introduction of numerous surface groups that impede its electrical conductivity. As a result, GO becomes electrically resistive, with a high resistivity of 1.64×10^4 Ω m [32,33]. Due to the high resistivity of GO, researchers have investigated methods to reduce GO and create rGO. Through the process of reduction, the electrical conductivity of GO can be significantly enhanced and adjusted over a wide range, with conductivities varying from approximately 0.1 S m^{-1} [34] to 2.98×10^4 S m^{-1} [35]. However, even after reduction, rGO still contains residual sp^3-bonded carbon to oxygen, which disrupts the flow of charge carriers within the sp^2 clusters. Consequently, electrical transport in rGO primarily occurs through a hopping mechanism, which differs from the electrical transport observed in mechanically exfoliated graphene [36].

To address the need for reducing GO and enhancing its electrical conductivity, researchers have been exploring different techniques for synthesizing conductive rGO. For instance, Stankovich et al. [37] achieved a significant improvement in conductivity by reducing a colloidal suspension of GO using hydrazine hydrate in water. This resulted in an impressive electrical conductivity of 2×10^2 S m^{-1}, representing a five-order-of-magnitude enhancement compared to GO. Voiry et al. [38] employed a conventional microwave at 1,000 W for 1–2 seconds, which effectively transformed GO into rGO and significantly increased the electron mobility in field effect transistors from approximately 1×10^{-4} to over 0.1 $m^2 V^{-1} s^{-1}$. Other successful methods for synthesizing highly conductive rGO include immersion in hydroiodic acid [39] and activation with KOH [40].

3.2.1.5.2 Optical Properties

GO exhibits high optical transmittance within the visible spectrum range, making it transparent in its usual form. However, the degree of transparency can be adjusted by controlling the thickness of the film and the level of reduction. For instance, a

GO film created from 0.5 mg/mL suspensions can have a wavelength of 550 nm and an optical transmittance of 96%. By reducing the wavelength to less than 30 nm, the material becomes semi-transparent. On the other hand, when GO is reduced to an atomically thin state, it becomes highly transparent [41]. This property makes GO well suited for transmitting information through optical processes.

3.2.1.5.3 Mechanical Strength

The exceptional mechanical properties of pristine monolayer graphene are widely documented. According to reports, graphene exhibits impressive characteristics such as a break strength of 42 N m^{-1}, a Young's modulus of 1.0 TPa, and an intrinsic tensile strength of 130.5 GPa [42]. The fracture toughness of graphene has also been investigated by Zhang et al. [43], revealing a relatively low value of 4.0 ± 0.6 MPa $m^{1/2}$. This confirms that graphene sheets have a propensity for low fracture toughness.

The properties of GO and rGO blends are influenced by various factors, including the presence of surface groups and defects resulting from oxidation or other treatment methods. In a study by Suk et al. [44], monolayer GO was produced using a modified Hummer's method and exhibited a Young's modulus of 207.6 ± 23.4 GPa. Although this represents a significant drop compared to pristine graphene, the value is still notably high. Similarly, Gomez-Navarro et al. [45] reported a monolayer of rGO produced through the original Hummer's method, followed by thermal annealing in a hydrogen environment. This rGO monolayer demonstrated a Young's modulus of 250 ± 150 TPa.

3.2.1.5.4 Thermal Conductivity

Thermal transport in graphene has become a prominent research area due to its potential applications in thermal management. The initial measurements of suspended graphene's thermal conductivity, in line with earlier predictions for graphene and CNTs, revealed an extraordinarily high thermal conductivity of up to 5,300 $W{\cdot}m^{-1}{\cdot}K^{-1}$ [46]. This value significantly surpasses the thermal conductivity of pyrolytic graphite, which is ~2,000 $W{\cdot}m^{-1}{\cdot}K^{-1}$ at room temperature (RT) [47]. The remarkable thermal conductivity of graphene has sparked considerable interest and exploration in the field of thermal transport.

Subsequent studies, focusing on graphene derived through chemical vapor deposition (CVD), which offers scalability but may introduce more defects, have been unable to replicate the exceptionally high thermal conductivity observed in earlier research. Instead, these studies have reported a wide range of thermal conductivities for suspended single-layer graphene, ranging from 1,500 to 2,500 $W{\cdot}m^{-1}{\cdot}K^{-1}$ [48–51]. The variation in reported values can be attributed to measurement uncertainties, as well as disparities in the quality of graphene and differences in processing conditions. These factors contribute to the significant range observed in the thermal conductivity measurements.

Furthermore, when single-layer graphene is reinforced with an amorphous material, it has been observed that the thermal conductivity decreases to around 500–600 $W{\cdot}m^{-1}{\cdot}K^{-1}$ at RT. This reduction is attributed to the scattering of graphene lattice waves by the substrate, as reported in studies [52,53]. For few-layer graphene encapsulated in amorphous oxide, the thermal conductivity can be even lower [54].

Additionally, the presence of polymeric residue can lead to a similar decrease in thermal conductivity for suspended bilayer graphene, bringing it to approximately 500–600 $W{\cdot}m^{-1}{\cdot}K^{-1}$ [55].

3.2.1.6 Functionalization of GO

Once the reduction of GO is complete, the resulting mixture can be modified to enhance the properties of GO for various applications. This modification process, known as functionalization, can be achieved through covalent or non-covalent methods [56]. Covalent functionalization involves attaching oxygenated groups, such as carboxylic groups at the edges and epoxy groups at the basal plane, to the surface of GO. On the other hand, non-covalent functionalization entails introducing non-covalent interactions between GO and organic compounds or polymers [56]. One approach for non-covalent functionalization is treating GO with isocyanates, among other attempts that have been made [31,57]. Treating GO with isocyanates is a method that reduces the hydrophilicity of GO and results in the formation of amine and carbamate esters. The carboxyl and epoxy functional groups are primarily responsible for the formation of these esters [13]. One advantage of using isocyanates to treat GO is that it leads to the formation of stable dispersions in polar aprotic solvents, which can be easily exfoliated to obtain single-layer graphene sheets. Additionally, these dispersions can create a uniform mixture of GO with the polymer matrix, ultimately leading to high-quality graphene-polymer composites [13]. Another way to modify GO is through chemical reduction using hydrazine as a functionalizing agent, which results in the production of rGO.

3.2.2 Overview of Carbon Nanotubes

3.2.2.1 Brief History of CNTs

Decades ago, in 1952, Russian scientists Lukyanovich and Radushkevich made a significant discovery of CNTs as carbon filaments. Unfortunately, due to the limitations imposed by the Cold War and restricted access to their article, this discovery went unnoticed by Western scientists. Many Russian papers were not included in the literature database [58]. In the 1970s, carbon nanofilaments started to emerge as by-products alongside carbon fibers during oxygen combustion. It was not until 1991, when Iijima utilized high-resolution transmission electron microscopy (HRTEM) to examine these nanofilaments, that the discovery of CNTs was officially acknowledged and credited to Iijima for introducing CNTs to the scientific community [59].

In 1993, both Iijima [60] and Bethune [61] independently reported the existence of single-walled carbon nanotubes (SWCNTs). This discovery of CNTs paved the way for new possibilities in electronics, thanks to their unique characteristics. Experimental evidence confirming these properties was obtained in 1998 [62].

3.2.2.2 Carbon Nanotubes

Carbon materials exhibit various structures, including graphite, amorphous, and diamond. Graphite and diamond are well-known allotropes of carbon [63]. Advancements in research methods have expanded our understanding of carbonaceous materials, leading to the identification of additional allotropes [63]. These include graphene, fullerene, and CNTs [64]. Among these allotropes, graphitic carbon has been the focus of extensive research in materials science. This is primarily due to its exceptional thermal [46,64,65], mechanical [64–66], and electrical [7–9] properties, which make it suitable for diverse applications such as electronics, communication, storage, medicine, and composite materials [65,67]. Research on nanotubes has experienced significant progress in the past decade due to their superior properties, leading to various commercial applications [64]. Following Iijima's discovery of CNTs [68], much of the research has focused on their synthesis, properties [64], and applications [67,69]. Synthesis and purification of CNTs involve a range of techniques such as arc discharge, laser ablation (LA), and CVD [70] for synthesis and oxidation, ultrasonication, acid treatment, annealing, microfiltering, and functionalization for purification [67].

3.2.2.3 Synthesis of CNTs

The arc discharge method involves the use of two graphite electrodes: a cathode and an anode. A high DC current of 100 A is passed through these electrodes, which are separated by 1–2 mm within a helium atmosphere at 400 mbar. As a result, carbon rods accumulate on the cathode. This technique is commonly employed to produce multi-walled carbon nanotubes (MWCNTs), and by introducing catalysts such as Mo, Ni, Co, or Fe, SWCNTs can be generated. The quality, quantity, purity, lengths, and diameters of the produced nanotubes are influenced by parameters such as the concentration of the catalyst metals, temperature, type of gas used, and inert gas pressure. Another method called laser ablation (LA), introduced by Smalley in 1996, has proven successful in producing large quantities of SWCNTs [71].

The LA technique involves the use of a high-temperature furnace to heat a mixture of graphite and metal (referred to as the target). The furnace can reach temperatures of up to 1,200°C. A laser beam is then directed at the target and moved across its surface to cause vaporization. This vaporization process generates some debris, which is subsequently removed by the flow of argon (Ar) gas through the system. The Ar gas moves from a region of high temperature to a copper collector outside the furnace, which is cooled by water. In the LA method, the target's composition typically includes 1.2% nickel or cobalt, which act as catalysts for the growth of SWCNTs. The remaining 98.8% of the target consists of other components [72].

Arc discharge and LA are two commonly used methods for creating high-quality CNTs. However, these methods have limitations, such as being expensive and having limited production capacity. Additionally, purifying the CNTs obtained through these methods can be difficult due to the presence of unwanted carbon impurities and metal elements. In contrast, CVD is a well-established technique that allows for large-scale production of CNTs while providing better control over their length, diameter, and structure. This technique shows significant promise for industrial applications [67,70]. In the CVD process, nitrogen and hydrocarbon gases such as methane, acetylene, and ethylene are broken down under atmospheric pressure on a hot metal substrate with catalytic properties, which is placed inside a reactor maintained at approximately 1,000°C. These conditions facilitate the growth of CNTs on the metallic catalyst. Different substrates, such as glass, SiO_2, or Si, can be used to deposit solid catalysts or introduce gases into the reactor [73,74]. A high magnetic field of 10 T can be utilized to control the morphology of CNTs [75]. Different types of CVD techniques include radiofrequency (RF-CVD), microwave plasma (MPECVD), catalytic chemical (CCVD) (both thermal and plasma-enhanced (PE) oxygen-assisted CVD), and hot-filament (HFCVD) or water-assisted CVD. Currently, CCVD is the most employed CVD technique for synthesizing CNTs. Compared to arc discharge and LA, CCVD offers economic benefits and the capability of large-scale production of pure CNTs [76].

3.2.2.4 Purification of CNTs

The process of oxidation involves the utilization of substances such as H_2O_2 and H_2SO_4 as oxidizing agents to eliminate carbonaceous impurities. However, a significant drawback of this method is that it not only removes the impurities but also affects the CNTs themselves. In the ultrasonication technique, purification of CNTs is achieved through the application of vibration, which leads to improved separation and dispersion of nanoparticles (NPs). The specific reagent, solvent, and surfactant used play a crucial role in determining the extent of particle separation. Acid treatment is a widely used technique for effectively removing impurities, including metal catalysts, from CNTs. Prior to being exposed to acid and solvent, it is necessary to oxidize or subject the metal surface to sonication to expose it. This treatment affects the metal catalyst but has minimal or no impact on the obtained pure CNTs. Another method, known as annealing, involves subjecting the CNTs to extremely high temperatures ranging from 873 to 1,873 K to achieve purity. The elevated temperature causes decomposition of the graphitic carbon. In the case of a high-temperature treatment at 1,873 K, the metal will melt and can be separated from the pure CNTs. Microfiltration is a purification method that relies on separating particles based on their size. In this technique, CNTs are trapped in a filter, while other impurities pass through, resulting in the isolation of pure CNTs. Another commonly employed technique is the functionalization of CNTs, where additional groups are attached to the CNT walls, rendering them more soluble compared to impurities. This attachment facilitates the filtration-based separation of CNTs from insoluble impurities, such as metals [67].

3.2.2.5 Structure of CNTs

CNTs are cylindrical structures composed of carbon atoms arranged in a hexagonal pattern, forming a single-layer graphene sheet [67,77]. These nanotubes can be categorized as SWCNTs, double-walled CNTs (DWCNTs), or MWCNTs [67]. The bonding between carbon atoms is covalent, with each carbon atom bonded to three other carbon atoms. The folding pattern of the graphene sheet determines the structure of the nanotube, which can be armchair, zigzag, or chiral [78]. Figures 3.2 and 3.3 depict the different structures of CNTs.

SWCNTs are cylindrical structures formed by wrapping a sheet of graphene, which is a one-atom-thick layer of graphite, into a cylinder. The ends of the nanotube can be either open or closed, and has a specific size and direction. When considering the symmetry of the cylinder, the graphene sheet contains multiple atoms, and two atoms are selected, with one serving as the reference point. The wrapping process of the graphene sheet continues until the two chosen atoms align or overlap.

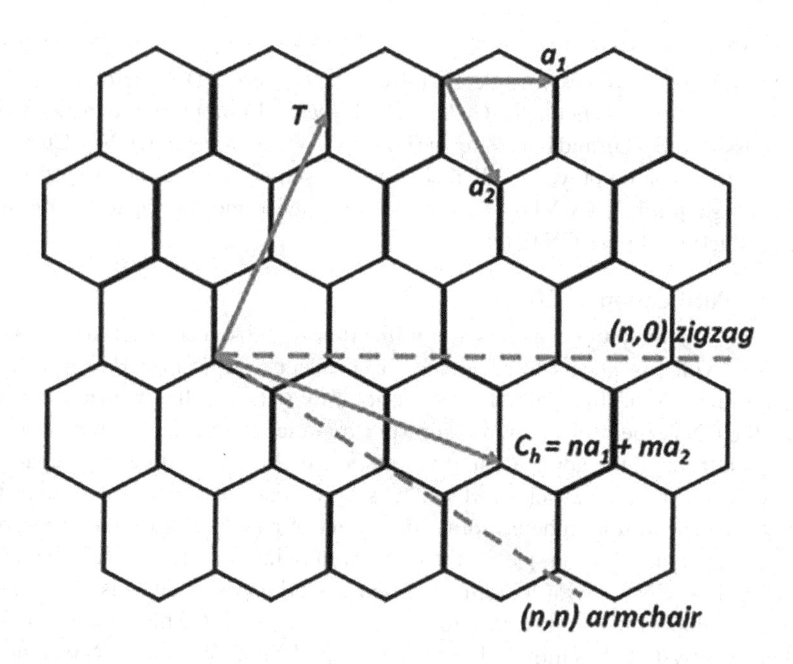

FIGURE 3.2 CNTs chiralities of vector direction.

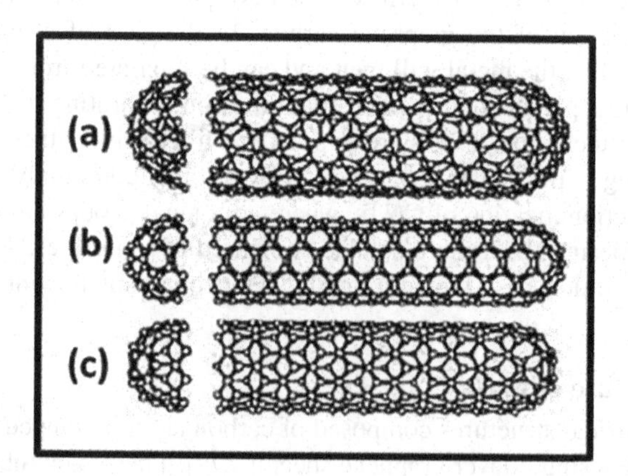

FIGURE 3.3 CNTs chiralities structures: (a) chiral, (b) armchair, and (c) zigzag.

This wrapping results in a tube with a circumference equivalent to the length of the vector connecting the two atoms, known as the chiral vector (see Figure 3.2). The alignment between the axis of the tube and its chiral vector is perpendicular. The diverse chiral vectors found in SWCNTs contribute to distinct properties such as electrical conductivity, mechanical strength, and optical characteristics. In contrast, DWCNTs comprise two SWCNTs, with one enveloping the other, exhibiting different diameters. Additionally, MWCNTs are composed of multiple concentric SWCNTs, with interlayers bonded by Van der Waal's forces [77]. The structures of MWCNTs can be described using two models: the Russian Doll model and the

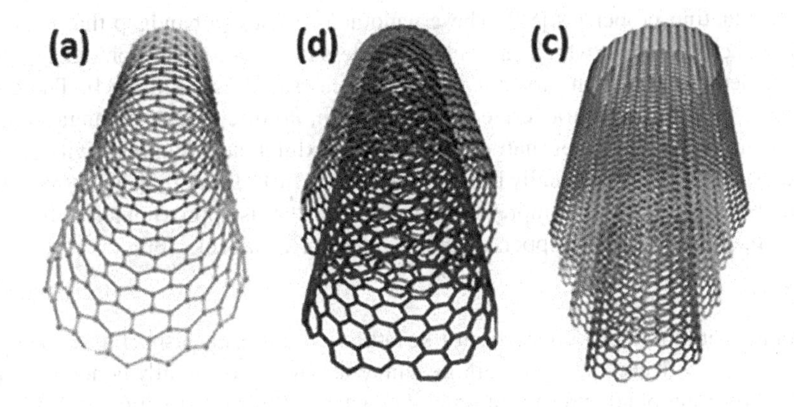

FIGURE 3.4 CNTs: (a) single wall (SW), (b) double wall (DW), and (c) multi-wall (MW).

Parchment model. In the Russian Doll model, the layers of graphite are arranged in concentric cylinders, resembling a set of nesting dolls. On the other hand, the Parchment model depicts a single sheet of graphite rolled upon itself, much like a tightly wrapped newspaper. The outermost layer of MWCNTs can vary in diameter, ranging from a few nanometers (nm) to several hundred nanometers, while their length can extend up to 100 μm. Figure 3.4 illustrates these different forms of CNTs.

MWCNTs offer certain advantages over SWCNTs due to their cost-effective production and abundant availability [79]. In order to enhance the physical and chemical properties for electronic applications, functionalization becomes essential for CNTs due to their inherent chemical inertness [79].

3.2.2.6 Properties of CNTs

The one-dimensional structure of CNT has led to extensive research on its electrical, thermal, and mechanical properties. The following are some of the noteworthy properties of CNTs:

3.2.2.6.1 Electrical Properties

The electrical characteristics of CNTs are determined by the specific wrapping pattern of the graphene sheet. Additionally, the electrical properties of CNTs are influenced by factors such as tube diameter and chirality (n, m). Depending on the wrapping direction, CNTs can exhibit either metallic or semiconductor behavior, with varying energy gap sizes ranging from large to small [80]. An interesting example is the distinction between different types of CNTs. CNTs with a zigzag structure exhibit semiconducting behavior, while those with an armchair structure display metallic characteristics [81]. The naming convention for these structures is based on the shape of their cross-sectional ring. In particular, the armchair structure stands out, as it possesses a chiral (n, n) direction at an angle of 30° and exhibits conductivity similar to that of metals [82]. Within the armchair structure, there exists a pseudo-forbidden energy gap with a range spanning from 2 to 50 meV. Furthermore, the diameter of these armchair CNTs can vary between 3.0 and 0.7 nm [83]. The zigzag structure of CNTs is characterized by a chiral (n, 0) direction at an angle of 0° and exhibits semiconducting properties [82]. These nanotubes possess a bandgap that is dependent on their diameter, which can range from less than 0.4 to 2 eV. For zigzag CNTs, the diameter varies from larger than 3.0 nm to as small as 0.4 nm [84]. The chiral structure of CNTs is characterized by a chiral (n, m) direction at an angle ranging from 0° to 30° [82]. In their natural state, CNTs exhibit electrical resistivity comparable to that of metals, typically ranging from 10^{-4} to 10^{-3} Ω cm [85]. Moreover, these nanotubes can support an impressive current flow that is approximately a thousand times higher than that of copper, reaching values of 10^9 A cm^{-2} [86].

3.2.2.6.2 Mechanical Strength

Graphite, composed of a honeycomb-like lattice structure, consists of multiple layers of graphene, with each layer containing numerous atoms chemically bonded together [87]. The strong bonds present in graphite contribute to its distinction as a material with the highest basal-plane elastic modulus. Taking advantage of this characteristic, CNTs have been envisioned as potential materials for high-strength fibers. For instance, SWCNTs are remarkably rigid, surpassing the hardness of steel, and exhibit resistance to physical forces that could cause damage. When pressure is applied to the tip of a CNT, it flexes and then readily returns to its original state once the force is released, without sustaining any significant damage. The distinct characteristics of CNTs render them valuable in high-resolution scanning microscopy applications [88]. CNTs possess exceptional mechanical properties, offering a combination of a low density of 1–2 g cm^{-3} and a large surface area of 200–900 $m^2 g^{-1}$. MWCNTs exhibit a high Young's modulus and tensile strength, with reported values of approximately 1.8 TPa and 150 GPa, respectively [86,89]. On the other hand, SWCNTs demonstrate slightly lower values with a Young's modulus of around 1.0 TPa and a tensile strength of approximately 100 GPa [86,89].

3.2.2.6.3 Thermal Conductivity

The compact dimensions of CNTs can be attributed to their wrapped graphene structure. These CNTs exhibit quantum effects that significantly impact their thermal conductivity and specific heat behavior, particularly at low temperatures [87]. The distinct structural characteristics and shapes of CNTs contribute to their significantly higher longitudinal thermal conductivity compared to graphite. Specifically, CNTs exhibit a range of 1,750–5,800 W m^{-1} K^{-1} at 300 K, which is equivalent to the thermal conductivity observed in pristine diamond [80]. CNTs possess exceptional thermal stability, demonstrating resistance to high temperatures of approximately 750°C in air and approximately 2,800°C in a vacuum environment [89,90].

3.2.2.7 Functionalization of CNTs

The process of functionalizing CNTs involves intentionally introducing impurity atoms or molecules into the lattice of CNTs through physical absorption or chemical bonding [81]. This functionalization aims to alter the electrical, chemical, physical, and magnetic properties of CNTs, enabling their utilization in a wide range of applications [91]. Functionalization provides a viable pathway to tailor the properties of materials according to specific application requirements [81]. By attaching or immobilizing chemical or biological atoms and molecules onto the surface of

CNTs, the development of biochemical sensors becomes feasible [81,92]. Through the process of functionalization, CNTs can undergo a transformation from diamagnetic to ferromagnetic behavior by incorporating ferromagnetic NPs or adsorbing oxygen atoms. When CNTs are wrapped in a zigzag chirality configuration, they exhibit semiconducting properties and possess diamagnetic characteristics. However, coating them with titanium enhances their magnetization and conductivity [81]. Functionalizing CNTs involves two primary methods: covalent and non-covalent modification [67,91]. Covalent modification entails attaching functional groups to the CNTs through chemical bonding, while non-covalent modification involves the attachment of functional groups without altering the CNT structure. These functional groups can be attached to either the ends or sidewalls of the CNTs [91]. Moreover, CNTs tend to form agglomerates due to their hydrophobic nature, which hinders their dispersion in solvents. To achieve CNT dispersion in a solvent, surface modification is required to reduce hydrophobicity and enhance the interfacial bonding with other materials through chemical attachment. This modification process enables improved compatibility and dispersion of CNTs in the desired solvent [70].

The distinctive characteristics of CNTs, combined with their surface modifiability, make them highly promising materials for a wide range of technological applications [77]. One prominent area where CNTs hold great potential is in nanoelectronic devices [93]. It is precisely these reasons that make CNTs a subject of extensive study and exploration.

3.2.2.7.1 Non-covalent Modification of CNTs

Non-covalent functionalization methods were employed to modify CNTs without compromising their inherent structure while sustaining stability [91]. Various methods, such as endohedral, metal, and polymer biological, were utilized to achieve non-covalent functionalization. In the process of endohedral functionalization, CNTs capture molecules due to the capillary effect within their structures [94]. Metal functionalization entails the decoration of CNT surfaces with NPs such as Ni, Au, or Ag [85]. Polymer functionalization of CNTs enhances the Van Der Waals forces, leading to the rolling of CNTs along polymer chains [94]. In biological functionalization, proteins and DNA are attached to the surface of CNTs, enabling the production of bioactive CNTs for applications such as drug delivery and biosensing [85,95].

3.2.2.7.2 Covalent Modification of CNTs

The covalent process ensures high stability, but it can disrupt the sp^2 hybridization through the formation of σ-bonds. Defect sites commonly arise in the σ-bonds formed during mechanical loading and production, leading to the disruption of local bonds [81]. The functionalization of CNTs with H_2SO_4/HNO_3 [96–98], H_2O_2/H_2SO_4 [99,100], or HNO_3 [101,102] serves as an example, as it triggers the creation of covalent bonds such as C=O, C–OH, and COOH [103,104]. These bonds are directly linked to the structure of CNTs, establishing a strong covalent connection. Additionally, the introduction of NPs can result in the direct formation of a covalent bond by converting the carbon atom from sp^2 to sp^3 hybridization. This approach

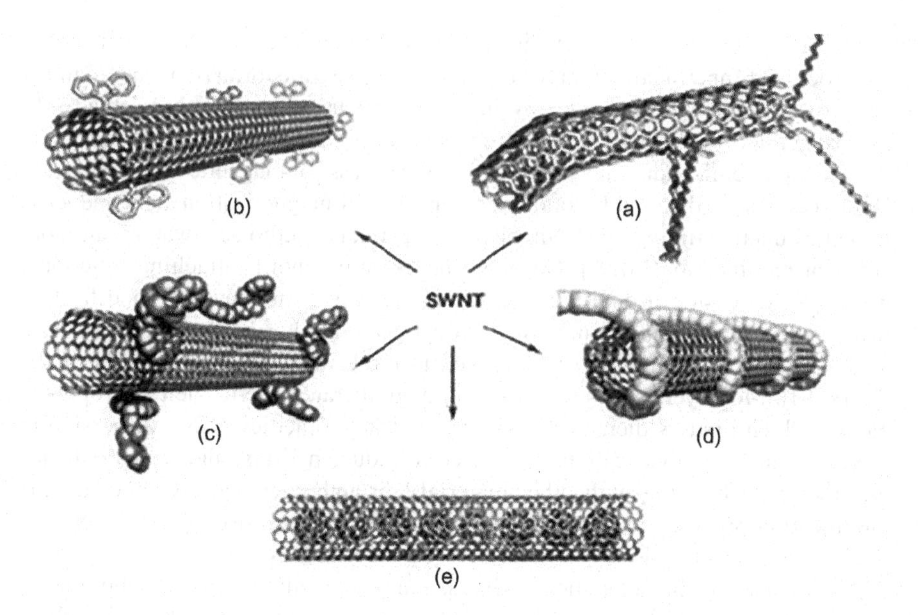

FIGURE 3.5 Potential CNTs functionalization: (a) defect, (b) wall-side, (c) bio-, (d) polymer, and (e) endohedral [79,105].

offers a robust covalent bond and significant structural alteration, leading to the creation of extensive defects [85,94]. Figure 3.5 illustrates the various types of covalent and non-covalent functionalization techniques.

3.2.3 Overview of Chalcogenides

3.2.3.1 History of Chalcogens

Sulfur has a long history dating back to ancient times. The ancient Greeks were familiar with it, and the ancient Romans mined it extensively. It was even utilized as an ingredient in Greek fire, a formidable incendiary weapon. During the Middle Ages, sulfur played a crucial role in alchemical experiments. During the 18th and 19th centuries, prominent scientists Joseph Louis Gay-Lussac and Louis-Jacques Thénard made significant contributions to definitively identifying sulfur as an individual chemical element. Their groundbreaking research and experiments solidified our understanding of sulfur's elemental nature [106].

In 1783, Franz Joseph Müller von Reichenstein made the initial discovery of tellurium. He encountered tellurium while examining a specimen that we now recognize as calaverite. Initially, Müller mistakenly believed the sample to be pure antimony, but subsequent tests contradicted this assumption. This led to the realization that the specimen contained a distinct element, later identified as tellurium. After considering that the sample might be bismuth sulfide, Müller's tests indicated otherwise, ruling out that possibility. Müller faced a perplexing challenge for several years. However, he eventually had a revelation—the sample consisted of gold bonded with an unidentified element. This realization marked a crucial breakthrough in understanding the composition of the sample. In 1796, Müller forwarded a portion of the sample to the

German chemist Martin Klaproth for further analysis. With his expertise, Klaproth successfully purified the previously unknown element. In recognition of its properties, Klaproth named the element "tellurium," derived from the Latin term for earth. This choice of name reflected the element's association with the Earth's crust and its geological significance [106].

In 1817, the Swedish chemist Jöns Jacob Berzelius made the discovery of selenium. While visiting a sulfuric acid manufacturing facility, Berzelius observed a distinctive reddish-brown deposit. This observation piqued his curiosity and led to the identification of a new element, which he named selenium. The chance encounter at the manufacturing plant played a crucial role in the recognition and subsequent characterization of this element. Initially, it was believed that the sediment in question contained arsenic. Berzelius, upon initial examination, mistakenly identified the substance as tellurium. However, further investigation led him to the realization that it contained an entirely new element. To honor the Greek moon goddess Selene, Berzelius decided to name this newly discovered element "selenium." This choice of name paid homage to its association with the celestial body and added another entry to the periodic table of elements [106–108].

3.2.3.2 Chalcogenides

The term "chalcogen" holds a literal significance as "ore" forming materials, derived from the combination of the Greek word "khalkos," denoting copper (ore or coin), and the Latinized "Greek" word "genes," meaning born or produced. Chalcogenides encompass the Group 16 elements in the periodic table, namely Sulfur (S), Selenium (Se), and Tellurium (Te). These elements represent the chalcogen family, each playing a distinctive role in various scientific and industrial applications. Chalcogens, including S, Se, and Te, find significant technical applications in industries such as electronics, photovoltaics (PVs), and optoelectronics. These chalcogen elements are commonly utilized in the form of chemical compounds, forming essential compositions with elements from other groups. Their unique properties make them invaluable in various technological advancements and industrial processes within these fields. [109].

Chalcogenides are chemical compounds composed of one or more electropositive elements and at least one chalcogen anion. They are often classified as members of the oxygen family, encompassing S, Se, and Te, and even the radioactive element Polonium (Po). Materials containing the element "S" exhibit noteworthy properties, such as a high refractive index, nonlinearity, significant Kerr effect, excellent infrared (IR) transmission beyond 1.5 μm, and notable chemical resistance. Additionally, they demonstrate a favorable response when subjected to direct patterning exposure near the band gap light, making them suitable for various applications. These favorable properties collectively render chalcogenides suitable for a wide range of technical applications. For instance, their utilization in fiber Bragg gratings, fiber-optic communication, and evanescent wave fiber sensors has proven highly beneficial. Chemical compositions incorporating the element "Se" find extensive application in rectifiers, solar cells, xerography, and photographic exposure meters, as well as in the development of anticancer agents. The versatility of chalcogenides enables their use in diverse fields, contributing to advancements in technology and various industries [109].

"Se" is valuable for its ability to eliminate bubbles in the glass industry and remove unwanted tints in iron production. Its high reactivity with specific chemicals allows for the conversion of selenium into functional materials. Amorphous materials based on selenium are utilized in imaging and biomedical applications. The incorporation of alloying elements such as Cu, In, and Ga, into crystalline Se finds extensive applications in PV and photo-detection fields. On the other hand, Te, which belongs to the chalcogen group, exhibits more metallic properties. Te-based materials are widely employed in data storage devices due to their ability to transform between amorphous and crystalline states. Se and Te are polymers that possess a divalent nature and exhibit chain-like structures. Their ability to create various bonds with varying strengths is attributed to the presence of cross-linking components. The materials formed through cross-linking of Se and Te display a substantial quantity of unbound lone pair electrons. These electrons can be readily stimulated when exposed to optical or electrical fields [109].

The transition from a crystalline phase to an amorphous phase happens when the amorphous phase lacks the energy required for lone pair excitation. In the case of Se/Te polymers with a coating of lone pairs, the polymeric structures exhibit vibrational properties that enable electronic transitions through the movement of the chains. The unique characteristics of these materials find significant application in phase change memory (PCM) technology. Their potential for PCM opens numerous possibilities for high-performance microelectronics applications. For instance, microprocessors can now incorporate transistors as small as 15 nm, pushing the boundaries of the well-established Moore's Law that governs optical limits in physical systems. The nanostructures exhibited by these materials position them at the forefront of diverse industrial applications. Their technical utility stems from the remarkable properties of the inorganic and organic compounds derived from Se and Te, highlighting their significance in various fields. In the realm of scientific exploration, the combination of Se chalcogen systems with organic additives presents a promising avenue for future advancements. These host systems hold great potential for a wide range of applications in diverse fields. Exciting possibilities lie ahead in areas such as biomedical sciences, sensor technologies, high-processing electronics, imaging, PVs, memory devices, and more. The fusion of Se chalcogen systems with organic compounds opens a world of possibilities for innovative solutions in various domains [109].

The group of elements known as chalcogens (Sulfur, Selenium, and Tellurium) is found in association with nearly every other group in the periodic table, except for noble gases and certain radioactive elements. Chalcogen-based alloys typically adhere to the established chemical valence patterns, as seen in compounds such as SeZn and In_2Se_3. Nevertheless, there exist certain chalcogen-containing alloys that deviate from this common behavior, such as P_4S_3. The crystallographic structures of the prominent chalcogen group demonstrate a directional nature primarily governed by covalent bonding with specific orientations. In most cases, chalcogens exhibit positive oxidation states when combined with oxides, halides, and nitrides [109].

3.2.3.3 Monochalcogenides

Metal monochalcogenides are compounds with the chemical formula ME, where M represents a transition metal and E denotes one of the chalcogens, namely Sulfur (S),

Selenium (Se), or Tellurium (Te). These compounds commonly adopt one of two distinct crystal structures, which are named after the corresponding arrangements found in zinc sulfide. The zinc blend structure observed in monochalcogenides such as zinc sulfide features a cubic arrangement where sulfide atoms are packed symmetrically and Zn^{2+} ions occupy half of the tetrahedral holes. This arrangement gives rise to a diamond-like framework. On the other hand, the wurtzite structure, an alternative configuration for monochalcogenides, exhibits similar tetrahedral atom connectivities but possesses a hexagonal crystal symmetry. Another structural motif for metal monochalcogenides is the nickel arsenide lattice, where the metal atoms and chalcogenide ions exhibit octahedral and trigonal prismatic coordination, respectively. However, this particular motif often experiences nonstoichiometry, meaning that the composition of the compound can deviate from the expected stoichiometric ratios [110]. Prominent examples of monochalcogenides encompass various pigments, with cadmium sulfide being particularly noteworthy. Furthermore, monosulfides constitute a significant portion of minerals and ores found in nature [111].

3.2.3.4 Dichalcogenides

Metal dichalcogenides, denoted as ME_2, with M representing a transition metal and E representing S, Se, or Te [112], are a class of compounds with sulfides being the most significant members. These compounds possess dark diamagnetic solid properties, rendering them insoluble in all solvents, and they exhibit semiconducting characteristics. Notably, certain metal dichalcogenides also exhibit superconducting properties [113]. When examining their electronic structures, these compounds are commonly perceived as derivatives of M^{4+} [114].

Where, $M^{4+} = Ti^{4+}$ (d^0 configuration), V^{4+} (d^1 configuration), and Mo^{4+} (d^2 configuration). The reversible intercalation ability of lithium in titanium disulfide was explored in prototype cathodes for secondary batteries. This characteristic of titanium disulfide was harnessed and studied for its potential applications in battery technology [115]. Molybdenum disulfide, commonly referred to as molybdenite, garners significant attention with numerous articles dedicated to its study. It serves as the primary ore of molybdenum and finds application as a solid lubricant and catalyst for hydrodesulfurization [116]. Additionally, analogous compounds such as titanium diselenide ($TiSe_2$), molybdenum diselenide ($MoSe_2$), and tungsten diselenide (WSe_2) are well-known examples within the same family, exhibiting similar properties.

3.2.3.5 Tri and Tetra Chalcogenides

Trichalcogenides, composed of metals primarily from the early transition metal groups such as Titanium (Ti), Vanadium (V), Chromium (Cr), and Manganese (Mn), are another category of compounds worth noting. Their chemical formula is typically expressed as $M^{4+}(E_2^{2-})(E^{2-})$, where E represents Sulfur (S), Selenium (Se), or Tellurium (Te). An example of a well-known trichalcogenide is niobium triselenide. Additionally, the amorphous form of molybdenum trisulfide (MoS_3) can be obtained by subjecting tetrathiomolybdate to acid treatment.

$$MoS_4^{2-} + 2H^+ \rightarrow MoS_3 + H_2S \tag{3.1}$$

Patrónite, a mineral with the chemical formula VS_4, serves as an illustration of a metal tetrachalcogenide. Through crystallographic analysis, it has been determined that the structure of this material can be described as a bis(persulfide), denoted as $V^{4+},(S_2^{2-})_2$ [117].

3.2.3.6 Synthesis Techniques

3.2.3.6.1 Solvothermal and Hydrothermal

Solvothermal and hydrothermal synthesis refer to techniques employed for the synthesis of chemical compounds. This method involves subjecting a solvent containing the necessary reagents to elevated pressure, usually ranging from 1 to 100 MPa, and elevated temperature, typically ranging from 373 to 1,273 K, within an autoclave. The high-pressure and high-temperature conditions facilitate the reaction and allow for the controlled formation of desired compounds. Under solvothermal and hydrothermal conditions, the solubility of numerous substances in a particular solvent is significantly enhanced compared to standard conditions. This unique property allows for reactions that would otherwise be improbable, resulting in the formation of novel compounds or polymorphs. Solvothermal synthesis closely resembles the hydrothermal approach, as both methods are commonly carried out using a stainless-steel autoclave. The primary distinction lies in the fact that solvothermal synthesis typically employs a non-aqueous precursor solution [118]. Researchers have utilized solvothermal and hydrothermal methods to synthesize chalcogenide materials [119].

In a conventional procedure, Cu_2FeSnS_4 was synthesized by Jiang et al. [120]. The process involved dissolving 1 mmol of iron (III) chloride trihydrate, 1 mmol of tin (IV) chloride tetrahydrate, 2 mmol of copper (II) chloride dihydrate, and 5 mmol of thiourea in 80 mL of dimethylformamide (DMF) under magnetic stirring. The resulting mixture was carefully transferred into a Teflon-lined stainless autoclave with a volume of 100 mL. The autoclave was securely sealed and subjected to a temperature of 250°C for a duration of 24 hours. After the reaction period, the autoclave was allowed to cool naturally to RT. The resulting precipitates were then separated by centrifugation and underwent six consecutive washing cycles using deionized water and absolute ethanol. The washing process was repeated until no traces of Cl^- ions could be detected. Finally, the obtained products were dried under vacuum conditions at 80°C for a duration of 4 hours.

In another study by Senthil et al. [121], Sb_2S_3 nanocrystals (NCs) were synthesized using both solvothermal and hydrothermal approaches. For the solvothermal method, 0.25 g of antimony chloride was dissolved in 50 mL of ethanol. After 30 minutes of continuous stirring, an equal amount of ethylenediamine and sodium sulfide (0.015 M) was added to the solution. The resulting solution was carefully transferred into a Teflon-lined stainless-steel autoclave and subjected to four distinct temperatures: 70°C, 100°C, 150°C, and 200°C. This temperature variation allowed for a period of 24 hours, enabling the nucleation and growth of NPs within the solution. Subsequently, the autoclave was allowed to cool down naturally to reach RT. In the hydrothermal synthesis approach, two solutions were prepared. The first solution involved dissolving 2.5 g of polyvinylpyrrolidone (PVP) in 25 mL of water at 40°C with continuous stirring. To this solution, 0.52 g of citric acid was added, followed by an additional 30 minutes of stirring. Subsequently, 0.5 g of antimony chloride ($SbCl_3$) was introduced to the solution. Another solution (Solution 2) was created by dissolving 0.15 g of thioacetamide in 20 mL of water in a separate beaker. Afterwards, solution 2 was slowly added, drop by drop, to another solution known as "solution 1." The combined solution was then carefully transferred into a Teflon-coated autoclave. The solution was subjected to four distinct temperatures: 200°C, 150°C, 100°C, and 70°C, and maintained at each temperature for a duration of 24 hours. This controlled heating process promoted the nucleation and growth of NPs within the solution. Subsequently, the solution was left to naturally cool down to RT. The resulting precipitate underwent a series of treatments, including filtration, washing, and centrifugation. The washing and centrifugation steps were iterated multiple times using both ethanol and water as cleaning agents. Ultimately, the obtained samples were subjected to drying under vacuum conditions at a temperature of 70°C.

3.2.3.6.2 *Hot Injection*

The hot injection technique encompasses the introduction of precursors into a solvent with a high boiling point at temperatures surpassing the breakdown point of the precursor [122]. The decomposition of individual molecular precursors offers practical and efficient pathways for synthesizing NCs of metal chalcogenides. A coordinating solvent is used to thermally decompose a molecular complex containing both metal and chalcogen elements [123]. The hot injection approach, pioneered by Murray et al. for the synthesis of CdSe NCs, involves rapid nucleation at a high temperature immediately after injection, followed by growth at a comparatively lower temperature [124]. In the usual procedure, a blend containing the metal precursor is initially concocted in a three-necked flask and heated to a predetermined temperature. At this stage, a cooled chalcogenide precursor is swiftly introduced into the heated solution, typically around 200°C–300°C, for a duration of 20–30 minutes. This rapid injection leads to a temperature decrease of 30°C–50°C, triggering burst nucleation and the formation of nanocrystallites. Subsequently, the solution is maintained at a lower temperature, allowing the reaction to proceed for a specific duration and enabling the growth of NCs within a targeted diameter range of approximately 10–100 Å. To obtain larger NCs, or elongated nanorods, a technique involving multiple injections is employed. This involves introducing additional chemical precursors into the reaction flask in a repetitive manner [125].

Iron chalcogenides have been synthesized by some researchers through the process of hot injection using $Fe(acac)_2$ [126], $FeCl_2$ [127], $Fe(acac)_3$ (acac = acectyl-acetonate) [128], or $Fe(CO)_5$ [129] as sources of iron in oxidation states. The selection of an iron source seems to influence both the shape control and nanoparticle phase obtained. The utilization of $FeCl_2$ promotes the formation of pyrite (FeS_2), wherein the iron oxidation state transitions from Fe(II) to Fe(IV) due to the oxidizing atmosphere generated by sulfur. In their study, Li et al. [130] discovered that by adjusting the concentration of $FeCl_2$ in their reaction, they could manipulate both the size and shape of the resulting products. By employing different concentrations of $FeCl_2$ in the presence of oleyl amine (OA), distinct outcomes in terms of NP morphology were observed. At low $FeCl_2$ concentrations, approximately 250 nm nanocubes were formed, whereas higher concentrations yielded ~10 nm nanodendrites. To generate

NPs with a cube shape, Puthussery et al. discovered that by substituting the OA ligands with octadexylxanthate, they were able to achieve more stable colloidal suspensions.

3.2.3.6.3 Solid Synthesis

The International Union of Pure and Applied Chemistry (IUPAC) defines a solid solution as a crystalline structure that incorporates a second component that seamlessly integrates into the lattice of the host crystal, emphasizing that it should not be mistaken for amorphous materials [131]. When it comes to semiconductor preparation, solid solutions are generally considered more advantageous than doped semiconductors. The presence of a hybridized valence band and/or conduction band, with their dispersed nature, enhances charge transfer rates [132], particularly in solid solution photocatalysts. When synthesizing solid solutions, it is crucial to consider other properties inherited from this process, such as electronegativity, valence state, and crystal structure. These factors also play a significant role in determining the overall performance of the solid solution [133]. To optimize the solid solution method, a recommended approach is to combine narrow band gap semiconductors with wide band gap semiconductors, such as utilizing sulfide-based solid solutions [134]. This combination allows for the achievement of enhanced properties and performance in the resulting solid solution.

In their work, Yu et al. effectively integrated $Cd_{1-x}Zn_xS$ into polyacrylonitrile nanofibers using solid solution synthesis, resulting in the development of a leaf-like structure [135]. On the other hand, Schlosser et al. employed the solid solution method to synthesize air-sensitive and colorless Li_6PS_5X (X=Cl, Br, I) at 550°C by combining LiX, Li_2S, and P_2S_5 [136]. In a study conducted by Kudo et al., they successfully synthesized $(AgIn)_xZn_{2(1-x)}S_2$ $(0 \leq x \leq 1)$ compounds by subjecting complex Ag-In-Zn sulfide solid solution precursors to heat treatment [137]. Various heat treatment strategies were explored to investigate the enhancement of the resulting compounds. Under visible light irradiation, $(AgIn)_{0.22}Zn_{1.58}S_2$ with a band gap of 2.33 eV was synthesized through heat treatment using aqueous solutions of K_2SO_3 and Na_2S, carried out under a N_2 flown [137].

3.2.3.6.4 Thermal Decomposition

In a typical thermal decomposition method, the substrate is heated to initiate thermolysis, leading to the nucleation and growth of particles to achieve the desired size and/or shape through controlled precursor reactions. This method is frequently employed for synthesizing CdS NPs, where the cadmium thiolate powder is heated to induce the nucleation of CdS particles [133,138].

In their study, Mokari et al. [139] developed a method for selectively growing semiconductor NCs onto the edges of elongated Cd-chalcogenide nanostructures. They successfully grew various materials, including PbS, Ag_2S, and $Cu_{22x}S$ nanostructures, onto the tips and vertices of Cd-chalcogenide nanostructures such as rods, tetrapods, and cubes. This innovative approach allowed for precise control over the spatial distribution of NCs on the nanostructure surfaces. To synthesize metal (Ag, Pb, Cu) NPs, single-source precursors (SSPs) in the form of metal bis-diethyldithiocarbamate compounds were introduced into a solution containing

Cd-chalcogenide (CdS or CdSe) in trioctylphosphine (TOP) or dodecanethiol (DDT) at RT. Subsequently, the mixture of Cd-chalcogenide nanostructures and the SSPs was subjected to heating at temperatures ranging from 218°C to 290°C, corresponding to the decomposition temperature of the SSPs, for several minutes. This thermal treatment facilitated the transformation of the SSPs into metal NPs, resulting in the integration of metal species into the Cd-chalcogenide nanostructures.

Afterward, the reactor was gradually cooled to RT, and the resulting product was separated from the unused precursors and free ligands through a centrifugation process. The heating process of the mixture comprising the single-source precursors (SSPs) and Cd-chalcogenide nanostructures, carried out at temperatures ranging from 218°C to 290°C, induced a distinct color change in the solution, transitioning from red or yellow to black. In a study by Nair et al. [140], CdS nanorods (NRs) were synthesized using a single-source approach through the thermal decomposition of an organometal complex in a TOPO solution. On the other hand, Mokurala et al. [139] developed chalcogenide NPs (CZTS, CFTS) by thermally decomposing precursor materials including metal acetates, nitrates, thiourea, and $SnCl_2$ for one hour at a temperature of 400°C [133].

3.2.3.6.5 Non-injection

Traditional synthetic techniques such as hydrothermal synthesis [141] and the hot injection route [142] have been associated with challenges such as polydispersion and limited scalability for large-scale production of NCs. To achieve precise control over morphology, size, and element ratio, alternative methods are required. Among these methods, non-injection synthesis has gained popularity due to its practicality and suitability for large-scale production of NCs, making it a preferred choice. The non-injection approach enables the synthesis of NCs at RT or even lower, utilizing precursors prepared through multiple reaction steps and allowing for precise control over their properties [143,144]. This method involves mixing the prepared precursors, providing greater flexibility and accuracy in tailoring the desired characteristics of the NCs.

In their study, Zhong and colleagues explored a non-injection synthesis route to produce $CuInS_2$ NCs with a distinctive triangular morphology and tunable band gap [145]. The researchers initiated the process by creating an intermediate complex through the reaction of $In(Ac)_3$, CuAc, and dodecanethiol. Subsequently, the mixture was heated until $CuInS_2$ NCs were formed, resulting in the desired triangular-shaped NCs with adjustable band gaps. By extending the reaction time from 20 to 120 minutes, the size of CuInS2 NCs can be increased to approximately 3–8 nm. The non-injection method primarily aims to reduce the reaction rate. To prevent rapid aggregation resulting from the fast reaction between metal ions in CuAc and dodecanethiol, the researchers opted to substitute CuAc with CuI, thus achieving better control over the synthesis process [145]. When examining the properties of the components involved, Ac is classified as a hard acid, while Cu is categorized as a soft acid. On the other hand, iodide and thiols (R-SH) are considered soft bases. It is well established that the combination of a soft acid with a soft base results in stronger binding compared to the pairing of a hard acid with a soft base [144]. By substituting CuAc with CuI in this case, the reaction rate between the Cu and S sources is reduced, which can be attributed to the altered interaction between the soft acid and soft base components.

3.2.3.7 Structure of Chalcogens

3.2.3.7.1 Crystalline Chalcogens

Sulfur, in its crystalline form, is in a solid state at standard RT and pressure of 1 atm. Typically, it appears as a yellow substance that is tasteless and nearly odorless. It occurs naturally in various manifestations, such as elemental sulfur, sulfides, sulfates, and organosulfur compounds. The renowned Frasch process is commonly employed for the extraction of sulfur from its natural sources. Unlike other chalcogen elements, sulfur possesses the remarkable capability of exhibiting a diverse array of allotropes. The most prevalent form is the solid S_8 ring, which represents the most thermodynamically stable state at RTRT. At RT, sulfur (S_a) adopts an orthorhombic crystalline structure (see Figure 3.6a), adding to its structural diversity and unique properties [146]. Sulfur exhibits a temperature-dependent crystallographic structure, as demonstrated by the presence of monoclinic sulfur (S_b) shown in Figure 3.6b [147]. As the temperature increases, the crystallographic structure of sulfur undergoes changes. At 95.5°C, it adopts the monoclinic sulfur (S_b) phase, while at 119°C and 160°C, it transitions into liquid sulfur (S_l) and liquid sulfur (S_m), respectively. When the temperature reaches 445°C, sulfur transforms into its gaseous state, forming sulfur vapor. In its gaseous form, sulfur can exist in five distinct molecular forms (S, S_2, S_4, S_6, and S_8), adding to the versatility and complexity of its chemical behavior.

In 1818, Berzelius made the initial discovery of crystalline selenium. The name of this element, selenium, originates from the Greek word “selene,” which translates to “moon.” Selenium can be found in the form of red or gray crystalline structures, as well as black amorphous solids. It exhibits multiple allotropes, which are distinct molecular forms of the element with diverse physical properties. The stable allotrope of selenium is characterized by a metallic gray color and a crystalline hexagonal structure. Selenium in its crystalline form can undergo various structural modifications, with the trigonal phase being the most stable under normal conditions, while the other phases are considered metastable. The crystal structure of trigonal selenium is characterized by interconnected spiral chains that are weakly bonded together. On the other hand, monoclinic selenium exhibits a structure composed of eight-membered rings, further adding to the structural diversity of this element [148].

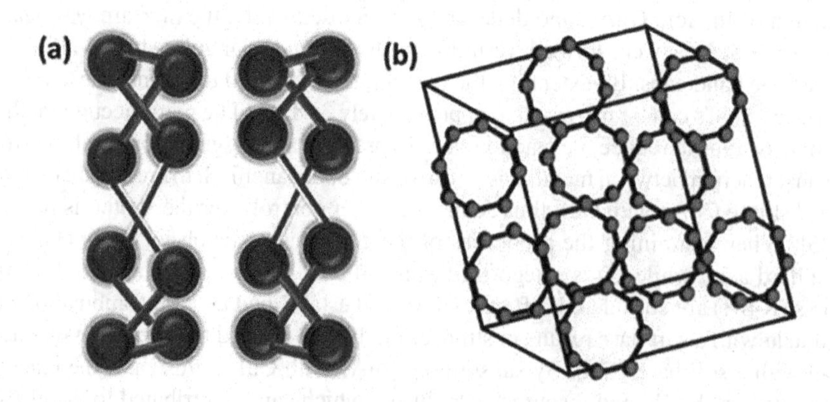

FIGURE 3.6 S_8 structures: (a) orthorhombic and (b) monoclinic crystalline.

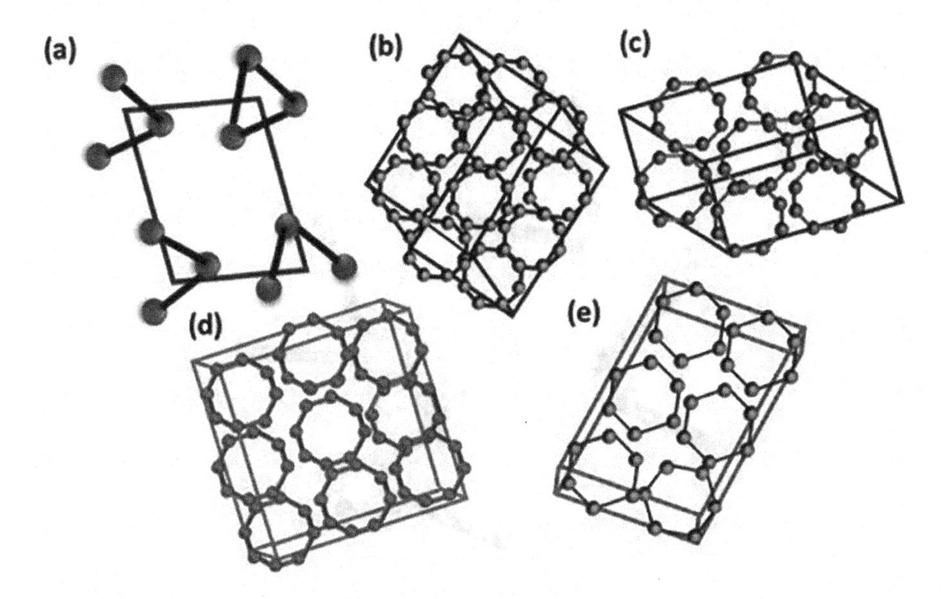

FIGURE 3.7 Rhombohedral crystallographic structures (a, b, c, d, and e) Se_8 trigonal, *a* Se_8 monoclinic, *b* Se_8 monoclinic, *g* Se_8 monoclinic and Se_6, respectively.

The extraction of selenium commonly takes place during the process of electrolytic copper refining. In addition to its crystalline forms, selenium can also exist as an amorphous allotrope, often appearing as a red powder. Selenium exhibits numerous isotopes, which share the same atomic number but differ in the number of neutrons. With over 20 known isotopes, only five of them are considered stable. These stable isotopes of selenium are ^{74}Se, ^{76}Se, ^{77}Se, ^{78}Se, and ^{80}Se [149]. The trigonal crystallographic structure of selenium, known as Se8, is considered the most stable. In contrast, the monoclinic structures, namely *a* Se_8, *b* Se_8, and *g* Se_8, are less stable. Additionally, selenium can also adopt a rohmohedral structure known as Se_6. These various crystallographic structures of selenium, including Se_8, *a* Se_8, *b* Se_8, *g* Se_8, and Se_6, are visually represented in Figure 3.7a–e.

The discovery of crystalline tellurium dates back to 1782, and its name is derived from the Latin word "tellus," meaning "earth." Tellurium possesses semi-metallic characteristics and displays a lustrous, crystalline, and brittle nature. At RT, it exhibits a silver-white color and is classified as a metalloid. In its common form, tellurium is typically found as a dark gray powder, displaying properties that lie between those of metals and non-metals. Tellurium is known to have eight isotopes, adding to its atomic variability. Tellurium exhibits six stable isotopes, namely ^{120}Te, ^{122}Te, ^{123}Te, ^{124}Te, ^{125}Te, and ^{126}Te [150]. The remaining two isotopes, ^{128}Te and ^{130}Te, are slightly radioactive. Figure 3.8 illustrates the stable trigonal crystalline structure of tellurium. Similar to sulfur and selenium, tellurium is capable of forming numerous compounds, particularly with sulfur and selenium. Furthermore, tellurium shares a photoconductivity property with selenium. Tellurium, an element of great rarity, is commonly found in the form of gold tellurides. It is frequently employed in metallurgy, where it is combined with copper, lead, and iron. Moreover, tellurium finds

FIGURE 3.8 Te_8 trigonal crystallographic structure.

application in solar panels and computer memory chips. While it is not toxic or carcinogenic, excessive exposure to tellurium can cause individuals to exhibit a garlic-like odor on their breath [151].

3.2.3.7.2 Polycrystalline Chalcogens

Solid-state polysulfides typically exhibit chain structures consisting of 2–8 sulfur atoms. However, in the case of chains containing five or more sulfur atoms, isolation is only possible in the form of $[NR_{4-x}H_x]^+$ (where $n=6$–8 and R represents an alkyl or substituent) configurations. Crystal structures such as A_2S_2, A_2S_3, and A_2S_5 are observed with alkali metals A ranging from potassium (K) to cesium (Cs). These crystal structures feature highly symmetrical cations, and long-chain polysulfides ($n>6$) adopt a helical all-trans conformation with gauche torsion angles consistently sharing the same sign. When cations are present, the packing forces in polysulfides can lead to the formation of alternative chains, especially in cases where lower symmetry is observed. For heptasulfide anions, there are six theoretically possible configurations. The octasulfide chain, denoted as $S_8{}^2$, adopts a trans–cis–cis–trans (++− −++) conformation. In $S_n{}^2$ chains, typical S–S–S bond angles range from 106 to 111 degrees, while the S–S distances measure between 201 and 208 pm. The localization of charge in polysulfides affects the first and last members of the chain, resulting in shorter distances typically observed in the terminal bonds. In polysulfides, the presence of short intermolecular secondary S....S bonds is often absent [151].

The $Se_n{}^2$ molecules form discrete polyselenide chains that exhibit negligible secondary interactions between the Se atoms. These chains are structurally characterized by Se...Se contacts, with the number of atoms involved in the range of 2–8 [152]. In crystal structures of binary alkali metal selenides, such as A_2Se_n, where n ranges from 2 to 5 (e.g., A_2Se_2 for A = Na, K, Rb), the chains are shorter. On the other hand, longer Se_n^{2} chains with $n = 6–8$ can only accommodate isolated alkali metal cations or large non-coordinating organic monocations within their structure.

The highly symmetric counter cations typically accompany the well-known octaselenides Se_8, Se_7, and Se_6 [153]. In the polyselenide chains, the terminal Se–Se bond lengths are generally shorter compared to the interchain bonds. The average bond distances for heptaselenide anions range from 228 to 233 pm, and the Se–Se–Se bond angles fall within the range of 104°–111°. Furthermore, the variation from 66° to 84° is the torsion angle within the all-trans chain.

As we move down Group 16, the hypervalent Se_n^{2} anions display structures with values of n ranging from 9 to 11, which are distinct from polysulfides and not parallel to them. The presence of the Se_9 anion results in additional bond stabilization, possibly through weak secondary interactions between Se_1 and Se_6, with a bond length of approximately 295 pm [152]. The presence of weak secondary interactions in Se results in a noticeable elongation of the Se_5–Se_6 bond, with a distance of approximately 247 pm and a narrow Se_4–Se_5–Se_6 angle of 93°. In all polyselenide Se_n^{2-} structures with n greater than 8, the internal bonding interactions and cyclization of the crystalline Se_9^{2-} and Se_{10}^{2-} anions are stabilized. The packing factors of the cations can also play a significant role in determining the degree of distortion in the coordination sphere. These factors are involved in the structural directions and can contribute to the overall distortion observed in the coordination environment [151].

Tellurium exhibits a pronounced inclination to engage in intra- and intermolecular bonding involving np^2–ns^* orbitals. These interactions result in the formation of distorted linear Te–Te...Te units found in numerous polytellurides. As a result, polytellurides are often classified as discrete structures or categorized as polymeric chains, sheets, or 3D frameworks, with the specific arrangement often being somewhat arbitrary. Isolated chains of Te_{2n} can be formed through the establishment of strong Te–Te bonds (with bond lengths $d < 313$ pm). However, dianions with values of n ranging from 2 to 6, 8, 12, and 13 are also capable of existing. Similar to selenium (Se), tellurium (Te) also exhibits discrete bicyclic dianions when n equals 7 or 8 [151,154,155].

Unlike polysulfides and most polyselenides, certain polytelluride anions exhibit total negative charges exceeding 2 [156,157]. The structural motif changes in alkali metal polytellurides governed by the empirical A_xTe_y formula occur at various ratios (x/y) such as 5/3, 1/1, 2/3, 2/5, 1/3, 1/4, 1/6, 2/13, 1/7, and 3/22.

For ratios x/y greater than or equal to 2/3, discrete anions are observed, whereas polymeric chains and lamellar anionic networks form when x/y is less than or equal to 2/5 [34,35]. Isolated polytelluride anions are exclusively found in binary alkali metal compounds of the A_xTe_y formula with x/y values greater than or equal to 2/3. The presence of hypervalent bonding gives rise to linear $TeTe^{4-}{}_2$, T-shaped $TeTe^{4-}{}_3$, and square-planar $TeTe^{6-}{}_4$ building units within polymeric anions when x/y is less than or equal to 2/5. Heavy main group elements exhibit finite-length linear chains characterized by electron-rich multicenter bonding [36,37]. For instance, in Te_5 linear units, the chains intersect at right angles, and their bonding involves substantial s-p orbital mixing with relatively less significant p-bonding. This results in the splitting of five px orbitals into bonding ($x2$), non-bonding ($x1$), and anti-bonding orbitals. Similarly, in Te_6, two multicenter bonds can exist, each corresponding to 2×6 electrons. The presence of additional lone pair electrons from s and pz orbitals in six Te atoms (6×4), combined with the four in-plane py lone pair electrons from the two central Te atoms (2×2), allows for the formation of the $Te^{4-}{}_6$ configuration. However, weak secondary interactions can also occur between the Te_8 rings. As a result, polytellurides exhibit a wide range of structures that are governed by well-defined rules, depending on the specific circumstances and conditions [151].

3.2.3.7.3 Amorphous Structure of Chalcogenide

In their natural state, all solids are typically found to have a degree of disorder. This is because solid materials are grown at finite temperatures, resulting in the presence of defects. Interestingly, even crystalline solids such as Si and Ge exhibit a certain level of disorder, despite the common belief that the crystalline state of these materials is defect-free and devoid of impurities. Amorphous materials display a type of disorder known as homogeneous disorder, where many atoms exhibit similar average (or bulk) properties such as specific heat, electrical and thermal conductivities, optical properties, and density, among others. The classification of materials as amorphous or vitreous is based on the presence of this homogeneous disorder [151]. Figure 3.9 provides a schematic representation of an amorphous solid.

Amorphous solids, also known as non-crystalline materials, are characterized by the absence of long-range periodic ordering in their atomic arrangement. Unlike crystalline solids, amorphous materials lack a well-defined and repeating pattern of atom arrangement over extended distances. These materials can be obtained by rapidly cooling or "freezing" the liquid state into a solid state, preventing the formation of a crystalline structure. When amorphous or non-crystalline solids undergo a rapid phase change, they experience significant alterations in their physical properties. Unlike crystalline solids, the fundamental characteristics of amorphous materials primarily depend on the electronic configuration and chemical bonding between neighboring atoms, rather than the long-range periodic arrangement found in crystalline solids. In crystalline solids, these properties are predominantly governed by the extended and ordered arrangement of constituent atoms. The absence of atomic periodicity in amorphous materials allows for a diverse range of material compositions, depending on their method of preparation. These non-crystalline or amorphous materials can exhibit a variety of electrical properties, including electrical insulation, semiconducting behavior, or metallic conductivity [151,158]. The distinctive electrical and optical properties of amorphous semiconductors have garnered significant interest, especially in the context of PCM applications. These materials have proven to be particularly promising in PCM due to their specific characteristics in terms of conductivity and light absorption [159,160]. Amorphous semiconductors have emerged as promising contenders for various commercial applications. It's worth noting that the cohesive energy between atoms in crystals and amorphous materials remains unchanged. Consequently, amorphous semiconductors can be categorized into two main groups based on their chemical bonding: covalently bonded amorphous semiconductors and semiconducting oxide glasses. This classification helps to differentiate between these two broad categories of amorphous semiconductors with distinct bonding characteristics [151]. Covalent semiconductors can be

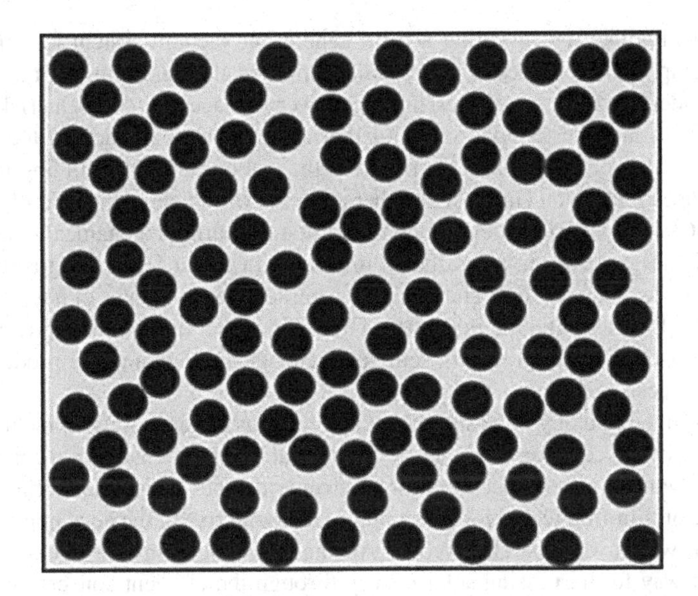

FIGURE 3.9 Structure of amorphous solid.

further categorized into two main categories [161,162]. The first category comprises tetrahedrally bonded amorphous solids such as amorphous silicon (a-Si), germanium (a-Ge), and others. The second category consists of chalcogen semiconductors, which involve combinations of chalcogen elements (sulfur, selenium, and tellurium) with elements from groups III, IV, or V of the periodic table [151]. This classification helps to differentiate between these distinct types of covalent semiconductors based on their bonding and elemental composition.

3.3 THIN FILMS

Thin films refer to layers of material that vary in thickness, ranging from a fraction of a nanometer (monolayer) to several micrometers. The technology behind thin films is a major driving force in our society, playing a crucial role in various industries. These materials find applications in diverse fields such as electronics, communication, and biotechnology, among others. The versatility and widespread use of thin-film technology make it a significant technological driver with substantial impact in multiple sectors.

3.3.1 History of Thin Films

The development of thin-film technology can be traced back to the discovery of vacuum technology in 1640, which coincided with the invention of the barometer. Vacuum technology played a pivotal role in enabling scientists to create cleaner surfaces necessary for the deposition of thin films. This technological advancement provided a crucial foundation for the progress of thin-film technology and its subsequent applications. In 1652, Otto von Guericke revolutionized the field of vacuum technology by inventing a third-generation vacuum system known as the piston pump. This remarkable development marked the beginning of the journey toward understanding and harnessing the power of vacuums. Von Guericke's pioneering work laid the foundation for further advancements in vacuum technology, paving the way for the remarkable progress that followed [163]. Building upon the groundwork laid by Von Guericke, further progress was made in the field of electricity. In 1663, Von Guericke devised a system that converted mechanical energy into electrical energy, marking a significant milestone in the understanding and utilization of electrical power [164]. The discovery sparked the emergence of physical vapor deposition (PVD) around the mid-1800s [165]. Since that time, thin-film technology has been extensively employed in semiconductor technology and the production of energy devices.

Thin-film technology finds significant applications in various areas, including the production of PV cells for solar energy and the fabrication of nanoscale components used in advanced microprocessors for electronic devices. It is noteworthy that the discovery of thin-film technology also played a crucial role in the development of the PV effect, which was first observed by Alexandre-Edmond in 1839. This discovery paved the way for harnessing solar energy through the efficient conversion of light into electricity, contributing to the advancement of PV technology [166–168].

In the realm of thin-film technology, significant advancements were propelled by the physicist's creation of the very first PV cell, marking a milestone in the field [169]. Subsequent breakthroughs in thin-film technology were also made possible by Herman Sprengel's contributions in 1865. Sprengel developed an improved vacuum pump, enhancing the capabilities of the vacuum system. This advancement enabled researchers to delve deeper into the study of gas discharges and explore the possibilities of sputter deposition, leading to further discoveries and advancements in thin-film technology.

In conclusion, the demand for thin-film technology has expanded in response to technological advancements. Moreover, significant progress has been achieved in the processing and fabrication of thin-film materials. Deposition techniques, both physical and chemical, have emerged as crucial methods to produce these materials. The progress made in thin-film deposition methods throughout the 20th century has paved the way for numerous technological advancements across various fields. These include light-emitting diodes (LEDs), magnetic recording media, integrated passive devices, electronic semiconductor devices, optical coatings, durable coatings for cutting tools, as well as thin-film solar cells and batteries for energy generation and storage, respectively. Additionally, the application of thin-film technology has extended to the field of pharmaceuticals, particularly through thin-film drug delivery [170]. The potential for thin-film materials and technologies is vast, and it necessitates extensive research and development to unlock its promising future.

3.3.2 Properties of Thin Films

Thin films, being nanostructured materials, possess distinct properties that set them apart from their bulk counterparts. Here are some commonly observed properties of thin-film materials.

3.3.2.1 Electrical Properties

Thin-film materials exhibit superior electrical properties when compared to bulk materials. Unlike bulk materials, where electrical properties are primarily determined by the material's composition, thin-film materials are influenced by scaling effects. The electrical transport in thin films is affected by various factors, such as film thickness, lattice dimensions, and surface roughness. These properties play a significant role in shaping the electrical characteristics of thin films.

3.3.2.2 Optical Properties

Thin-film materials surpass bulk materials in terms of their optical properties, particularly in terms of reflectivity and absorption across different energy ranges.

3.3.2.3 Magnetic Properties

Thin films can exhibit three distinct ordered magnetic states: ferromagnetism, antiferromagnetism, and ferrimagnetism. These states represent different arrangements of magnetic moments within the thin-film structure.

3.3.2.4 Mechanical Properties

Due to their nanostructured nature, thin films demonstrate exceptional mechanical properties. They possess high yield strengths, enabling them to withstand substantial residual stresses. Moreover, thin films display favorable elastic and plastic characteristics, making them valuable for a wide range of applications.

3.3.2.5 Chemical Compositions

By carefully controlling the deposition processes, the chemical compositions of thin-film materials can be precisely managed. This level of control enables the enhancement of their chemical behavior, opening possibilities for tailored chemical properties and improved performance.

3.3.2.6 Microstructural Stability

Certain thin-film materials demonstrate exceptional structural stability even under extreme conditions, including high temperatures and cryogenic environments. Their ability to maintain their structural integrity in such challenging conditions makes them highly desirable for applications that require stability across a wide range of temperature conditions.

3.3.2.7 Surface Protection Properties

Thin-film materials showcase outstanding mechanical and chemical properties, making them ideal for surface protection through deposition. Their notable attributes include superior wear and corrosion resistance, which effectively safeguard a variety of surfaces against degradation and damage.

3.3.2.8 Surface Roughness and Film Thickness

The properties of thin-film materials hold significant importance. Typically, thin films exhibit minimal surface roughness compared to bulk materials, often at micro- and nanoscale dimensions. The film thickness plays a crucial role in determining both surface protection and other functional characteristics of the thin-film materials. However, it is essential to acknowledge that these properties are general in nature, and each deposition process aims to achieve specific performance properties or a combination thereof. The desired characteristics may vary depending on the intended application and the specific goals of the deposition process.

3.3.3 Deposition and Characterization of Thin Films

Thin films exhibit distinct characteristics that set them apart from their bulk counterparts. When at least one dimension of a material is within the nanometer scale (less than 100 nm), it is referred to as a nanostructured material. Due to the imposed geometrical restrictions, nanostructured materials display novel physical and chemical properties that differ significantly from their bulk counterparts. Thin films, regardless of whether they are polycrystalline, single crystal (epitaxial), or amorphous, exhibit distinct material, physical, and chemical properties compared to their bulk counterparts. This is primarily attributed to their large surface-to-volume ratio, resulting in higher surface energy. The unique surface characteristics of thin films, combined with their reduced thickness (typically in the range of 1 μm), contribute to their altered properties and behavior [171]. The ability to customize and manipulate materials to achieve specific physical properties has sparked extensive research endeavors in various fields, including sensors, microelectronics, and optoelectronics [172]. To achieve the desired outcomes in these devices, the accurate deposition of multiple layers of films is crucial. These films can be in the form of polymers, metals, semiconductors, or insulators and may exist in amorphous, polycrystalline, or epitaxial forms. The precise growth and arrangement of these layers play a pivotal role in determining the functionality and performance of the devices. By carefully controlling the deposition process, scientists and engineers can tailor the properties of each layer to achieve the desired outcome and optimize the overall device performance.

Chapter 4 provides an overview of various techniques employed for depositing these films, including molecular beam epitaxy (MBE), CVD, pulsed laser deposition (PLD), sputtering, thermal evaporation, e-beam evaporation, electroplating, spin coating, and numerous other methods. These techniques are essential for precisely fabricating and depositing the films, allowing researchers and engineers to harness their unique properties and tailor them for specific applications.

The properties of thin-film materials often deviate from those observed at the bulk level. As the application of thin films spans various disciplines, there is a growing need for interdisciplinary approaches to characterize and measure the properties of both individual films and multilayer films [173]. This evolving demand underscores the importance of developing advanced techniques and methodologies to accurately assess and understand the unique characteristics of thin films, enabling their effective utilization in diverse fields. These techniques encompass a wide range of parameters, such as film thickness, microstructure, composition, mechanical properties, chemical properties, optical properties, electrical properties, magnetic properties, and many more. These various characterization techniques are also described in Chapter 5.

3.3.4 Applications of Thin Films

Over the course of history, thin-film materials have transitioned from being solely decorative to becoming integral components in advanced composites and functional elements due to their exceptional properties. Here are some of the prevalent commercial and industrial applications where thin-film materials find extensive use:

3.3.4.1 Semiconductors

Thin-film materials play a crucial role in the production of integrated circuits (ICs) used in semiconductor devices for electronics. The application of thin-film technology extends to the fabrication of various components such as transistors, resistors, rectifiers, LEDs, capacitors, and numerous other elements used in semiconductor devices. One of the notable advantages of thin-film semiconductors is their cost-effectiveness, making them a viable option for large-scale production. These semiconductors can be readily fabricated on large areas, even with complex structures and geometries. Furthermore, by adjusting the process parameters of thin-film deposition methods [174,175], it becomes relatively straightforward to create multi-functional semiconductor structures with diverse complex configurations.

3.3.4.2 Solar Cell

In the realm of solar energy, second-generation solar cells employ thin layers of PV materials. This technology holds significant importance as it enables the generation of renewable and clean energy through the utilization of solar cells. PVs not only play a crucial role in terrestrial applications, providing a sustainable energy source, but they also hold immense potential for space systems, where solar energy can serve as a primary source of electrical power [175]. For outdoor applications, three prevalent types of thin-film solar cells are commonly utilized: amorphous silicon (a-Si), cadmium telluride (CdTe), and copper indium gallium selenide (CIGS) [174].

3.3.4.3 Optical Devices

Thin-film materials are commonly applied to the surfaces of lenses or mirrors to enhance their reflection or transmission properties. By depositing thin-film coatings, the reflecting characteristics of mirrors can be significantly improved, catering to a wide range of applications. For instance, in laser mirrors, where high reflectivity is essential, thin-film coatings can achieve reflectivity levels as high as 99.99%. In optical devices designed to minimize reflection, thin-film coatings are employed to decrease undesired reflections from their surfaces. Several examples of such anti-reflective devices comprise solar panels, eyeglasses, and camera lenses.

3.3.4.4 Data Storage

The recording head and the recording medium are the two essential components in most storage technologies, and advancements in one coincide with advancements in the other. Therefore, enhancements in read/write head technology are closely linked to improvements in storage media. To cater to the ever-increasing need for information storage, thin films have emerged as the prevailing choice of medium, and their implementation has extended to read/write heads. The realm of materials science assumes a crucial role in establishing the relationship between the structure, properties, and manufacturing techniques of thin-film magnetic materials, serving both purposes effectively.

3.3.4.5 Sensor

In a diverse array of applications, sensor devices incorporate thin-film technology to create active layers. These thin-film sensors excel at monitoring environmental conditions by precisely measuring a multitude of physical parameters. Leveraging thin-film technology ensures precise measurements while maintaining stability and reliability. Additionally, the utilization of this technology enables the cost-effective production of sensor devices, making them more accessible. Thin-film sensors possess a multitude of advantages compared to conventional sensors, making them an attractive choice. These sensors are known for their precision, stability, reliability, and cost-effectiveness. Engineers particularly benefit from their compact size and versatile housing options, which are especially advantageous in applications that demand precision.

Thin-film technology holds tremendous potential for diverse applications in both commercial and industrial sectors, offering a wide range of possibilities and opportunities. As the world embraces Industry 4.0 and digital manufacturing, the need for compact and high-performance devices will only escalate, further fueling the demand for thin-film technology. The significance of thin-film technologies in enabling the Internet of Things (IoT) and other digital platforms crucial to the Fourth Industrial Revolution is set to persist. The ongoing innovation within the realm of thin-film technology will be fueled by the continued advancement of intelligent devices and sensors. These cutting-edge technologies play a crucial role in enabling optimization, industrial processes, and facilitating data analysis. In response to the growing need for health monitoring, particularly in the realm of comprehending and combating complex diseases such as cancer, there is a pressing requirement to create sensor structures that are exceptionally precise, innovative, and compact [174].

3.4 SUMMARY

Smart nanomaterials and thin films made from CNTs, graphene, and chalcogenides have unique properties that make them suitable for various applications. CNTs are cylindrical tubes of graphene with a diameter of only a few nanometers and a length that can be many times their diameter. They have exceptional mechanical, electrical, and thermal properties, making them attractive for a wide range of applications, including electronics, energy storage, and sensing. CNT-based thin films can be used to create flexible, transparent conductive films, which are ideal for use in touch screens and other electronic devices. Graphene is a single layer of carbon atoms arranged in a hexagonal lattice. It has high electrical and thermal conductivity, high mechanical strength, and is also transparent, making it a versatile material for many applications. Graphene-based thin films can be used for flexible and transparent electronics, energy storage, and sensing applications. Chalcogenides are materials made from elements from the chalcogen group, including sulfur, selenium, and tellurium. They have unique electronic and optical properties and are commonly used in photonics, energy storage, and sensing applications. Chalcogenide-based thin films can be used for IR optics, thermoelectrics, and PVs. These materials have the potential to revolutionize many fields, including electronics, energy, and biomedicine.

REFERENCES

[1] R. Bogue, Smart materials: a review of recent developments, *Assembly Automation*. 32 (2012) 3–7. https://doi.org/10.1108/01445151211198674.
[2] S. Thangudu, Next generation nanomaterials: smart nanomaterials, significance, and biomedical applications, in: *Applications of Nanomaterials in Human Health*, Springer, 2020: pp. 287–312.
[3] D.R. Dreyer, S. Park, C.W. Bielawski, R.S. Ruoff, The chemistry of graphene oxide, *Chemical Society Reviews*. 39 (2010) 228–240.
[4] A.K. Geim, Graphene prehistory, *Physica Scripta*. 2012 (2012) 14003.
[5] P.R. Wallace, The band theory of graphite, *Physical Review*. 71 (1947) 622.
[6] G. Ruess, F. Vogt, Höchstlamellarer Kohlenstoff aus Graphitoxyhydroxyd, *Monatshefte Für Chemie Und Verwandte Teile Anderer Wissenschaften*. 78 (1948) 222–242.
[7] H.-P. Boehm, A. Clauss, G.O. Fischer, U. Hofmann, Das adsorptionsverhalten sehr dünner kohlenstoff-folien, *Zeitschrift Für Anorganische Und Allgemeine Chemie*. 316 (1962) 119–127.
[8] C. Oshima, A. Nagashima, Ultra-thin epitaxial films of graphite and hexagonal boron nitride on solid surfaces, *Journal of Physics: Condensed Matter*. 9 (1997) 1.
[9] A.K. Geim, P. Kim, Carbon wonderland, *Scientific American*. 298 (2008) 90–97.
[10] H. Shioyama, Cleavage of graphite to graphene, *Journal of Materials Science Letters*. 20 (2001) 499–500.
[11] S. Horiuchi, T. Gotou, M. Fujiwara, T. Asaka, T. Yokosawa, Y. Matsui, Single graphene sheet detected in a carbon nanofilm, *Applied Physics Letters*. 84 (2004) 2403–2405.
[12] H.P. Boehm, R. Setton, E. Stumpp, Nomenclature and terminology of graphite intercalation compounds, *Carbon*. 24 (1986) 241–245.
[13] S. Ray, *Applications of Graphene and Graphene-Oxide Based Nanomaterials*, William Andrew, 2015.
[14] A.K. Geim, K.S. Novoselov, The rise of graphene, *Nature Material*. 6 (2007) 183–191.
[15] S.-H. Liao, P.-L. Liu, M.-C. Hsiao, C.-C. Teng, C.-A. Wang, M.-D. Ger, C.-L. Chiang, One-step reduction and functionalization of graphene oxide with phosphorus-based compound to produce flame-retardant epoxy nanocomposite, *Industrial & Engineering Chemistry Research*. 51 (2012) 4573–4581.
[16] F. Zhuge, B. Hu, C. He, X. Zhou, Z. Liu, R.-W. Li, Mechanism of nonvolatile resistive switching in graphene oxide thin films, *Carbon*. 49 (2011) 3796–3802.
[17] J. Zhang, X.S. Zhao, Conducting polymers directly coated on reduced graphene oxide sheets as high-performance supercapacitor electrodes, *The Journal of Physical Chemistry C*. 116 (2012) 5420–5426.
[18] M. Inagaki, F. Kang, *Materials Science and Engineering of Carbon: Fundamentals*, Butterworth-Heinemann, 2014.
[19] D.D.L. Chung, Exfoliation of graphite, *Journal of Materials Science*. 22 (1987) 4190–4198.
[20] B.C. Brodie, XIII. On the atomic weight of graphite, *Philosophical Transactions of the Royal Society of London*. 149 (1859) 249–259.
[21] N.I. Zaaba, K.L. Foo, U. Hashim, S.J. Tan, W.-W. Liu, C.H. Voon, Synthesis of graphene oxide using modified hummers method: solvent influence, *Procedia Engineering*. 184 (2017) 469–477.

[22] W.S. Hummers Jr, R.E. Offeman, Preparation of graphitic oxide, *Journal of the American Chemical Society*. 80 (1958) 1339.
[23] J. Chen, B. Yao, C. Li, G. Shi, An improved Hummers method for eco-friendly synthesis of graphene oxide, *Carbon*. 64 (2013) 225–229.
[24] L. Shahriary, A.A. Athawale, Graphene oxide synthesized by using modified hummers approach, *International Journal of Renewable Energy and Environmental Engineering*. 2 (2014) 58–63.
[25] T. Szabó, O. Berkesi, P. Forgó, K. Josepovits, Y. Sanakis, D. Petridis, I. Dékány, Evolution of surface functional groups in a series of progressively oxidized graphite oxides, Chemistry of materials. 18 (2006) 2740-2749.
[26] W. Gao, The chemistry of graphene oxide, in: W. Gao, *Graphene Oxide: reduction recipes, spectroscopy, and applications*, Springer, 2015: pp. 61–95.
[27] A. Lerf, H. He, M. Forster, J. Klinowski, Structure of graphite oxide revisited, *The Journal of Physical Chemistry B*. 102 (1998) 4477–4482.
[28] V. Gupta, N. Sharma, U. Singh, M. Arif, A. Singh, Higher oxidation level in graphene oxide, *Optik*. 143 (2017) 115–124.
[29] S. Park, R.S. Ruoff, Chemical methods for the production of graphenes, *Nature Nanotechnology*. 4 (2009) 217–224. https://doi.org/10.1038/nnano.2009.58.
[30] K.S. Novoselov, L. Colombo, P.R. Gellert, M.G. Schwab, K. Kim, A roadmap for graphene, *Nature*. 490 (2012) 192–200.
[31] S. Stankovich, D.A. Dikin, G.H.B. Dommett, K.M. Kohlhaas, E.J. Zimney, E.A. Stach, R.D. Piner, S.T. Nguyen, R.S. Ruoff, Graphene-based composite materials, *Nature*. 442 (2006) 282–286.
[32] Z. Wang, J.K. Nelson, H. Hillborg, S. Zhao, L.S. Schadler, Graphene oxide filled nanocomposite with novel electrical and dielectric properties, *Advanced Materials*. 24 (2012) 3134–3137.
[33] L. Tang, X. Li, R. Ji, K.S. Teng, G. Tai, J. Ye, C. Wei, S.P. Lau, Bottom-up synthesis of large-scale graphene oxide nanosheets, *Journal of Materials Chemistry*. 22 (2012) 5676–5683.
[34] G. Eda, G. Fanchini, M. Chhowalla, Large-area ultrathin films of reduced graphene oxide as a transparent and flexible electronic material, *Nature Nanotechnology*. 3 (2008) 270–274.
[35] S. Pei, J. Zhao, J. Du, W. Ren, H.-M. Cheng, Direct reduction of graphene oxide films into highly conductive and flexible graphene films by hydrohalic acids, *Carbon*. 48 (2010) 4466–4474.
[36] A. Bagri, C. Mattevi, M. Acik, Y.J. Chabal, M. Chhowalla, V.B. Shenoy, Structural evolution during the reduction of chemically derived graphene oxide, *Nature Chemistry*. 2 (2010) 581–587.
[37] S. Stankovich, D.A. Dikin, R.D. Piner, K.A. Kohlhaas, A. Kleinhammes, Y. Jia, Y. Wu, S.T. Nguyen, R.S. Ruoff, Synthesis of graphene-based nanosheets via chemical reduction of exfoliated graphite oxide, *Carbon*. 45 (2007) 1558–1565.
[38] D. Voiry, J. Yang, J. Kupferberg, R. Fullon, C. Lee, H.Y. Jeong, H.S. Shin, M. Chhowalla, High-quality graphene via microwave reduction of solution-exfoliated graphene oxide, *Science*. 353 (2016) 1413–1416.
[39] P. Kumar, F. Shahzad, S. Yu, S.M. Hong, Y.-H. Kim, C.M. Koo, Large-area reduced graphene oxide thin film with excellent thermal conductivity and electromagnetic interference shielding effectiveness, *Carbon*. 94 (2015) 494–500.
[40] L.L. Zhang, X. Zhao, M.D. Stoller, Y. Zhu, H. Ji, S. Murali, Y. Wu, S. Perales, B. Clevenger, R.S. Ruoff, Highly conductive and porous activated reduced graphene oxide films for high-power supercapacitors, *Nano Letters*. 12 (2012) 1806–1812.
[41] J. Zhao, L. Liu, F. Li, *Graphene Oxide: Physics and Applications*, Springer, 2015.

[42] C. Lee, X. Wei, J.W. Kysar, J. Hone, Measurement of the elastic properties and intrinsic strength of monolayer graphene, *Science.* 321 (2008) 385–388.
[43] P. Zhang, L. Ma, F. Fan, Z. Zeng, C. Peng, P.E. Loya, Z. Liu, Y. Gong, J. Zhang, X. Zhang, Fracture toughness of graphene, *Nature Communications.* 5 (2014) 1–7.
[44] J.W. Suk, R.D. Piner, J. An, R.S. Ruoff, Mechanical properties of monolayer graphene oxide, *ACS Nano.* 4 (2010) 6557–6564.
[45] C. Gómez-Navarro, M. Burghard, K. Kern, Elastic properties of chemically derived single graphene sheets, *Nano Letters.* 8 (2008) 2045–2049.
[46] A.A. Balandin, S. Ghosh, W. Bao, I. Calizo, D. Teweldebrhan, F. Miao, C.N. Lau, Superior thermal conductivity of single-layer graphene, *Nano Letters.* 8 (2008) 902–907. https://doi.org/10.1021/nl0731872.
[47] Y.S. Touloukian, T. Makita, *Thermophysical Properties of Matter – The TPRC Data Series. Volume 6. Specific Heat-Nonmetallic Liquids and Gases*, Thermophysical and Electronic Properties Information Analysis Center, 1970.
[48] W. Cai, A.L. Moore, Y. Zhu, X. Li, S. Chen, L. Shi, R.S. Ruoff, Thermal transport in suspended and supported monolayer graphene grown by chemical vapor deposition, *Nano Letters.* 10 (2010) 1645–1651.
[49] X. Xu, L.F.C. Pereira, Y. Wang, J. Wu, K. Zhang, X. Zhao, S. Bae, C. Tinh Bui, R. Xie, J.T.L. Thong, Length-dependent thermal conductivity in suspended single-layer graphene, *Nature Communications.* 5 (2014) 1–6.
[50] C. Faugeras, B. Faugeras, M. Orlita, M. Potemski, R.R. Nair, A.K. Geim, Thermal conductivity of graphene in corbino membrane geometry, *ACS Nano.* 4 (2010) 1889–1892.
[51] J.-U. Lee, D. Yoon, H. Kim, S.W. Lee, H. Cheong, Thermal conductivity of suspended pristine graphene measured by Raman spectroscopy, *Physical Review B.* 83 (2011) 81419.
[52] J.H. Seol, I. Jo, A.L. Moore, L. Lindsay, Z.H. Aitken, M.T. Pettes, X. Li, Z. Yao, R. Huang, D. Broido, Two-dimensional phonon transport in supported graphene, *Science.* 328 (2010) 213–216.
[53] P.G. Klemens, Theory of thermal conduction in thin ceramic films, *International Journal of Thermophysics.* 22 (2001) 265–275.
[54] W. Jang, Z. Chen, W. Bao, C.N. Lau, C. Dames, Thickness-dependent thermal conductivity of encased graphene and ultrathin graphite, *Nano Letters.* 10 (2010) 3909–3913.
[55] M.T. Pettes, I. Jo, Z. Yao, L. Shi, Influence of polymeric residue on the thermal conductivity of suspended bilayer graphene, *Nano Letters.* 11 (2011) 1195–1200.
[56] V. Georgakilas, M. Otyepka, A.B. Bourlinos, V. Chandra, N. Kim, K.C. Kemp, P. Hobza, R. Zboril, K.S. Kim, Functionalization of graphene: covalent and non-covalent approaches, derivatives and applications, *Chemical Reviews.* 112 (2012) 6156–6214.
[57] S. Stankovich, R.D. Piner, S.T. Nguyen, R.S. Ruoff, Synthesis and exfoliation of isocyanate-treated graphene oxide nanoplatelets, *Carbon.* 44 (2006) 3342–3347.
[58] P.J.F. Harris, Transmission electron microscopy of carbon: a brief history, *C – Journal of Carbon Research.* 4 (2018) 4. https://doi.org/10.3390/c4010004.
[59] S. Iijima, Helical microtubules of graphitic carbon, *Nature.* 354 (1991) 56–58. https://doi.org/10.1038/354056a0.
[60] T.I.S. Iijima, Single-shell carbon nanotubes of 1-nm diameter, *Nature.* 362 (1993) 603. https://doi.org/10.1111/j.1753-4887.1971.tb07294.x.
[61] D.S. Bethune, C.H. Klang, M.S. De Vries, G. Gorman, R. Savoy, J. Vazquez, R. Beyers, Cobalt-catalysed growth of carbon nanotubes with single-atomic-layer walls, *Nature.* 363 (1993) 605–607. https://doi.org/10.1038/363605a0.
[62] S.B. Sinnott, R. Andrews, Carbon nanotubes: synthesis, properties, and applications carbon nanotubes, *Critical Reviews in Solid State and Materials Sciences.* 26 (2001) 145–249. https://doi.org/10.1080/20014091104189.

[63] M.I. Katsnelson, *Graphene: Carbon in Two Dimensions*, Cambridge University Press, 2012: pp. 1–351. https://doi.org/10.1017/CBO9781139031080.

[64] S.K. Tiwari, V. Kumar, A. Huczko, R. Oraon, A. De Adhikari, G.C. Nayak, Magical allotropes of carbon: prospects and applications, *Critical Reviews in Solid State and Materials Sciences*. 41 (2016) 257–317. https://doi.org/10.1080/10408436.2015.1127206.

[65] G. Lalwani, B. Sitharaman, Multifunctional fullerene - and metallofullerene-based nanobiomaterials, *Nano LIFE*. 03 (2013) 1342003. https://doi.org/10.1142/S1793984413420038.

[66] I.W. Frank, D.M. Tanenbaum, A.M. van der Zande, P.L. McEuen, Mechanical properties of suspended graphene sheets, *Journal of Vacuum Science & Technology B: Microelectronics and Nanometer Structures*. 25 (2007) 2558. https://doi.org/10.1116/1.2789446.

[67] M. Daenen, R.D. de Fouw, B. Hamers, P.G.A. Janssen, K. Schouteden, M.A.J. Veld, *The Wondrous World of Carbon Nanotubes*, Eindhoven University of Technology, 2003: pp. 1–35. https://doi.org/10.1126/science.1078727.

[68] S. Iijima, P.M. Ajayan, T. Ichihashi, Growth model for carbon nanotubes, *Physical Review Letters*. 69 (1992) 3100–3103. https://doi.org/10.1103/PhysRevLett.69.3100.

[69] M. Bansal, C. Lal, R. Srivastava, M.N. Kamalasanan, L.S. Tanwar, Comparison of structure and yield of multiwall carbon nanotubes produced by the CVD technique and a water assisted method, *Physica B: Condensed Matter*. 405 (2010) 1745–1749. https://doi.org/10.1016/j.physb.2010.01.031.

[70] N. Karousis, N. Tagmatarchis, D. Tasis, Current progress on the chemical modification of carbon nanotubes, *Chemical Reviews*. 110 (2010) 5366–5397. https://doi.org/10.1021/cr100018g.

[71] H. Dai, A.G. Rinzler, P. Nikolaev, A. Thess, D.T. Colbert, R.E. Smalley, Single-wall nanotubes produced by metal-catalyzed disproportionation of carbon monoxide, *Chemical Physics Letters*. 260 (1996) 471–475.

[72] A. Mostofizadeh, Y. Li, B. Song, Y. Huang, Synthesis, properties, and applications of low-dimensional carbon-related nanomaterials, *Journal of Nanomaterials*. 2011 (2011) 1–21. https://doi.org/10.1155/2011/685081.

[73] J. Prasek, J. Drbohlavova, J. Chomoucka, J. Hubalek, O. Jasek, V. Adam, R. Kizek, Chemical vapor depositions for carbon nanotubes synthesis, in: A. K. Mishra, *Carbon Nanotubes*, Nova Science, 2013: pp. 87–106.

[74] M.L. Terranova, V. Sessa, M. Rossi, The world of carbon nanotubes: an overview of CVD growth methodologies, *Chemical Vapor Deposition*. 12 (2006) 315–325. https://doi.org/10.1002/cvde.200600030.

[75] H. Yokomichi, M. Ichihara, S. Nimori, N. Kishimoto, Morphology of carbon nanotubes synthesized by thermal CVD under high magnetic field up to 10 T, *Vacuum*. 83 (2008) 625–628. https://doi.org/10.1016/j.vacuum.2008.04.037.

[76] A. Eatemadi, H. Daraee, H. Karimkhanloo, M. Kouhi, N. Zarghami, Carbon nanotubes: properties, synthesis, purification, and medical applications, *Nanoscale Research Letters*. 9 (2014) 1–13. https://doi.org/10.1186/1556-276X-9-393.

[77] L. Stobinski, B. Lesiak, L. Kövér, J. Tóth, S. Biniak, G. Trykowski, J. Judek, Multiwall carbon nanotubes purification and oxidation by nitric acid studied by the FTIR and electron spectroscopy methods, *Journal of Alloys and Compounds*. 501 (2010) 77–84. https://doi.org/10.1016/j.jallcom.2010.04.032.

[78] A. Aqel, K.M.M.A. El-Nour, R.A.A. Ammar, A. Al-Warthan, Carbon nanotubes, science and technology part (I) structure, synthesis and characterisation, *Arabian Journal of Chemistry*. 5 (2012) 1–23. https://doi.org/10.1016/j.arabjc.2010.08.022.

[79] A. Hirsch, Functionalization of single-walled carbon nanotubes, *Angewandte Chemie - International Edition*. 41 (2002) 1853–1859. https://doi.org/10.1002/1521-3773(20020603)41:11<1853::AID-ANIE1853>3.0.CO;2-N.

[80] J.M. Wernik, S.A. Meguid, Recent developments in multifunctional nanocomposites using carbon nanotubes, *Applied Mechanics Reviews*. 63 (2010) 1–40. https://doi.org/10.1115/1.4003503.

[81] S. Ciraci, S. Dag, T. Yildirim, O. Gülseren, R.T. Senger, Functionalized carbon nanotubes and device applications, *Journal of Physics: Condensed Matter*. 16 (2004) R901. https://doi.org/10.1088/0953-8984/16/29/R01.

[82] A. Ahmed, Synthesis and characterization of magnetic carbon nanotubes, Doctoral Dissertation, McMaster University, 2017: pp. 1–181.

[83] C. Zhou, J. Kong, H. Dai, Intrinsic electrical properties of individual single-walled carbon nanotubes with small band gaps, *Physical Review Letters*. 84 (2000) 5604–5607. https://doi.org/10.1103/PhysRevLett.84.5604.

[84] G.R.A. Jamal, M.S. Arefin, S.M. Mominuzzaman, Empirical prediction of bandgap in semiconducting single-wall carbon nanotubes, in: *2012 7th International Conference on Electrical and Computer Engineering, ICECE 2012*, 2012: pp. 221–224. https://doi.org/10.1109/ICECE.2012.6471525.

[85] L. Meng, C. Fu, Q. Lu, Advanced technology for functionalization of carbon nanotubes, *Progress in Natural Science*. 19 (2009) 801–810. https://doi.org/10.1016/j.pnsc.2008.08.011.

[86] M.F.L. De Volder, S.H. Tawfick, R.H. Baughman, A.J. Hart, Carbon nanotubes: present and future commercial applications, *Science*. 339 (2013) 535–540.

[87] K.S. Ibrahim, Carbon nanotubes-properties and applications: a review, *Carbon Letters*. 14 (2013) 131–144. https://doi.org/10.5714/cl.2013.14.3.131.

[88] L. Ebert, Science of fullerenes and carbon nanotubes: by M. S. Dresselhaus, G. Dresselhaus and P. C. Eklund, Academic Press, 1996, 965 pages, $130.00. ISBN 0-12-221820-5, *Carbon*. 35 (1997) 437–438. https://doi.org/10.1016/S0008-6223(97)89618-2.

[89] K. Varshney, Carbon nanotubes: a review on synthesis, properties and applications, *International Journal of Engineering Research and General Science*. 2 (2014) 660–677.

[90] C. Wang, J. Meyer, N. Teichert, A. Auge, E. Rausch, B. Balke, A. Hütten, G.H. Fecher, C. Felser, Heusler nanoparticles for spintronics and ferromagnetic shape memory alloys, *Journal of Vacuum Science & Technology B*. 32 (2014) 020802. https://doi.org/10.1116/1.4866418.

[91] N. Karousis, N. Tagmatarchis, D. Tasis, Current progress on the chemical modification of carbon nanotubes, *Chemical Reviews*. 110 (2010) 5366–5397. https://doi.org/10.1021/cr100018g.

[92] R.J. Chen, S. Bangsaruntip, K.A. Drouvalakis, N. Wong, S. Kam, M. Shim, Y. Li, W. Kim, P.J. Utz, H. Dai, Noncovalent functionalization of carbon nanotubes for highly specific electronic biosensors, *Proceedings of the National Academy of Sciences*. 100 (2003) 4984–4989.

[93] S.C. Ray, S. Sitha, P. Papakonstantinou, Change of magnetic behaviour of nitrogenated carbon nanotubes on chlorination/oxidation, *International Journal of Nanotechnology*. 14 (2017) 356–366. https://doi.org/10.1504/IJNT.2017.082456.

[94] P. Ma, N.A. Siddiqui, G. Marom, J. Kim, Dispersion and functionalization of carbon nanotubes for polymer-based nanocomposites : a review, *Composites Part A*. 41 (2010) 1345–1367. https://doi.org/10.1016/j.compositesa.2010.07.003.

[95] F. Yang, C. Jin, D. Yang, Y. Jiang, J. Li, Y. Di, Magnetic functionalised carbon nanotubes as drug vehicles for cancer lymph node metastasis treatment, *European Journal of Cancer*. 47 (2011) 1873–1882. https://doi.org/10.1016/j.ejca.2011.03.018.

[96] H. Zhang, H. Guo, X. Deng, P. Gu, Functionalization of multi-walled carbon nanotubes via surface unpaired electrons, *Nanotechnology*. 21 (2010) 085706. https://doi.org/10.1088/0957-4484/21/8/085706.

[97] A. Baykal, M. Senel, B. Unal, E. Karaog, H. So, Acid functionalized multiwall carbon nanotube/magnetite (MWCNT)-COOH/Fe3O4 hybrid: synthesis, characterization and conductivity evaluation, *Journal of Inorganic and Organometallic Polymers and Materials*. 23 (2013) 726–735. https://doi.org/10.1007/s10904-013-9839-4.
[98] A. Sánchez, R.C. Sampedro, L. Peña-parás, E. Palacios-aguilar, Functionalization of carbon nanotubes and polymer compatibility studies, *Journal of Materials Science Research*. 3 (2014) 1–12. https://doi.org/10.5539/jmsr.v3n1p1.
[99] D.S. Ahmed, A.J. Haider, M.R. Mohammad, Comparesion of functionalization of multi Ğ walled carbon nanotubes treated by oil olive and nitric acid and their characterization, *Energy Procedia*. 36 (2013) 1111–1118. https://doi.org/10.1016/j.egypro.2013.07.126.
[100] V. Datsyuk, M. Kalyva, K. Papagelis, J. Parthenios, D. Tasis, A. Siokou, I. Kallitsis, C. Galiotis, Chemical oxidation of multiwalled carbon nanotubes, *Carbon*. 46 (2008) 833–840. https://doi.org/10.1016/j.carbon.2008.02.012.
[101] M.A. Atieh, Effect of functionalized carbon nanotubes with carboxylic functional group on the mechanical and thermal properties of styrene butadiene rubber, *Fullerenes, Nanotubes and Carbon Nanostructures*. 19 (2011) 617–627. https://doi.org/10.1080/1536383X.2010.504953.
[102] Y. Kanai, V.R. Khalap, P.G. Collins, J.C. Grossman, Atomistic oxidation mechanism of a carbon nanotube in nitric acid, *Physical Review Letters*. 104 (2010) 066401. https://doi.org/10.1103/PhysRevLett.104.066401.
[103] D. Tasis, N. Tagmatarchis, A. Bianco, M. Prato, Chemistry of carbon nanotubes, *Chemical Reviews*. 106 (2006) 1105–1136. https://doi.org/10.1021/cr050569o.
[104] B.S. Banerjee, T. Hemraj-benny, S.S. Wong, Covalent surface chemistry of single-walled carbon nanotubes, *Advanced Materials*. 17 (2005) 17–29. https://doi.org/10.1002/adma.200401340.
[105] A.M. Díez-Pascual, Chemical functionalization of carbon nanotubes with polymers: a brief overview, *Macromol*. 1 (2021) 64–83.
[106] J. Emsley, *Nature's Building Blocks: An AZ Guide to the Elements*, Oxford University Press, 2011.
[107] J. Trofast, Berzelius' discovery of selenium, *Chemistry International*. 33 (2011) 16.
[108] V. Nayak, K.R.B. Singh, A.K. Singh, R.P. Singh, Potentialities of selenium nanoparticles in biomedical science, *New Journal of Chemistry*. 45 (2021) 2849–2878.
[109] G.K. Ahluwalia, *Applications of Chalcogenides: S, Se, and Te*, Springer, 2017.
[110] P.H. Ribbe, *Sulfide Mineralogy: Mineralogical Society of America Short Course Notes*, Southern Printing Company, 1974.
[111] N.N. Greenwood, A. Earnshaw, 30-The lanthanide elements (Z= 58-71), in: *Chemistry of the Elements*, 2nd ed., Butterworth-Heinemann, Oxford, 1997: pp. 1227–1249.
[112] A.F. Wells, *Structural Inorganic Chemistry*, 5th ed., Clarendon Press, 1984.
[113] M. Bouroushian, Electrochemistry of the chalcogens, in: *Electrochemistry of Metal Chalcogenides*, Springer, 2010: pp. 57–75.
[114] E. Ridolfi, Electronic structure, transport and optical properties of MoS2 monolayers and nanoribbons, Doctoral Dissertation, Universidade Federal Fluminense, 2017.
[115] J. Ding, W. Hu, E. Paek, D. Mitlin, Review of hybrid ion capacitors: from aqueous to lithium to sodium, *Chemical Reviews*. 118 (2018) 6457–6498.
[116] I. Song, C. Park, H.C. Choi, Synthesis and properties of molybdenum disulphide: from bulk to atomic layers, *RSC Advances*. 5 (2015) 7495–7514.
[117] D.J. Vaughan, J. R. Craig, *Mineral Chemistry of Metal Sulfides, Cambridge Earth Science Series*, Cambridge University Press, 1978: p. 493.
[118] G. Demazeau, Solvothermal reactions: an original route for the synthesis of novel materials, *Journal of Materials Science*. 43 (2008) 2104–2114.

[119] J. Li, Z. Chen, R.-J. Wang, D.M. Proserpio, Low temperature route towards new materials: solvothermal synthesis of metal chalcogenides in ethylenediamine, *Coordination Chemistry Reviews*. 190 (1999) 707–735.
[120] X. Jiang, W. Xu, R. Tan, W. Song, J. Chen, Solvothermal synthesis of highly crystallized quaternary chalcogenide Cu2FeSnS4 particles, *Materials Letters*. 102 (2013) 39–42.
[121] T.S. Senthil, N. Muthukumarasamy, M. Kang, Study of various Sb2S3 nanostructures synthesized by simple solvothermal and hydrothermal methods, *Materials Characterization*. 95 (2014) 164–170.
[122] P.D. Matthews, M. Akhtar, M.A. Malik, N. Revaprasadu, P. O'Brien, Synthetic routes to iron chalcogenide nanoparticles and thin films, *Dalton Transactions*. 45 (2016) 18803–18812.
[123] P.K. Bajpai, S. Yadav, A. Tiwari, H.S. Virk, Recent advances in the synthesis and characterization of chalcogenide nanoparticles, *Solid State Phenomena*. 222 (2015) 187–233.
[124] Cb. Murray, D.J. Norris, M.G. Bawendi, Synthesis and characterization of nearly monodisperse CdE (E= sulfur, selenium, tellurium) semiconductor nanocrystallites, *Journal of the American Chemical Society*. 115 (1993) 8706–8715.
[125] D.R. Alfonso, Computational investigation of FeS2 surfaces and prediction of effects of sulfur environment on stabilities, *The Journal of Physical Chemistry C*. 114 (2010) 8971–8980.
[126] J.H.L. Beal, S. Prabakar, N. Gaston, G.B. Teh, P.G. Etchegoin, G. Williams, R.D. Tilley, Synthesis and comparison of the magnetic properties of iron sulfide spinel and iron oxide spinel nanocrystals, *Chemistry of Materials*. 23 (2011) 2514–2517.
[127] C. Steinhagen, T.B. Harvey, C.J. Stolle, J. Harris, B.A. Korgel, Pyrite nanocrystal solar cells: promising, or fool's gold?, *The Journal of Physical Chemistry Letters*. 3 (2012) 2352–2356.
[128] C. Xu, Y. Zeng, X. Rui, N. Xiao, J. Zhu, W. Zhang, J. Chen, W. Liu, H. Tan, H.H. Hng, Controlled soft-template synthesis of ultrathin C@ FeS nanosheets with high-Li-storage performance, *ACS Nano*. 6 (2012) 4713–4721.
[129] A. Kirkeminde, B.A. Ruzicka, R. Wang, S. Puna, H. Zhao, S. Ren, Synthesis and optoelectronic properties of two-dimensional FeS2 nanoplates, *ACS Applied Materials & Interfaces*. 4 (2012) 1174–1177.
[130] J. Puthussery, S. Seefeld, N. Berry, M. Gibbs, M. Law, Colloidal iron pyrite (FeS2) nanocrystal inks for thin-film photovoltaics, *Journal of the American Chemical Society*. 133 (2011) 716–719.
[131] IUPAC, Isotopomer, in: *IUPAC Compendium of Chemical Terminology*, 2008.
[132] C. Xu, P.R. Anusuyadevi, C. Aymonier, R. Luque, S. Marre, Nanostructured materials for photocatalysis, *Chemical Society Reviews*. 48 (2019) 3868–3902.
[133] A. Ridhova, V. Puspasari, M.I. Amal, Synthesis methods for chalcogenides and chalcogenides-based nanomaterials for photocatalysis, in: M. M. Khan, *Chalcogenide-Based Nanomaterials as Photocatalysts*, Elsevier, 2021: pp. 105–134.
[134] Z. Mei, M. Zhang, J. Schneider, W. Wang, N. Zhang, Y. Su, B. Chen, S. Wang, A.L. Rogach, F. Pan, Hexagonal Zn 1– x Cd x S (0.2≤ x≤ 1) solid solution photocatalysts for H 2 generation from water, *Catalysis Science & Technology*. 7 (2017) 982–987.
[135] J. Fu, B. Zhu, W. You, M. Jaroniec, J. Yu, A flexible bio-inspired H2-production photocatalyst, *Applied Catalysis B: Environmental*. 220 (2018) 148–160.
[136] H. Deiseroth, S. Kong, H. Eckert, J. Vannahme, C. Reiner, T. Zaiß, M. Schlosser, Li6PS5X: a class of crystalline Li-rich solids with an unusually high Li+ mobility, *Angewandte Chemie International Edition*. 47 (2008) 755–758.

[137] I. Tsuji, H. Kato, H. Kobayashi, A. Kudo, Photocatalytic H2 evolution reaction from aqueous solutions over band structure-controlled (AgIn) x Zn2 (1-x) S2 solid solution photocatalysts with visible-light response and their surface nanostructures, *Journal of the American Chemical Society*. 126 (2004) 13406–13413.
[138] Y. Lee, Y. Lo, Highly efficient quantum-dot-sensitized solar cell based on co-sensitization of CdS/CdSe, *Advanced Functional Materials*. 19 (2009) 604–609.
[139] T. Mokari, E. Rothenberg, I. Popov, R. Costi, U. Banin, Selective growth of metal tips onto semiconductor quantum rods and tetrapods, *Science*. 304 (2004) 1787–1790.
[140] P.S. Nair, T. Radhakrishnan, N. Revaprasadu, G.A. Kolawole, P. O'Brien, A single-source route to CdS nanorods, *Chemical Communications*. (2002) 564–565.
[141] J.Q. Hu, B. Deng, C.R. Wang, K.B. Tang, Y.T. Qian, Hydrothermal preparation of CuGaS2 crystallites with different morphologies, *Solid State Communications*. 121 (2002) 493–496.
[142] J. Zhong, Y. Zhao, H. Yang, J. Wang, X. Liang, W. Xiang, Sphere-like CuGaS2 nanoparticles synthesized by a simple biomolecule-assisted solvothermal route, *Applied Surface Science*. 257 (2011) 10188–10194.
[143] Z. Liu, J. Liu, Y. Huang, J. Li, Y. Yuan, H. Ye, D. Zhu, Z. Wang, A. Tang, From one-dimensional to two-dimensional wurtzite CuGaS 2 nanocrystals: non-injection synthesis and photocatalytic evolution, *Nanoscale*. 11 (2019) 158–169.
[144] S. Li, X. Tang, Z. Zang, Y. Yao, Z. Yao, H. Zhong, B. Chen, I-III-VI chalcogenide semiconductor nanocrystals: synthesis, properties, and applications, *Chinese Journal of Catalysis*. 39 (2018) 590–605.
[145] H. Zhong, S.S. Lo, T. Mirkovic, Y. Li, Y. Ding, Y. Li, G.D. Scholes, Noninjection gram-scale synthesis of monodisperse pyramidal CuInS2 nanocrystals and their size-dependent properties, *ACS Nano*. 4 (2010) 5253–5262.
[146] O. Diéguez, N. Marzari, First-principles characterization of the structure and electronic structure of α–S and Rh-S chalcogenides, *Physical Review B*. 80 (2009) 214115.
[147] L.K. Templeton, D.H. Templeton, A. Zalkin, Crystal structure of monoclinic sulfur, *Inorganic Chemistry*. 15 (1976) 1999–2001.
[148] I.V. Golosovsky, O.P. Smirnov, R.G. Delaplane, A. Wannberg, Y.A. Kibalin, A.A. Naberezhnov, S.B. Vakhrushev, Atomic motion in Se nanoparticles embedded into a porous glass matrix, *The European Physical Journal B-Condensed Matter and Complex Systems*. 54 (2006) 211–216.
[149] O. Degtyareva, E. Gregoryanz, H.K. Mao, R.J. Hemley, Crystal structure of sulfur and selenium at pressures up to 160 GPa, *High Pressure Research*. 25 (2005) 17–33.
[150] F.A. Devillanova, W.-W. Du Mont, *Handbook of Chalcogen Chemistry: New Perspectives in Sulfur, Selenium and Tellurium*, Royal Society of Chemistry, 2013.
[151] A.K. Singh, T.-C. Jen, *Chalcogenide: Carbon Nanotubes and Graphene Composites*, CRC Press, 2021.
[152] V. Müller, C. Grebe, U. Müller, K. Dehnicke, Synthese und Kristallstruktur des Nonaselenids [Sr (15-Krone-5) 2] Se9, *Zeitschrift Für Anorganische Und Allgemeine Chemie*. 619 (1993) 416–420.
[153] R. Staffel, U. Müller, A. Ahle, K. Dehnicke, Die Kristallstruktur von [Na (12-Krone-4) 2+] 2Se82-·(Se6, Se7) [The crystal structure of [Na (12-Crown-4) 2+] 2Se82-·(Se6, Se7)], *Zeitschrift Für Naturforschung B*. 46 (1991) 1287–1292.
[154] F. Klaiber, W. Petter, F. Hulliger, The structure type of Re2Te5, a new [M6X14] cluster compound, *Journal of Solid State Chemistry*. 46 (1983) 112–120.
[155] B. Schreiner, K. Dehnicke, K. Maczek, D. Fenske, [K (15-Krone-5) 2] 2Te8–ein bicyclisches Polytellurid, *Zeitschrift Für Anorganische Und Allgemeine Chemie*. 619 (1993) 1414–1418.
[156] W.S. Sheldrick, M. Wachhold, Discrete crown-shaped Te8 rings in Cs3 Te22, *Angewandte Chemie International Edition in English*. 34 (1995) 450–451.

[157] P. Böttcher, R. Keller, The crystal structure of NaTe and its relationship to tellurium-rich tellurides, *Journal of the Less Common Metals.* 109 (1985) 311–321.
[158] G.D. Wilk, R.M. Wallace, J. Anthony, High-κ gate dielectrics: current status and materials properties considerations, *Journal of Applied Physics.* 89 (2001) 5243–5275.
[159] L.F. Kiss, I. Bakonyi, A. Lovas, M. Baran, J. Kadlecová, Magnetic properties of amorphous Ni 81.5− x Fe x B 18.5 alloys (x= 1, 2, 3): a further key to understand the magnetism of amorphous Ni 81.5B 18.5, *Physical Review B.* 64 (2001) 64417.
[160] T. Egami, Low-field magnetic properties of amorphous alloys, *Journal of the American Ceramic Society.* 60 (1977) 128–133.
[161] A. Madan, M.P. Shaw, *The Physics and Applications of Amorphous Semiconductors*, Elsevier, 2012.
[162] J. Tauc, *Amorphous and Liquid Semiconductors*, Springer Science & Business Media, 2012.
[163] A. Khadher, M. Farooqui, M. Mohsin, G. Rabbani, Metal oxide thin films: a mini review, *Journal of Advanced Scientific Research.* 7 (2016) 1–8.
[164] J.E. Greene, Tracing the 5000-year recorded history of inorganic thin films from ~3000 BC to the early 1900s AD, *Applied Physics Reviews.* 1 (2014) 041302.
[165] K.K. Schuegraf, *Handbook of Thin Film Deposition: Processes and Technologies*, Elsevier Science & Technology, 2002.
[166] J. Datta, C. Bhattacharya, S. Bandyopadhyay, Cathodic deposition of CdSe films from dimethyl formamide solution at optimized temperature, *Applied Surface Science.* 253 (2006) 2289–2295.
[167] T. Unold, H.-W. Schock, Nonconventional (non-silicon-based) photovoltaic materials, *Annual Review of Materials Research.* 41 (2011) 297–321.
[168] N.J. Vickers, Animal communication: when i'm calling you, will you answer too?, *Current Biology.* 27 (2017) R713–R715.
[169] O.O. Abegunde, E.T. Akinlabi, O.P. Oladijo, S. Akinlabi, A.U. Ude, Overview of thin film deposition techniques, *AIMS Materials Science.* 6 (2019) 174–199.
[170] A. Jilani, M.S. Abdel-Wahab, A.H. Hammad, Advance deposition techniques for thin film and coating, *Modern Technologies for Creating the Thin-Film Systems and Coatings.* 2 (2017) 137–149.
[171] S.C. Tjong, H. Chen, Nanocrystalline materials and coatings, *Materials Science and Engineering: R: Reports.* 45 (2004) 1–88.
[172] P. Muralt, Ferroelectric thin films for micro-sensors and actuators: a review, *Journal of Micromechanics and Microengineering.* 10 (2000) 136.
[173] P. Waters, *Stress Analysis and Mechanical Characterization of Thin Films for Microelectronics and MEMS Applications*, University of South Florida, 2008.
[174] F.M. Mwema, T.-C. Jen, L. Zhu, *Thin Film Coatings: Properties, Deposition, and Applications*, CRC Press, 2022.
[175] O. Sancakoglu, Technological background and properties of thin film semiconductors, in: P. Pham, P. Goel, S. Kumar, K. Yadav, *21st Century Surface Science – A Handbook*, IntechOpen, 2020.

4 Fabrication Techniques of Smart Thin Films

4.1 INTRODUCTION

Thin films have a great impact on the modern era of technology. They are considered the backbone for advanced applications in various fields such as optical devices, telecommunications devices, energy storage devices, memory storage devices, and so on. The crucial issue for all applications of thin films depends on their morphology and stability. The morphology of the thin films strongly hinges on deposition techniques. Thus, thin-film deposition techniques are very crucial. Films can be deposited by the physical and chemical routes. In this chapter, the principles, advantages, and disadvantages of various deposition techniques of thin films were discussed.

4.2 PHYSICAL DEPOSITION TECHNIQUE

4.2.1 Physical Vapor Deposition

4.2.1.1 Evaporation

Among the various deposition methods for thin-layer films, these techniques are widely employed. In general, the mechanism of these techniques involves the transformation of material from a solid state to a vapor phase, followed by its reversion back to a solid state on a specific substrate. The deposition process can occur under both atmospheric and vacuum conditions, depending on the specific requirements and desired film properties.

4.2.1.1.1 Electron Beam Evaporation Process

Electron beam evaporation is a technique of PVD that involves the use of a concentrated electron beam generated by an electron gun (filament) within a vacuum chamber, aided by a magnetic field and an electric field. In this process, the focused electron beam is directed toward a target material, leading to its vaporization. The resulting vapor is then deposited onto a chosen substrate, as depicted in Figure 4.1. EB evaporation enables the deposition of various materials, including crystalline and amorphous semiconductors, molecular substances, metals, and oxides. Notably, this technique yields thin films of high purity and excellent quality [1].

To facilitate the passage of electrons from the electron gun to the evaporation material, an EB system requires the deposition chamber to be evacuated to a minimum pressure of 7.5×10^{-5} Torr (10^{-2} Pa). The evaporation material, which can be in the form of an ingot or rod, is thus subjected to this low-pressure environment [2]. In some contemporary EB systems, an arc-suppression mechanism is employed, enabling their operation at remarkably low vacuum levels, reaching as low as

DOI: 10.1201/9781003331940-4

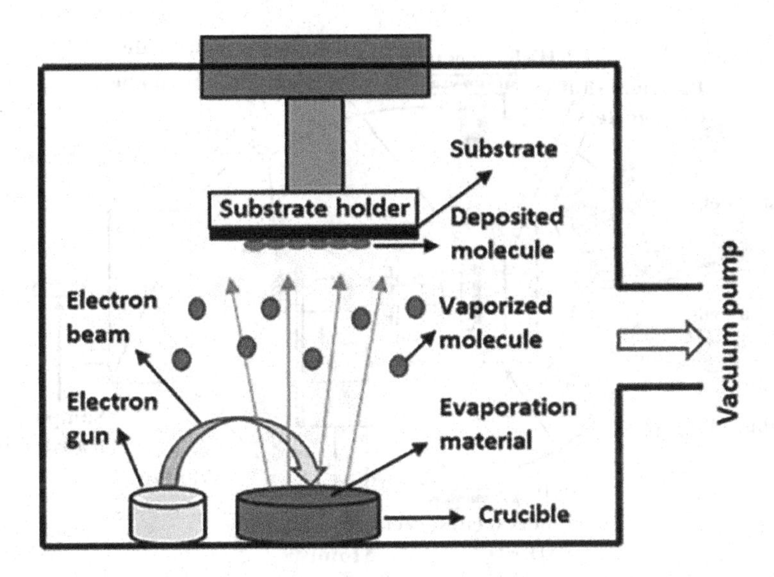

FIGURE 4.1 Schematic of electron beam evaporation process.

5.0×10^{-3} Torr. This feature proves advantageous in scenarios where parallel usage with magnetron sputtering is required [3]. In a single EB system, it is possible to utilize various types of evaporation materials and electron guns simultaneously. These components can possess power ranging from tens to hundreds of kilowatts.

EB within a system can be produced through three methods: thermionic emission, field electron emission, or the anodic arc method. Once generated, the EB is accelerated to attain significant kinetic energy and then directed toward the evaporation material. When the electrons collide with the evaporation material, they promptly dissipate their energy [3]. As the electrons interact with the evaporation material, their kinetic energy is transformed into different forms of energy. This energy conversion process primarily generates thermal energy, which in turn heats up the evaporation material, causing it to either melt or sublimate. When the temperature and vacuum level reach adequate levels, the melted or solid material transforms into vapor. This vapor can subsequently be utilized for surface coatings [3].

4.2.1.1.1.1 Advantages High deposition rate, high-purity films, good adhesion, and wide range of materials.

4.2.1.1.1.2 Disadvantages Limited material size, high equipment cost, susceptibility to electron beam deflection, and substrate heating.

4.2.1.1.2 Molecular Beam Epitaxy Process

MBE, or molecular beam epitaxy, is a method utilized for depositing thin films. It entails the creation of an epitaxial film by subjecting a pristine single substrate to thermal atom beams. In this technique, films are deposited at extremely high temperatures within an environment of ultrahigh vacuum [4,5]. The film growth process of MBE is illustrated in Figure 4.2. To enable the growth of epitaxial films with

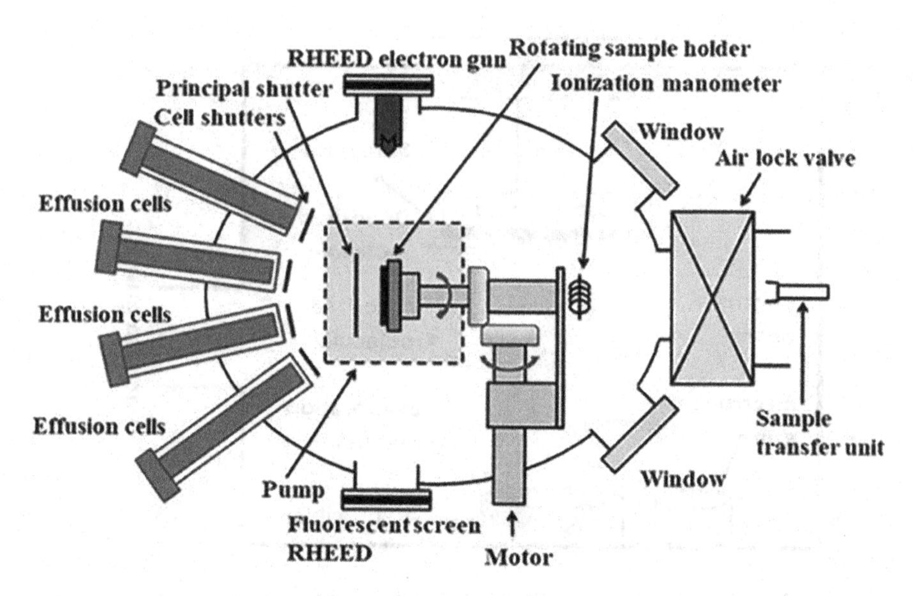

FIGURE 4.2 Schematic of MBE growth chamber [5].

excellent crystal properties and high purity, MBE utilizes a vacuum process typically operating at pressures below 10^{-10} torr. This allows for relatively low substrate temperatures during the film deposition [4]. Moreover, the ultrahigh vacuum environment provided by MBE allows for real-time investigation of the interface, bulk, and surface properties of the deposited thin film. This capability facilitates in-depth analysis and understanding of the film's characteristics.

MBE stands out as a flexible method for fabricating thin epitaxial structures comprising semiconductors, metals, or insulators. Its distinguishing feature from earlier vacuum deposition techniques lies in the remarkable precision it offers in controlling beam fluxes and growth conditions. Operating under vacuum deposition, MBE takes place in conditions far removed from thermodynamic equilibrium, with the surface processes predominantly governed by the kinetics of reactions between the impinging beams and the outermost atomic layers of the substrate crystal. In contrast to epitaxial growth techniques such as liquid phase epitaxy (LPE) or atmospheric pressure (AP) vapor phase epitaxy (VPE), which operate under conditions close to thermodynamic equilibrium and are primarily governed by diffusion processes within the crystallizing phase surrounding the substrate crystal, MBE offers a distinct advantage. It provides a unique edge over other epitaxial growth methods. As mentioned earlier, the ultrahigh vacuum environment of MBE allows for in situ control through surface diagnostic methods such as reflection high-energy electron diffraction, reflection mass spectrometry, and optical techniques such as reflectance anisotropy spectroscopy, spectroscopic ellipsometry, and laser interferometry. These advanced tools for control and analysis significantly reduce uncertainties in MBE processes and facilitate the fabrication of complex device structures with precision [6].

4.2.1.1.2.1 Advantages Precise control and atomic-scale precision, high-purity films, low-temperature growth, and versatility in material selection

4.2.1.1.2.2 Disadvantages Slow deposition rate, limited substrate size, complexity and equipment cost, and sensitivity to environmental conditions.

4.2.1.1.3 Pulsed Laser Deposition

Pulsed laser deposition (PLD) is a widely utilized technique for depositing thin films onto a substrate. The PLD process occurs within a vacuum chamber and employs a laser beam to ablate the material, enabling the deposition of films. Multiple laser sources can be employed for target ablation, with common examples including XeCl (308 nm), Nd-YAG, and KrF (248 nm) lasers [1]. A visual representation of the laser beam ablation process can be observed in Figure 4.3.

When the material (target) is exposed to the laser beam, it produces a plume that can be applied to different substrates. This plume comprises a mixture of ground-state atoms, ionized species, and neutral atoms. Oxygen is commonly utilized in the deposition of metal oxide films to enable their formation. The quality of the deposited films in PLD is influenced by several factors, including the duration of the laser pulse, the distance between the target and substrate, the wavelength of the laser, the pressure of the ambient gas, and the energy employed during the process. To control and monitor the process of ablation in PLD, techniques such as optical emission, laser-induced fluorescence, and laser ablation molecular isotopic spectroscopy can be employed. Additionally, the temperature of the substrate plays a crucial role in determining the morphology of the thin films. The compatibility

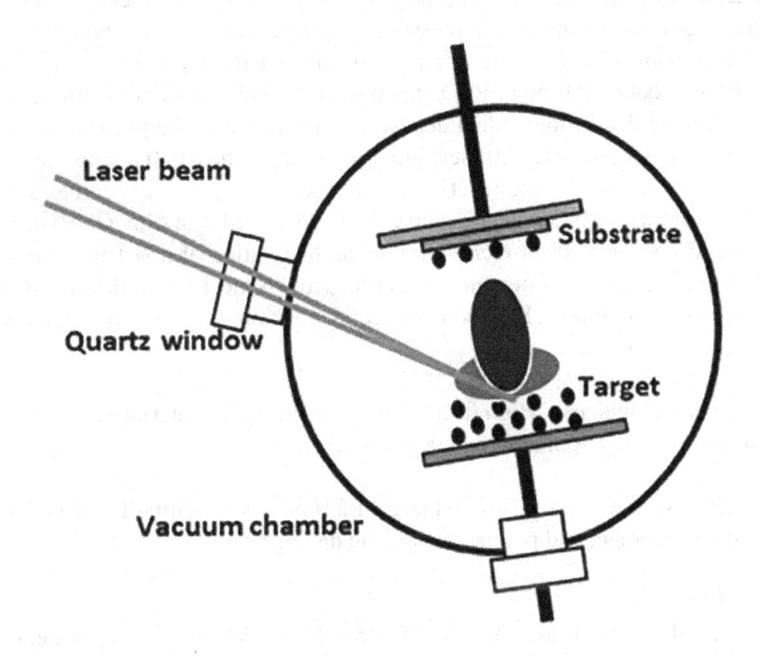

FIGURE 4.3 Schematic of PLD process.

of oxygen with the PLD technique, along with its rapid deposition time, makes it a preferred choice over other deposition methods [1].

The deposition process is conducted either in a vacuum or in a low-pressure environment, which is established using vacuum pumps situated near the deposition chamber. By manipulating the gas environment with the desired pressure and composition, the properties of the deposited film can be precisely adjusted and tuned. The deposition of target material in PLD is influenced by five key factors. These factors encompass the absorption of the laser by the target material, the process of laser ablation and plasma formation, the dynamics of the plasma, the deposition of material onto the substrate, and the growth of the desired film [7]. Attaining superior film coatings necessitates meticulous deliberation of the following elements. The quality of the film, including attributes such as crystallinity, composition, thickness, roughness, conductivity, optical properties, and mechanical properties, relies on diverse deposition factors. These factors encompass target material, target-sample distance, temperature, pulse length and rate, radiation intensity, as well as laser wavelength. Increasing the separation between the target material and substrate reduces the deposition of material due to an extended particle travel time, causing more particles to veer away from the substrate. Additionally, employing a laser with short pulses induces electrostatic ablation, wherein the rapid excitation of electrons holding the material results in small particle explosions emanating from the surface of the target. When longer pulses are utilized, a significant portion of the incident pulse is absorbed by the plasma plume [8,9].

The temperature of the substrate also influences film deposition. Films deposited at higher temperatures tend to be crystalline, whereas those deposited on cooler substrates tend to be amorphous. Consequently, the process of PLD can be conducted in different gas environments and pressures, as these factors impact both the growth rate of the film and its crystalline structure. To enhance the crystalline characteristics of thin films produced through PLD, it is possible to subject them to annealing processes following deposition. When depositing metal oxides, the presence of oxygen gas is necessary to facilitate sufficient bonding between the metal and oxygen atoms. In such cases, the ratio of the reactive gas (oxygen) to the metal becomes a crucial factor to ensure the desired stoichiometry of the deposited thin film. Over time, PLD has emerged as a significant technique for the fabrication of novel materials, even though there is a common perception that its growth mechanism deviates to some extent from the continuous PD and chemical deposition (CD) methods employed in pulsed deposition [8,9].

4.2.1.1.3.1 Advantages High film quality, versatility in material selection, versatility in material selection, and high deposition rate.

4.2.1.1.3.2 Disadvantages Limited target lifetime, uneven film thickness distribution, limited coverage, and potential substrate damage.

4.2.1.1.4 Ion Plating

Ion plating, also known as ion-assisted deposition (IAD) or ion vapor deposition (IVD), is a technique employed in physical vapor deposition (PVD) processes.

This modified form of vacuum deposition involves the simultaneous bombardment of both substrates and target material with highly energetic particles at the atomic level. Initially proposed in 1963, ion plating aimed to enhance film adhesion and optimize surface coverage by utilizing the effects of energetic particle interactions [10]. Since its inception, ion plating has undergone significant advancements and is now being widely applied in various fields such as tribology, electrical contacts, and corrosion protection. In ion plating, the targeted utilization of atomic-sized energetic particles allows for the simultaneous bombardment of both the substrate and the target material [11]. This bombardment of the substrate prior to deposition plays a crucial role in surface cleaning by the process of sputtering.

In the ion-plating process, several critical processing variables come into play, including the energy, flux, and mass of the bombarding species, as well as the ratio of bombarding particles to depositing particles. The material intended for deposition can be vaporized through various methods such as evaporation, sputtering (bias sputtering), arc vaporization, or the decomposition of a chemical vapor precursor in chemical vapor deposition (CVD). The energetic particles employed for bombardment typically consist of ions derived from inert or reactive gases. In specific cases, ions derived from the film material itself, referred to as "film ions," can also be employed in ion plating. The ion-plating process can be carried out in two different settings: a plasma environment, where ions used for bombardment are extracted from the plasma, or a vacuum environment, where ions for bombardment are generated using a separate ion gun. The alternative configuration of ion plating is commonly referred to as ion-beam-assisted deposition (IBAD). This method involves the utilization of a reactive gas or vapor in the plasma, enabling the deposition of compound material films [12]. Figure 4.4 illustrates the diagram of the ion-plating process.

An ion-plating apparatus consists of several key components, including a vacuum chamber and a pumping system, an inert gas inlet with a needle valve to control the gas flow, and an insulated high-tension electrode where the specimen is positioned. For vapor deposition, an electron beam gun, boat, or resistance-heated filament is utilized as the source of vapor. Ion plating can be performed under two distinct conditions: plasma ion plating and vacuum ion plating. The process of ion plating encompasses three primary stages for the deposition of thin films: (i) surface preparation, involving the elimination of contaminants such as oxide layers that may hinder ion adhesion; (ii) nucleation, surface coverage, and interface formation of the deposited materials; and (iii) the subsequent growth and formation of the film [12,13].

According to [14], the general procedure for an ion-plating coating sequence involves securing the specimen onto the electrode. The coating material is loaded onto the filament, and the pressure within the chamber is gradually reduced to reach near-vacuum levels. Simultaneously, argon gas is introduced into the chamber while the vacuum pumps maintain the desired pressure conditions. Next, a bias voltage of approximately 5 kV is applied to the specimen holder, resulting in a glow discharge between the specimen and the grounded components. This discharge leads to the bombardment of the specimen by high-energy argon ions, effectively cleaning the surface through sputtering. Consequently, the vapor source is energized, causing the coating material to vaporize and subsequently deposit onto the surface of the specimen. One significant advancement in ion-plating technology

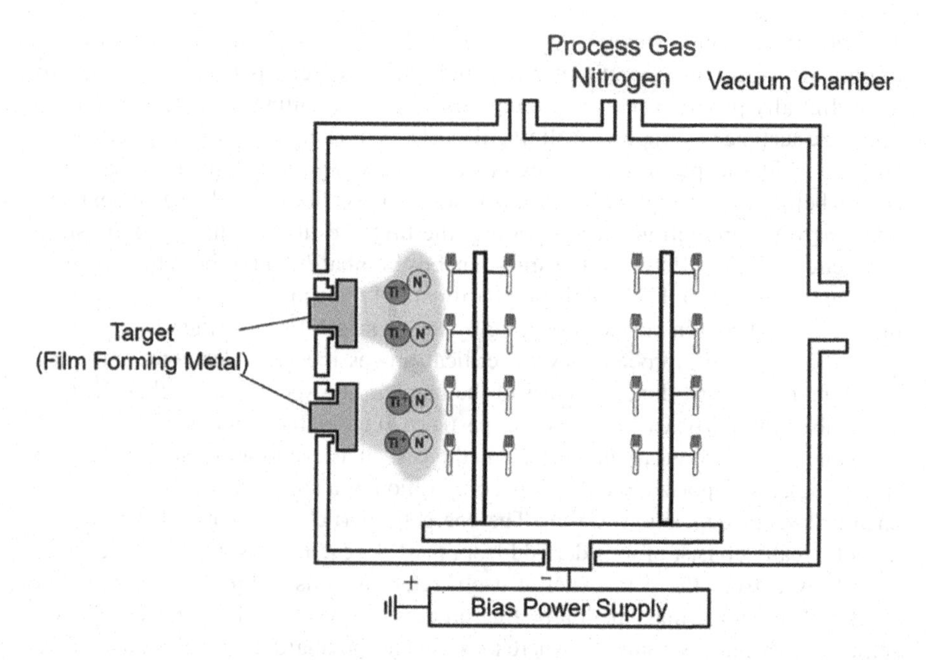

FIGURE 4.4 Schematic of ion-plating technique.

has been the adoption of an electron beam gun as the energy source for vaporizing the target material. This development has greatly expanded the capabilities of ion plating by enabling the deposition of coatings from high-melting-point metals such as molybdenum and tungsten [14]. Furthermore, the utilization of an electron beam gun instead of a resistance-heated source allows for faster deposition rates, facilitating the production of thicker coatings.

Ion plating finds applications in various areas, such as the deposition of durable compound coatings on tools, the formation of adherent metal coatings, the creation of high-density optical coatings, and the application of conformal coatings on intricate surfaces.

4.2.1.1.4.1 Advantages Improved adhesion, enhanced film quality, precise film thickness control, wide range of materials, uniform coating, and temperature control.

4.2.1.1.4.2 Disadvantages High equipment cost, complex operation, limited substrate size, ion beam damage, limited film thickness, and line-of-sight deposition.

4.2.2 Sputtering

Sputtering is a process that involves bombarding a target, which serves as the cathode plate, with energetic ions generated in a glow discharge plasma located on the opposite side of the target. This bombardment causes atoms to be dislodged from the target and subsequently deposited onto a substrate through the condensation process [15]. The ion bombardment during the sputtering process results in the emission of

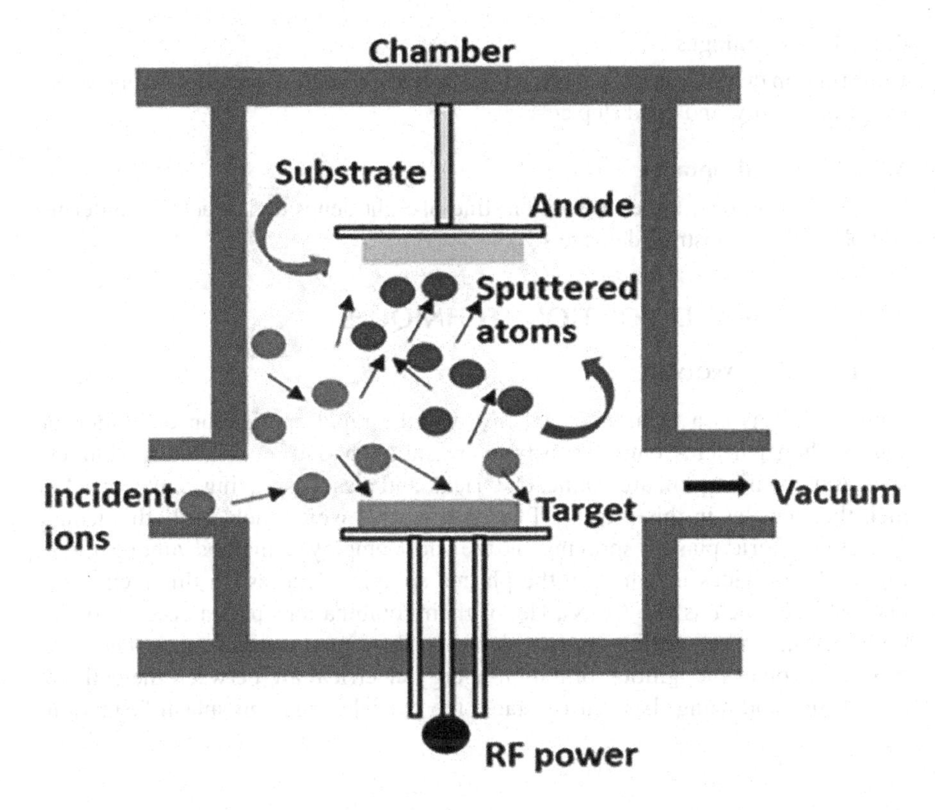

FIGURE 4.5 Schematic of sputtering system.

secondary electrons from the surface of the target, which contribute to the maintenance of the plasma [15]. Figure 4.5 provides a visual representation of the sputtering technique.

Over the years, the sputtering technique has been widely employed by researchers for depositing various thin-film materials [15]. However, the sputtering process does have limitations, including high substrate heating, a low deposition rate, and low ionization efficiencies. To address these challenges, magnetron sputtering was developed as an innovative solution. Another advancement in sputtering technology is the development of unbalanced magnetron sputtering. In this process, the target surface and a magnetic field are arranged parallel to each other, directing the motion of secondary electrons toward the target region. The magnets are strategically positioned, with the first pole located at the central axis of the target, while magnet rings form a second pole at the outer edge of the target. By implementing this configuration, there is a notable enhancement in electron ionization, attributed to the increased likelihood of atom collisions. Consequently, dense plasma forms near the target, facilitating more efficient ion bombardment of the target material. This heightened ion bombardment leads to an accelerated sputtering rate and, in turn, a higher deposition rate on the substrate. The growth of the film within the dense plasma region, where simultaneous ion bombardment occurs, has a significant impact on the structure and properties of the deposited film [15].

4.2.2.1 Advantages

Uniform film deposition, wide material compatibility, control over film composition, high film density, and good step coverage.

4.2.2.2 Disadvantages

Low deposition rate, target utilization, line-of-sight deposition, reactive sputtering complexity, and substrate damage.

4.3 CHEMICAL DEPOSITION TECHNIQUES

4.3.1 Spray Pyrolysis

Spray pyrolysis is a technique that involves the rapid application of molten or semi-molten particles onto a substrate by spraying. Different heating sources, such as detonation-gun, arc, flame, electrical, and plasma spraying, can be used to melt the particles in this process. Plasma spraying, which includes both vacuum and atmospheric plasma spraying, is the most employed method among them. Figure 4.6 provides insight into the plasma spraying process. In this technique, plasma gases such as H_2, Ar, N_2, He, or their combinations play a crucial role in both melting and propelling particles onto a substrate at a significant speed. The process involves the ignition of a high-energy electrical arc between the cathode and anode. Following the melting stage, the particles undergo splash formation,

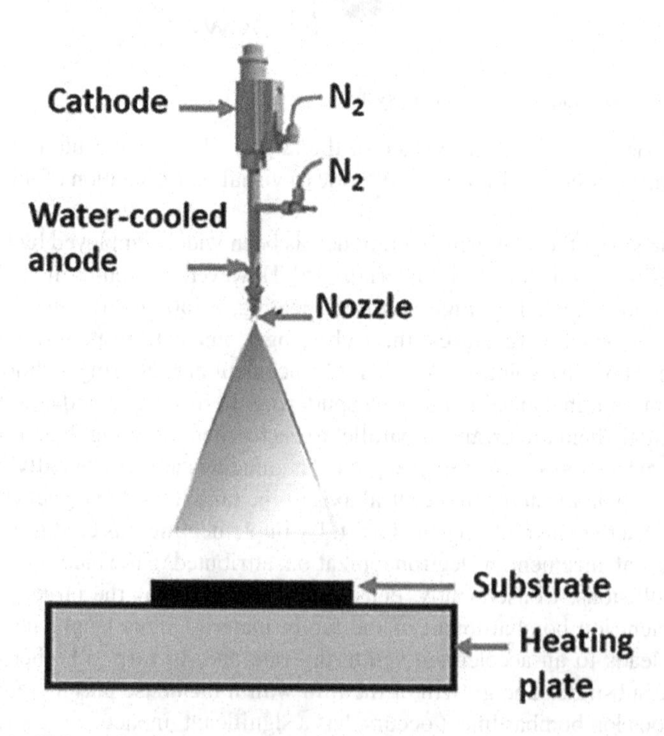

FIGURE 4.6 Schematic of spray pyrolysis process.

splash layering, and coating formation. Eventually, on the substrate surface, the film got flattened and solidified, resulting in the coating formation [16].

Spray pyrolysis is a technique that achieves the formation of a thin film by applying a solution to a surface that has been heated. When this process occurs, a chemical reaction takes place among the constituents, ultimately resulting in the creation of a compound [17]. The chemical spray deposition process can be categorized into three groups, depending on the specific type of reaction that occurs [18]. In the first category, the solution droplets remain on the heated surface as the solvent gradually evaporates. This enables the components to react in a dry state without the presence of a liquid medium. The second category involves the solvent evaporating before the droplets reach the heated surface. Consequently, a solid material is deposited through the process of decomposition, where the solvent transforms into a gas and leaves behind a dry residue. The third category encompasses processes in which the solvent vaporizes as the droplets approach the substrate. This results in a heterogeneous reaction among the solution components, as they react in the presence of both gas and solid phases.

In all these processes, several crucial parameters need to be carefully controlled. These include the substrate temperature, carrier gas flow rate, nozzle-to-substrate distance, as well as the content and concentration of the solution. Among these variables, the substrate temperature is widely recognized as the most critical factor in the production of thin films through spray pyrolysis. This is primarily due to the significant influence it has on the drying, decomposition, crystallization, and grain growth of the droplets [19]. The central component of the spray pyrolysis deposition apparatus is the atomizer, which is responsible for generating an aerosol from the precursor solution. The design of this equipment can vary significantly, ranging from inexpensive atomizers used for cosmetic or perfume applications [20,21], to Pyrex glass or metallic atomizers available in individual or commercial designs and even more complex ultrasonic devices [22]. Although the spray pyrolysis technique may seem straightforward, achieving optimal results requires a careful understanding of the deposition parameters and their interrelationships. In order to comprehend this correlation and optimize the deposition process, theoretical modeling of spray pyrolysis deposition has been undertaken [23]. This modeling aims to investigate how the film's topography evolves, treating the droplets as a continuous flux rather than individual drops. It also considers scenarios where the droplets evaporate in close proximity to the surface before coming into full contact with the substrate in the form of liquid. A conventional spray deposition system comprises several key components, such as the spray atomizer housing the precursor solution, a substrate heater, pressurized air, liquid flow, and temperature control mechanisms. Due to the inherent nature of the spray process, the resulting film thickness can often be nonuniform. To achieve a more uniform deposition, introducing random motion to either the substrate or the spray nozzle can prove beneficial [17].

4.3.1.1 Advantages

Cost-effectiveness, scalability, uniformity, versatility, and complex film formation.

4.3.1.2 Disadvantages

Limited thickness control, substrate compatibility, post-processing requirements, limited film thickness, and precursor limitations.

4.3.2 Sol-Gel Process

Extensive research has been conducted on the sol-gel method for synthesizing oxide materials, establishing it as a highly effective wet chemical technique. This approach offers numerous benefits, such as the capability to achieve uniformity in compositions containing multiple materials and the ability to operate at lower temperatures. In the context of this method, "sol" denotes the formation of a colloidal suspension, while "gel" signifies the transition of the "sol" into a viscous gel or solid substance [1]. Heat treatments are employed to achieve the desired physical, mechanical, and chemical properties of the formulated material [24]. This technique is commonly employed to create thin oxide coatings. Furthermore, it requires only a small quantity of the precursor solution, making it cost-effective and minimizing the significance of the initial cost of the organic metal compound utilized [25]. It has widespread applications in various industries, such as abrasive powder production and the optical industry.

This method utilizes a colloidal solution as a precursor for various polymers, discrete molecules, or materials, facilitating the production of materials. It takes advantage of easily accessible materials, resulting in cost reductions. The materials produced through this approach are commonly utilized across diverse industries, including electronics, energy, and space. Furthermore, this technique has been employed to fabricate sensors in the biomedical field [26]. The precursor solution (sol) undergoes a sequence of reactions, including hydrolysis and poly-condensation, resulting in the formation of nanoparticles consisting of hydroxides. This process leads to the creation of a monolithic gel that possesses characteristics of both liquid and solid phases [27]. The gel formed through this technique can take different forms, such as ambigels, xerogels, cryogels, and aerogels, each finding diverse applications in the field of engineering [28]. The sol-gel process is illustrated in Figure 4.7.

In the sol-gel process, precursor solutions (organic and inorganic materials) are used to develop thin-film coatings [28]. These coatings can be applied to a substrate through two methods: spin and dip coating [29] (refer to Figure 4.7). Several factors influence the gel formation process, such as pH, changes in material mass, and the external environment, such as solvent substitution. The thin-film coating provides protection against corrosion and abrasion, reduces catalytic activity, and enhances the substrate's conductive and ferroelectric properties [30].

4.3.2.1 Spin Coating

The spin-coating process is a fast method of depositing a substance onto a mostly flat surface. It involves using a specialized equipment with a chamber that securely holds the surface in place using a vacuum. The coating solution is applied to the surface, and then the surface is made to spin, causing the solution to spread evenly across the surface due to the spinning action. Researchers have conducted extensive studies

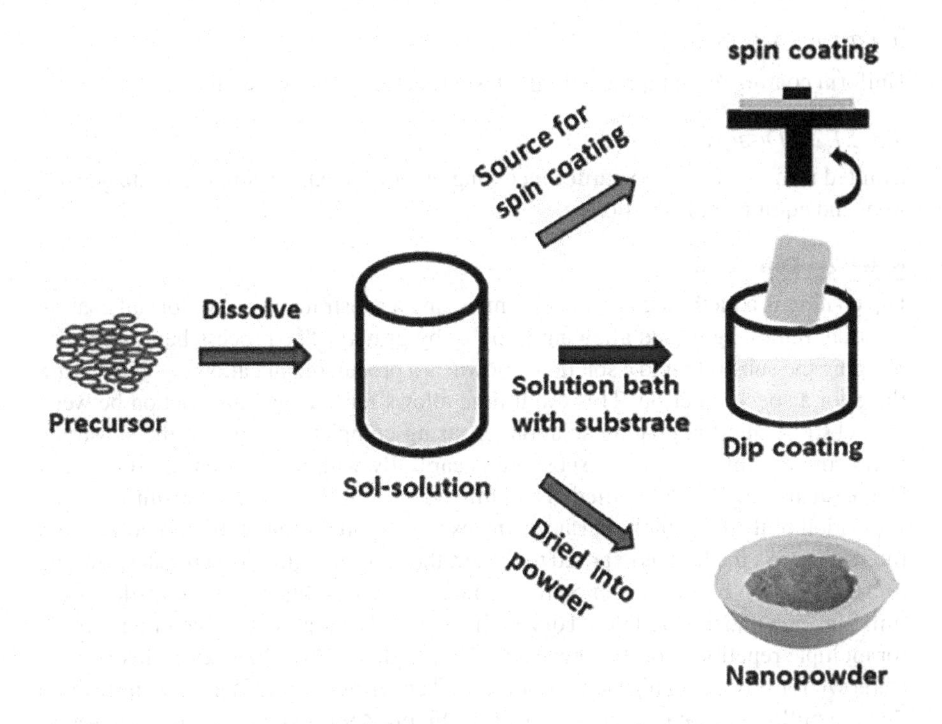

FIGURE 4.7 Schematic of sol-gel process leading to spin and dip coatings.

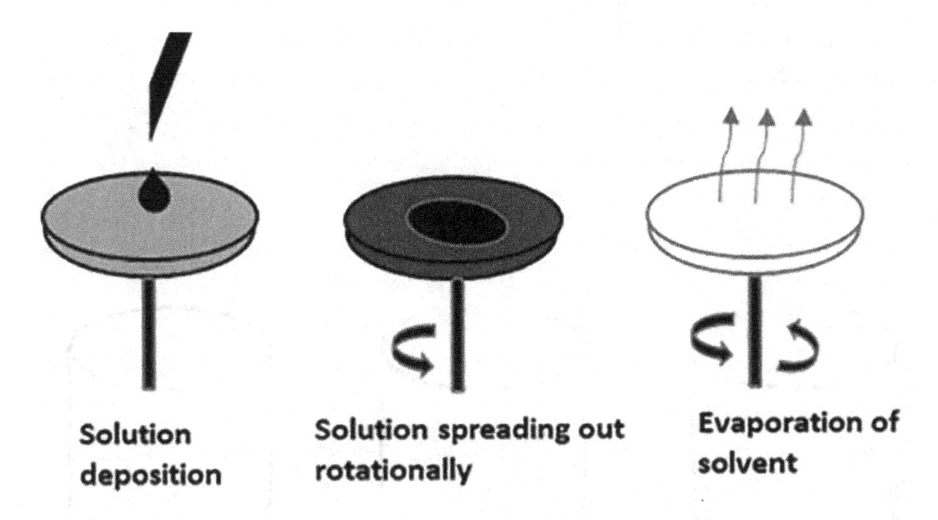

FIGURE 4.8 Schematic of spin-coating process.

on this coating process, investigating the various factors that affect the deposition process and the resulting thickness [1]. A visual representation of the spin-coating process can be seen in Figure 4.8.

4.3.2.1.1 Advantages

Uniform coating, high reproducibility, fast processing, and versatility.

4.3.2.1.2 Disadvantages

Limited film thickness, nonuniform coating on non-planar substrates, waste generation, and equipment limitations.

4.3.2.2 Dip Coating

Dip coating is a method that involves immersing a substrate into a conformal coating solution, removing it, and allowing it to dry by gravity. The process begins by submerging the substrate into a solution, known as a precursor, and allowing it to remain there for a specific period. This dwell time allows for adequate interaction between the substrate and the coating solution, ensuring complete wetting of the substrate. Following the immersion, the substrate is carefully withdrawn from the solution at a consistent speed. As it is lifted, a wet film adheres to the substrate, resulting in the deposition of the film on the surface. Any excess liquid present will drain away from the surface. As the fluid is exposed to the air, the solvent within it evaporates, leading to the formation of a thin film on the surface. This film, known as the as-deposited thin film, can undergo additional drying if needed. The dip-coating technique allows for multiple repetitions of this process to achieve the desired thickness. This method is known for producing high-quality films and ensuring uniformity, even on surfaces that are bulky or have complex shapes [31]. Figure 4.9 provides a visual representation of the dip-coating process.

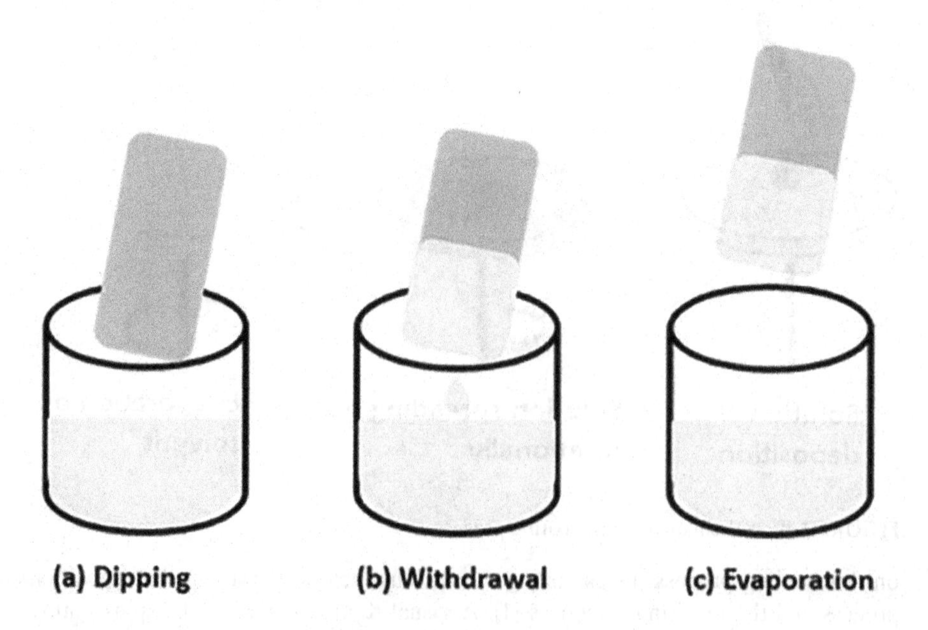

FIGURE 4.9 Schematic of dip-coating process.

4.3.2.2.1 *Advantages*

Simple and cost-effective, uniform coating thickness, versatile and adaptable, high material utilization, and ability to coat complex shapes.

4.3.2.2.2 *Disadvantages*

Limited coating thickness, surface tension effects, limited control over coating properties, slow production speed, and waste disposal.

4.3.3 Electroplating

Electroplating, also referred to as electrochemical deposition, is a technique that employs direct electrical current to coat a solid surface with a layer of metal. In this process, the surface to be coated functions as the negative electrode (cathode) within an electrolytic cell. The electrolyte used is a solution containing metal salt, while the positive electrode (anode) is usually made of the same metal or an inert conductive material. By utilizing this method, it becomes feasible to achieve the deposition of a metal coating onto the substrate. An external power supply is used to provide the necessary current. Electroplating finds extensive applications in industries and decorative arts to enhance the surface properties of various objects, including resistance to corrosion and abrasion, lubrication, reflectivity, electrical conductivity, and visual appearance [32]. Figure 4.10 depicts the electroplating process, showcasing its steps and components.

The electroplating process, as shown in Figure 4.10, requires an electrolyte containing cations (positive ions) of the metal to be deposited. These cations undergo a reduction reaction at the cathode, resulting in the formation of the metal in its neutral state. For example, in the case of copper plating, the electrolyte may consist of a solution containing copper (II) sulfate, which dissociates into Cu^{2+} cations and $SO^{2-}{}_4$ anions. At the cathode, Cu^{2+} is reduced through the gain of two electrons, leading to the creation of metallic copper. If the anode is composed of the same metal as the coating, a reverse reaction can occur, causing it to dissolve into cations. For instance, in the case of copper, it would be oxidized at the anode, resulting in the formation of Cu^{2+} ions through the loss of two electrons. In this scenario, the dissolution rate of the anode will match the plating rate at the cathode, ensuring a continuous replenishment of ions in the electrolyte solution by the anode. As a result, there is an efficient transfer of metal from the anode to the cathode, facilitating the electroplating process. Alternatively, the anode can be constructed from a material that is resistant to electrochemical oxidation, such as lead or carbon. In this scenario, substances such as oxygen, hydrogen peroxide, or other by-products are generated at the anode instead. Consequently, the metal ions required for plating must be regularly replenished in the electrolyte solution as they are depleted from the bath [32]. This approach has been successfully employed for depositing metal/graphene (Gr) composites, such as Cu/Gr [33,34] and Ni/Gr [35].

4.3.3.1 Advantages

Enhanced appearance, corrosion resistance, improved durability, and versatility.

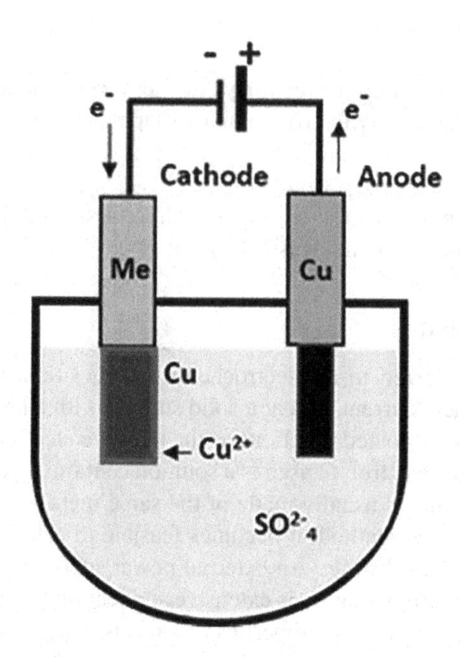

FIGURE 4.10 Schematic of electroplating process.

4.3.3.2 Disadvantages

Environmental impact, cost, limitations in thickness and size, surface preparation requirements, and limited choice of materials.

4.3.4 Chemical Vapor Deposition

CVD methods have emerged as a more favorable option because they generate high-quality thin films at a remarkably affordable cost, making them widely adopted worldwide. The effectiveness of CVD relies significantly on factors such as solution chemistry, viscosity, and pH value. The following CVD techniques, as explained below, possess the capability to produce superior films.

4.3.4.1 Atmospheric Pressure Chemical Vapor Deposition

Atmospheric Pressure Chemical Vapor Deposition (APCVD) stands as one of the initial techniques within the realm of CVD. As its name implies, this method typically takes place under AP conditions [36]. To achieve the desired operating pressures, the precursors used are often mixed with other inert gaseous substances (gases that do not engage or interfere with the intended precursor during adsorption), such as oxygen or a combination of nitrogen and ozone [37]. Nevertheless, the operational temperatures of APCVD differ based on the specific substrate and precursor gas utilized in the reaction system. Consequently, APCVD can be categorized into two subtypes: low-temperature APCVD and high-temperature APCVD. Additionally, the classification of reactors as hot wall or cold wall is determined by whether the reaction is endothermic or exothermic, respectively, in order to eliminate by-products.

CVD typically comprises three primary components: precursor delivery, the CVD reactor, and exhaust management. Here are the details for each component:

- **Precursor Delivery:** This component is responsible for delivering the appropriate composition and quantity of gas reactants into the CVD reactor for the deposition process. Depending on the state of the reactants used, the precursor must be capable of converting liquids and solids into gaseous phase reactants before delivering them to the reactor.
- **CVD Reactor:** The CVD reactor is where the actual deposition of the material occurs. The gas reactants interact with the substrate, leading to the formation of the desired thin-layer coating. To ensure precise control over deposition rate, film uniformity, and precursor utilization efficiencies, the reactor design should facilitate effective interaction between the gas reactants and the extensive surface area of the substrate.
- **Exhaust Management:** Exhaust management is a crucial component of the system, encompassing vacuum pumps, sensors, and control systems necessary for maintaining the desired total pressure within the system. It ensures that the by-products of the reaction are efficiently released from the reactor chamber.

In the case of APCVD, the reactive gases are introduced into the chamber and subsequently heated. This heating process activates the gases, causing their decomposition. The substrates absorb the gases, leading to reactions taking place on the surface of the substrate, resulting in the formation of a thin film. Throughout this process, the transportation of by-products away from the substrate occurs [38].

4.3.4.1.1 Advantages

Simplicity, wide substrate compatibility, large-area deposition, and better film quality.

4.3.4.1.2 Disadvantages

Lower deposition rates, limited process control, thermal stress, and potential chemical reactions.

4.3.4.2 Plasma-Enhanced Chemical Vapor Deposition

PECVD, or Plasma-Enhanced Chemical Vapor Deposition, is a technique used to deposit thin films onto various substrates by transforming gas into solid state. This process involves injecting gas reactants between an RF-energized electrode and parallel-grounded electrodes, facilitating film deposition. By employing capacitive coupling between the electrodes, the gas reactant is converted into plasma, triggering a chemical reaction that leads to the desired deposition of reaction products onto the chosen substrate. The PECVD process, depicted in Figure 4.11, supports the placement of the substrate. To achieve successful film deposition, a substrate heating temperature ranging from 250°C to 350°C is typically necessary, varying based on the specific film type being deposited. This PECVD technique outperforms traditional CVD methods, which typically demand higher temperatures ranging from 600°C to

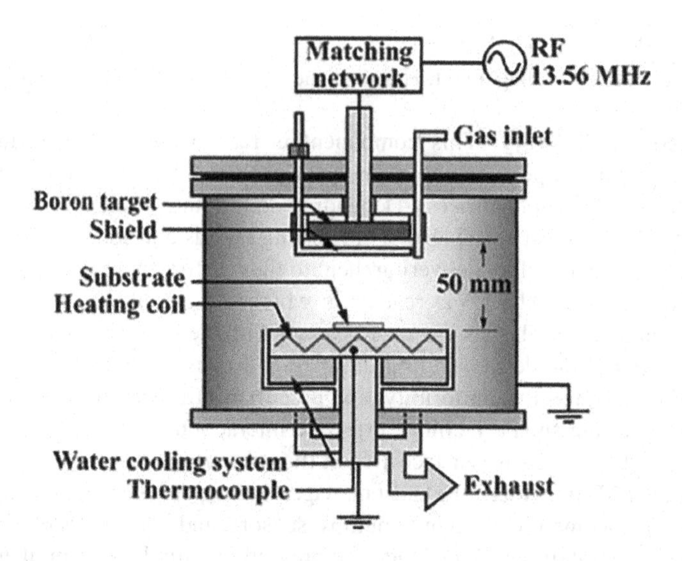

FIGURE 4.11 Schematic of PECVD process [40].

800°C. This lower temperature requirement is crucial in various applications where deposition at higher temperatures could potentially harm the fabricated devices [39]. PECVD can be classified according to the plasma type used. Two primary categorizations are remote systems and direct systems. The direct process involves placing the substrate between two parallel electrodes, and the cathode electrode is directly exposed to the energized plasma. Conversely, remote systems ensure that the substrate does not come into direct contact with the plasma.

4.3.4.2.1 Advantages

Film uniformity, wide material range, low-temperature deposition, high deposition rate, and controllable film properties.

4.3.4.2.2 Disadvantages

Complex equipment, limited deposition thickness, susceptibility to contamination, plasma damage, and process complexity.

4.3.4.3 Low-Pressure Chemical Vapor Deposition

LPCVD, or Low-Pressure Chemical Vapor Deposition, is a deposition technique that utilizes heat to induce a reaction between a precursor gas and a solid substrate. This reaction occurs at the surface and leads to the formation of the desired solid material. The application of low pressure helps minimize undesired gas-phase reactions and enhances uniformity throughout the substrate. LPCVD can be performed in a quartz tube reactor, which can be either cold-walled or hot-walled. Hot-walled furnaces enable batch processing, allowing for high throughput. Hot wall systems offer excellent thermal uniformity, ensuring the deposition of uniform films. However, a drawback of these systems is that the deposition can also happen on the walls of the furnace, necessitating regular cleaning or eventual replacement of the tube to prevent the flaking of the deposited material and potential particle contamination.

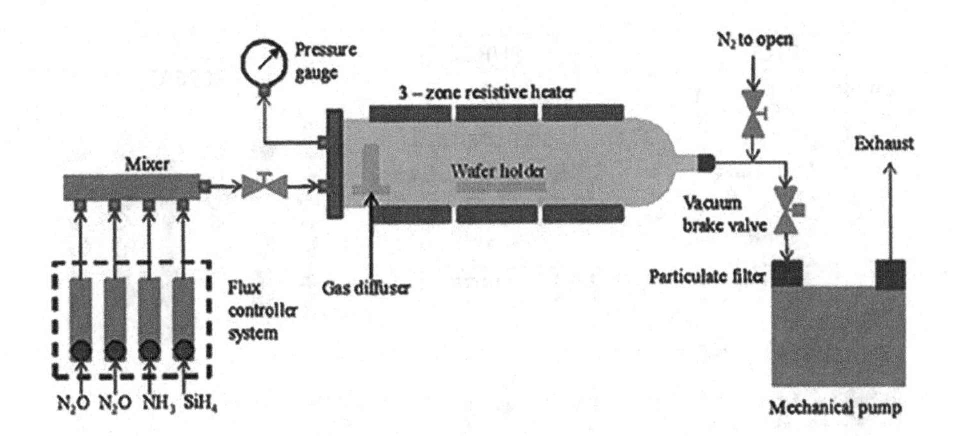

FIGURE 4.12 Schematic of LPCVD process [42].

On the other hand, cold-walled reactors require less maintenance since there is no film deposition on the walls of the reactor [41]. Refer to Figure 4.12 for a schematic representation of the LPCVD process.

LPCVD involves evacuating the tube to low pressures (10 mTorr to 1 Torr) and heating it to deposition temperature. This temperature (425°C–900°C) causes the precursor gas to decompose. The injected gas diffuses and reacts with the substrate surface, forming the solid material. LPCVD involves removing excess gas through an abatement system. It yields films with improved uniformity, fewer defects, and better step coverage compared to PECVD and PVD techniques. However, LPCVD's drawback is its requirement for higher temperatures, which restricts the choice of substrates and other materials used in the process [41].

4.3.4.3.1 Advantages

High-quality film deposition, uniform films, wide range of materials, and good step coverage.

4.3.4.3.2 Disadvantages

Slow deposition rate, high equipment cost, limited scalability, hazardous gases and safety concerns, and complex process control.

4.3.4.4 Atomic Layer Deposition

Atomic Layer Deposition (ALD) focuses on precisely controlling chemical reactions on the substrate surface, resulting in ultra-thin deposition. By using inert gas purges to separate precursors, ALD is similar to CVD but offers distinct advantages. ALD benefits from sequential vapor phase reactions, self-limiting behavior, and transient separation, making it an advanced technique for depositing atomic layers of thin films. Traditional PVD and CVD methods may not meet the requirements of 60 nm generation technology, highlighting the potential of ALD in this context [43].

ALD follows a bottom-up approach, involving multiple processes and purge steps. In the first phase, the substrate is exposed to a precursor or reactant (e.g., reactant 1),

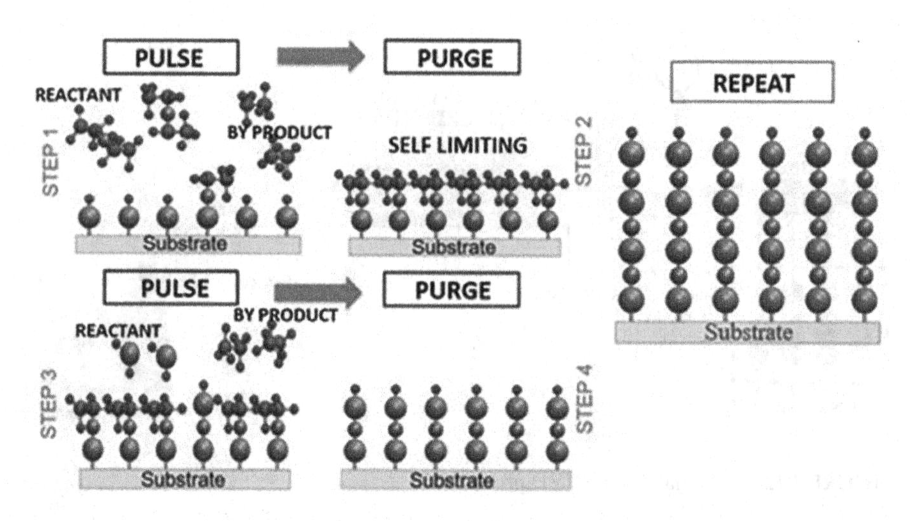

FIGURE 4.13 Schematic of general technique of ALD process [45].

leading to chemisorption on the substrate surface and surface interactions with the functional groups. This process continues until all functional groups are consumed and by-products are released due to surface interactions. The second phase includes a purge step to remove these by-products and any unreacted reactant 1. The third phase involves reactant 1 chemisorbing onto the substrate surface and undergoes a self-saturating reaction with the co-reactant in the third step. The fourth phase involves purging to remove any remaining by-products and unreacted co-reactants. This results in the creation of new active sites with functional groups on the substrate surface. By repeating this cycle, layer-by-layer deposition occurs, allowing for the desired thickness to be achieved. ALD has gained popularity in various fields, such as catalysis, fuel cells, and electronics, due to its ability to produce conformal coatings, uniform deposition, and precise thickness control [44]. A general overview of the ALD process can be found in Figure 4.13.

4.3.4.4.1 Thermal Atomic Layer Deposition

Thermal Atomic Layer Deposition (TALD) is a deposition technique that enables precise thickness control and conformal coating on the surface of a substrate. It is particularly suitable for complex structures with high aspect ratios. However, TALD typically requires high process temperatures ranging from 150°C to 350°C, which limits its applications. To overcome this limitation, PEALD was developed [45].

An example of TALD film deposition involves the fabrication of Al_2O_3 using trimethylaluminum (TMA) and water (H_2O) as precursor and reactant, respectively. In this process, TMA undergoes dissociative chemisorption on the substrate surface, while the excess TMA is pumped out. This allows the surface to be covered with $AlCH_3$. Then, H_2O vapor is introduced, reacting with the surface and forming a by-product of CH_4, resulting in a hydroxylated Al_2O_3 surface [45]. The reaction between $Al(CH_3)_3$ and H_2O in ALD is depicted in Figure 4.14.

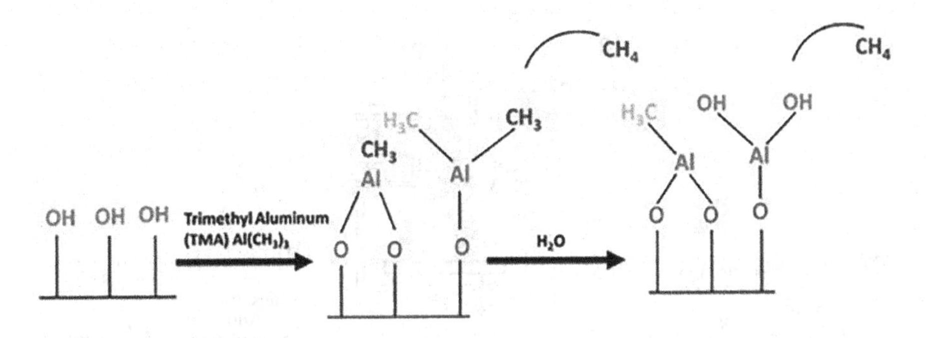

FIGURE 4.14 TALD of $Al(CH_3)_3$ and H_2O reaction [46].

4.3.4.4.1.1 Advantages Wide material compatibility, conformal coating, high film quality, precise film thickness control, and scalability.

4.3.4.4.1.2 Disadvantages Slow deposition rate, high temperature required, limited precursor options, sensitivity to substrate temperature, complex equipment setup, and high cost.

4.3.4.4.2 Plasma-Enhanced ALD

Plasma-Enhanced ALD (PEALD) is an alternative mode of ALD that involves the use of plasma species with high reactivity. This allows for the deposition of films at lower temperatures while maintaining the quality of the deposited material. PEALD serves to enhance the properties and quality of materials. Additionally, PEALD offers a broader range of precursor options that can be facilitated by plasma, enabling the deposition of materials that may be challenging with traditional TALD methods [47].

PEALD offers the advantage of enabling deposition at lower temperatures compared to TALD, resulting in enhanced film properties [48]. The use of highly reactive catalysts or species in PEALD further enhances the reaction process. This allows for a wider range of precursor and substrate options for deposition, making it possible to grow films on heat-sensitive substrates [45]. Additionally, PEALD allows to produce highly pure films, surpassing the purity achieved with TALD [45]. Another advantage of PEALD is its higher growth rate, making it a versatile and efficient process [49].

While PEALD offers numerous advantages, it also has several limitations. These limitations encompass challenges in achieving conformal coatings on structures with high aspect ratios or large surface areas, complex reaction chemistry, intricate reactor designs, low throughput, potential issues with achieving adequate conformality, and the need for additional growth parameters [45]. Low-pressure PEALD can be classified into four main configurations: radical-enhanced, direct plasma, remote, and direct plasma with mesh, as shown in Figure 4.15. The radical-enhanced configuration stands out from the other configurations (direct and remote plasma) because it exclusively employs neutral plasma species during deposition, protecting the substrate and film growth from potential damage [45]. On the other hand, the direct plasma with mesh configuration was designed to precisely control the interaction between the plasma and the substrate surface [50].

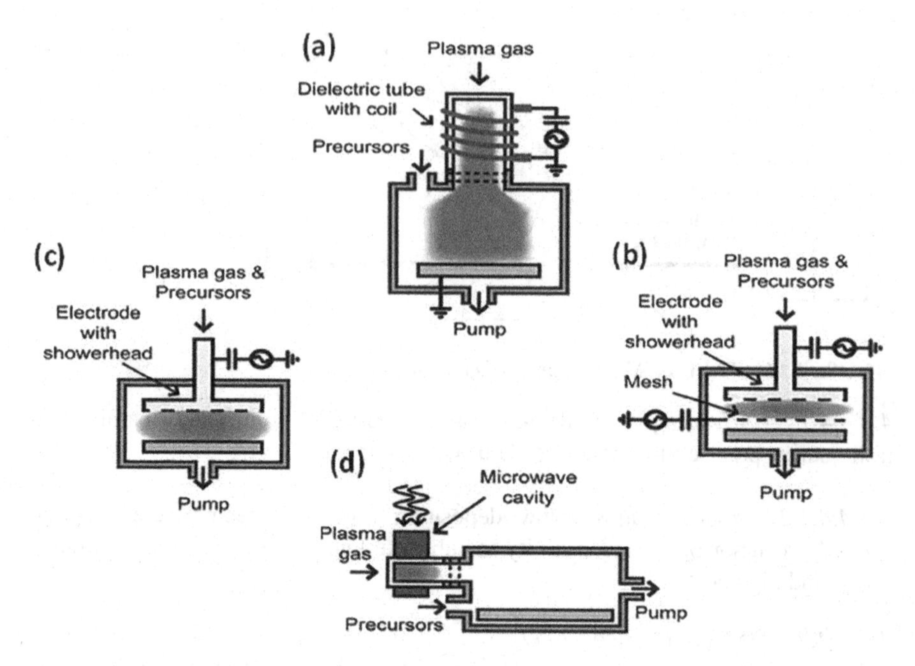

FIGURE 4.15 Schematic of plasma configurations: (a) remote, (b) direct with mesh, (c) direct, and (d) radical [49].

PEALD has demonstrated successful application in the deposition of various nitride compounds, metal films, and metal oxide films [51].

4.3.4.4.2.1 Advantages Precise thickness control, conformal coating, high film quality, wide material compatibility, and low-temperature deposition.

4.3.4.4.2.2 Disadvantages Slow deposition rate, equipment complexity and cost, process sensitivity, limited scalability, and plasma-induced damage.

4.3.4.4.3 Spatial Atomic Layer Deposition

When the size and surface area of an ALD reactor are increased, it leads to a decrease in the efficiency of purge time and a longer pulse duration. However, the use of spatial ALD (SALD) can help overcome this limitation. SALD resolves the issue by precisely directing a specific precursor to the substrate based on its position, eliminating the need for pulse-purge compartments [52]. This innovative SALD method enables a deposition rate of approximately 3,600 nm h^{-1} to be achieved [53].

The operational principle of SALD distinguishes it from thermal ALD (TALD) and PEALD. Unlike TALD and PEALD, which utilize sequential pulses and purge steps, creating a time-separated process, SALD utilizes separate inlets in the reactor to introduce precursors (see Figure 4.16). This methodology involves the substrate moving toward different inlets for film deposition. SALD achieves surface-limited and self-terminating reactions similar to those found in traditional ALD, making it equivalent to standard ALD processes. Hence, SALD ensures the production of

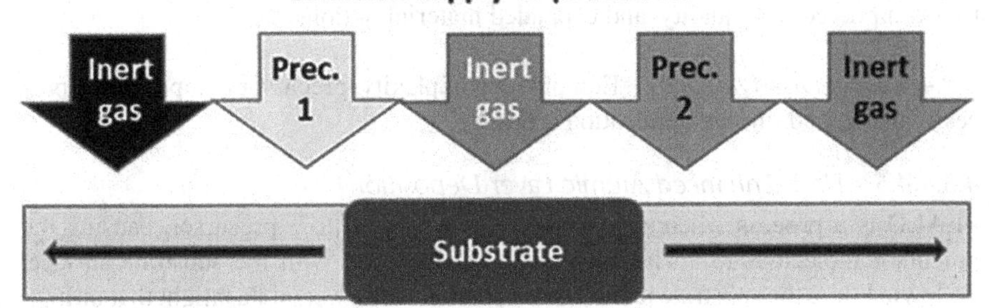

FIGURE 4.16 Schematic of SALD process [46].

high-quality films at lower temperatures while providing precise control overgrowth per cycle and enabling conformal coatings with high aspect ratios. It retains the unique characteristics of traditional ALD while offering notable benefits. SALD can be remarkably faster, up to 10^2 times, compared to conventional methods. It is also cost-effective, as thin films can be fabricated at AP (AP-SALD), eliminating the need for expensive and complex vacuum processes. Moreover, SALD is easily scalable. Nonetheless, SALD encounters challenges such as a limited range of available precursors [52], as well as the need to address impurities and undesired gas-phase reactions during the process.

4.3.4.4.3.1 Advantages High throughput, large-area deposition, conformal coating, precise film control, and low-temperature deposition.

4.3.4.4.3.2 Disadvantages Equipment complexity and cost, reduced film thickness control, surface contamination, process complexity, and limited material compatibility.

4.3.4.4.4 Photo-Assisted ALD

Photo-assisted ALD (PA-ALD) is also known as ultraviolet (UV)-enhanced ALD. Unlike PEALD, UV-enhanced ALD offers a milder activation, but it grants enhanced control through adjustments in UV light intensity, timing of illumination, and wavelength [45]. The presence of UV light expedites surface reactions on the substrate, leading to an improvement in film quality [45,47]. Essentially, UV energy facilitates the deposition reactions. In recent times, a similar variation of PA-ALD called flash-enhanced atomic layer deposition (FEALD) has been developed. The primary distinction is that FEALD employs light to deliver short pulses of heat to the substrate surface, driving the surface reactions [61]. PA-ALD has been successfully used in depositing various materials, including Al_2O_3, ZrO_2, TiO_2, ZnO, BN, Ta_2O_5, and GaAs, utilizing alkoxide precursors [54]. However, research on PA-ALD remains limited due to the challenges associated with constructing photo-ALD reactors [54].

4.3.4.4.4.1 Advantages Enhanced deposition rate, lower deposition temperatures, improved film quality, and expanded material options.

4.3.4.4.4.2 Disadvantages Equipment complexity, precursor compatibility, process control, and limited application scope.

4.3.4.4.5 Flash-Enhanced Atomic Layer Deposition

FEALD is a process whereby a substrate is subjected to a precursor, causing the precursor molecules to chemically or physically bond with the substrate surface. Following this, the substrate undergoes exposure to a powerful flashlight that utilizes xenon flash lamps. The emitted light from these lamps is predominantly visible and near-infrared, with a small portion reaching near-UV [55]. Because the precursor molecules have limited direct interactions with light, the photolytic effects are insignificant. Rather, the flash treatment briefly heats the illuminated surface, thereby initiating the chemical reaction between the adsorbed precursor molecules and the second reactant. Alternatively, it can thermally decompose the adsorbed precursor molecules. These reactions lead to the deposition of atoms from the precursors or compounds generated during the reactions onto the substrate surface. Subsequently, the volatile by-products are purged from the reaction compartment. The process proceeds in a step-by-step fashion, utilizing a deposition technique that adds layers one by one. Furthermore, the periodic annealing of the previously deposited film, achieved through flash heating in each cycle, offers the possibility of improving the film's quality [56]. This technique has been specifically used for depositing materials such as Si, Ru, and Al_2O_3 [54]. The configuration process of FEALD is illustrated in Figure 4.17.

4.3.4.4.5.1 Advantages Improved film quality, faster deposition rates, precise thickness control, and reduced temperature requirements.

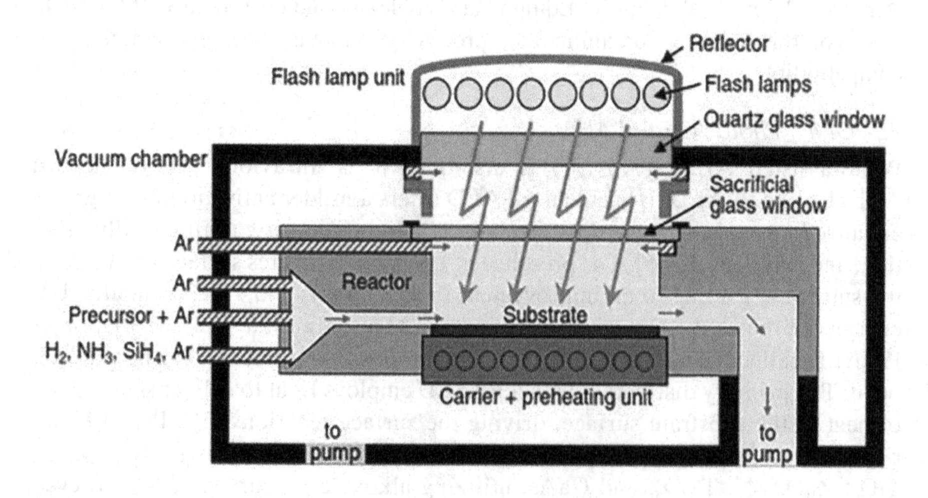

FIGURE 4.17 Schematic of FEALD deposition reactor [56].

4.3.4.4.5.2 Disadvantages Equipment complexity and cost, process optimization challenges, limited material compatibility, and increased energy consumption.

It's worth noting that the advantages and disadvantages mentioned above are based on the general characteristics of FALD. The specific benefits and drawbacks may vary depending on the material system, equipment setup, and application requirements.

4.4 SUMMARY

In this chapter, we have explored the physical and chemical deposition techniques utilized for fabricating thin films. The selection of a suitable method for thin-film fabrication relies on various factors, such as desired film properties, substrate material, and intended application. Researchers and engineers must carefully evaluate these considerations when deciding on the appropriate deposition technique to fulfill their specific requirements. Each method possesses its own strengths and limitations, and often a combination of techniques is employed to attain the anticipated film characteristics. The field of thin-film fabrication continues to evolve with the development of novel techniques and materials to meet the demands of modern applications. It is important to acknowledge that the realm of thin-film technologies is vast, and while this book does not extensively cover all the available technologies, it aims to provide a comprehensive understanding of the underlying principles and concepts.

REFERENCES

[1] A. Jilani, M.S. Abdel-Wahab, A.H. Hammad, Advance deposition techniques for thin film and coating, *Modern Technologies for Creating the Thin-Film Systems and Coatings*. 2 (2017) 137–149.

[2] K.S.S. Harsha, *Principles of Vapor Deposition of Thin Films*, Elsevier, 2005.

[3] S. Ganesan, S. Padmanabhan, J. Hemanandh, S.P. Venkatesan, Influence of substrate temperature on coated engine piston head using multi-response optimisation techniques, *International Journal of Ambient Energy*. 43 (2022) 610–617.

[4] T.D. Moustakas, Molecular beam epitaxy: thin film growth and surface studies, *MRS Bulletin*. 13 (1988) 29–36.

[5] J. Puebla, *Spin Phenomena in Semiconductor Quantum Dots*, Jorge Luis Puebla Nunez, 2015.

[6] M.A. Herman, W. Richter, H. Sitter, Molecular beam epitaxy, in: *Epitaxy*, Springer, 2004: pp. 131–170.

[7] C. Gumiel, D.G. Calatayud, Thin film processing of multiferroic BiFeO3: from sophistication to simplicity. A review, *Boletín de La Sociedad Española de Cerámica y Vidrio*. 61 (2022) 708–732.

[8] V.D. Okunev, Z.A. Samoilenko, N.N. Pafomov, A.L. Plehov, R. Szymczak, M. Baran, H. Szymczak, S.J. Lewandowski, P. Gierłowski, A. Abal'oshev, Effect of cluster structure on the transition from spin-dependent tunneling to percolation mechanism of conductivity in LaSr (Ca) MnO thin films, *Physics Letters A*. 332 (2004) 275–285.

[9] Y. Ahn, J.Y. Son, Mixed grains and orientation-dependent piezoelectricity of polycrystalline Nd-substituted Bi4Ti3O12 thin films, *Ceramics International*. 42 (2016) 13061–13064.

[10] D.M. Mattox, Fundamentals of ion plating, *Journal of Vacuum Science and Technology.* 10 (1973) 47–52.
[11] M.T. Taschuk, M.M. Hawkeye, M.J. Brett, *Handbook of Deposition Technologies for Films and Coatings*, Elsevier, 2010: pp. 621–678.
[12] D.M. Mattox, Ion plating-past, present and future, *Surface and Coatings Technology.* 133 (2000) 517–521.
[13] K. Friedrich, Polymer composites for tribological applications, *Advanced Industrial and Engineering Polymer Research.* 1 (2018) 3–39.
[14] D.G. Teer, Ion plating, *Vacuum.* 24 (1974) 482.
[15] P.J. Kelly, R.D. Arnell, Magnetron sputtering : a review of recent developments and applications, *Vacuum.* 56 (2000) 159–172.
[16] P.H. Li, P.K. Chu, H. Kong, Thin Film Deposition Technologies and Processing of Biomaterials, In: H. J. Griesser, *Thin Film Coatings for Biomaterials and Biomedical Applications*, Elsevier, 2016: pp. 3–28. https://doi.org/10.1016/B978-1-78242-453-6.00001-8.
[17] J.B. Mooney, S.B. Radding, Spray pyrolysis processing, *Annual Review of Materials Science.* 12 (1982) 81–101.
[18] J.C. Viguie, J. Spitz, Chemical vapor deposition at low temperatures, *Journal of the Electrochemical Society.* 122 (1975) 585–588.
[19] G.J. Exarhos, X.-D. Zhou, Discovery-based design of transparent conducting oxide films, *Thin Solid Films.* 515 (2007) 7025–7052.
[20] Y. Sawada, C. Kobayashi, S. Seki, H. Funakubo, Highly-conducting indium tin-oxide transparent films fabricated by spray CVD using ethanol solution of indium (III) chloride and tin (II) chloride, *Thin Solid Films.* 409 (2002) 46–50.
[21] T. Fukano, T. Motohiro, Low-temperature growth of highly crystallized transparent conductive fluorine-doped tin oxide films by intermittent spray pyrolysis deposition, *Solar Energy Materials and Solar Cells.* 82 (2004) 567–575.
[22] B. Benhaoua, S. Abbas, A. Rahal, A. Benhaoua, M.S. Aida, Effect of film thickness on the structural, optical and electrical properties of SnO2: F thin films prepared by spray ultrasonic for solar cells applications, *Superlattices and Microstructures.* 83 (2015) 78–88.
[23] L. Filipovic, S. Selberherr, G.C. Mutinati, E. Brunet, S. Steinhauer, A. Köck, J. Teva, J. Kraft, J. Siegert, F. Schrank, Modeling spray pyrolysis deposition, in: *Proceedings of the World Congress on Engineering*, 2013: pp. 987–992.
[24] L. Chen, G. Jianming Zhanga, X. Pangb, In-situ coating of MWNTs with solgel TiO2 nanoparticles, *Advanced Materials Letters.* 1 (2010) 75–78.
[25] E. Celik, M. Ozgul, E. Avci, Y.S. Hascicek, Adhesion properties of CeO2 films produced from different precursors using sol-gel process, *Materials Science and Engineering: B.* 261 (2020) 114774.
[26] M. Garcia-Rios, P. Gouze, Experimental assessment of the sealing potential of hydrated solgel for the remediation of leaky reservoirs, *Geosciences.* 8 (2018) 290.
[27] S. Sakka, *Handbook of Advanced Ceramics, Sol-Gel Process and Applications*, Elsevier, 2013.
[28] W. Chebil, A. Fouzri, A. Fargi, B. Azeza, Z. Zaaboub, V. Sallet, Characterization of ZnO thin films grown on different p-Si substrate elaborated by solgel spin-coating method, *Materials Research Bulletin.* 70 (2015) 719–727.
[39] S. Thiagarajan, A. Sanmugam, D. Vikraman, Facile methodology of sol-gel synthesis for metal oxide nanostructures, In: U. Chandra, ed., *Recent Applications in Sol-Gel Synthesis*, IntechOpen, 2017, pp. 1–17.
[30] J.M. Köhler, A. Knauer, The mixed-electrode concept for understanding growth and aggregation behavior of metal nanoparticles in colloidal solution, *Applied Sciences.* 8 (2018) 1343.

[31] J.E. ten Elshof, Chemical solution deposition techniques for epitaxial growth of complex oxides, in: *Epitaxial Growth of Complex Metal Oxides*, Elsevier, 2015: pp. 69–93.
[32] J. Dufour, *An Introduction to Metallurgy*, Cameron, New York, 2006: p. 12.
[33] R. Cui, Y. Han, Z. Zhu, B. Chen, Y. Ding, Q. Zhang, Q. Wang, G. Ma, F. Pei, Z. Ye, Investigation of the structure and properties of electrodeposited Cu/graphene composite coatings for the electrical contact materials of an ultrahigh voltage circuit breaker, *Journal of Alloys and Compounds*. 777 (2019) 1159–1167.
[34] R.T. Mathew, S. Singam, P. Kollu, S. Bohm, M. Prasad, Achieving exceptional tensile strength in electrodeposited copper through grain refinement and reinforcement effect by co-deposition of few layered graphene, *Journal of Alloys and Compounds*. 840 (2020) 155725.
[35] J. Li, Z. An, Z. Wang, M. Toda, T. Ono, Pulse-reverse electrodeposition and micromachining of graphene-nickel composite: an efficient strategy toward high-performance microsystem application, *ACS Applied Materials & Interfaces*. 8 (2016) 3969–3976.
[36] K. Maeda, Atmospheric pressure/low-pressure CVD, in: T. Hattori, ed., *Ultraclean Surface Processing of Silicon Wafers*, Springer, 1998: pp. 317–330, Part IV.
[37] R.G. Gordon, S. Barry, J.T. Barton, R.N.R. Broomhall-Dillard, Atmospheric pressure chemical vapor deposition of electrochromic tungsten oxide films, *Thin Solid Films*. 392 (2001) 231–235.
[38] A. de Graaf, J. van Deelen, P. Poodt, T. van Mol, K. Spee, F. Grob, A. Kuypers, Development of atmospheric pressure CVD processes for highquality transparent conductive oxides, *Energy Procedia*. 2 (2010) 41–48.
[39] N.K. Jain, M.S. Sawant, S.H. Nikam, S. Jhavar, L. Shohet, Metal deposition: plasma-based processes, in: J. L. Shohet, ed., *Encyclopedia of Plasma Technology*, CRC Press, 2016: pp. 722–740.
[40] T.-S. Chen, S.-E. Chiou, S.-T. Shiue, The effect of different radio-frequency powers on characteristics of amorphous boron carbon thin film alloys prepared by reactive radio-frequency plasma enhanced chemical vapor deposition, *Thin Solid Films*. 528 (2013) 86–92.
[41] K.K. Schuegraf, *Handbook of Thin Film Deposition: Processes and Technologies, Elsevier Science & Technology*, 2002.
[42] J. Alarcón-Salazar, R. López-Estopier, E. Quiroga-González, A. Morales-Sánchez, J. Pedraza-Chávez, I.E. Zaldívar-Huerta, M. Aceves-Mijares, Silicon-rich oxide obtained by low-pressure chemical vapor deposition to develop silicon light sources, in: S. Neralla, *Chemical Vapor Deposition-Recent Advances and Applications in Optical, Solar Cells and Solid State Devices*, IntechOpen, (2016) 159–181.
[43] O. Sneh, R.B. Clark-phelps, A.R. Londergan, J. Winkler, T.E. Seidel, Thin film atomic layer deposition equipment for semiconductor processing, *Thin Solid Films*. 402 (2002) 248–261.
[44] B.C. Mallick, C. Hsieh, K. Huang, K. Yin, Y.A. Gandomi, Review - on atomic layer deposition : current progress and future challenges, *ECS Journal of Solid State Science and Technology*. 8 (2019) N55. https://doi.org/10.1149/2.0201903jss.
[45] P.O. Oviroh, R. Akbarzadeh, D. Pan, A.M. Coetzee, T. Jen, New development of atomic layer deposition : processes, methods and applications, *Science and Technology of Advanced Materials*. 20 (2019) 465–496. https://doi.org/10.1080/14686996.2019.1599694.
[46] J.A. Oke, T.-C. Jen, Atomic layer deposition thin film techniques and its bibliometric perspective, *The International Journal of Advanced Manufacturing Technology*. 126 (2023) 4811–4825. https://doi.org/10.1007/s00170-023-11478-y.
[47] M. Napari, *Low-Temperature Thermal and Plasma-Enhanced Atomic Layer Deposition of Metal Oxide Thin Films*, Research Report/Department of Physics, University of Jyväskylä, 2017.

[48] J. Musschoot, *Advantages and Challenges of Plasma Enhanced Atomic Layer Deposition*, Doctoral Dissertation, Ghent University, 2011.
[49] S.B.S. Heil, J.L. Van Hemmen, C.J. Hodson, N. Singh, J.H. Klootwijk, F. Roozeboom, M.C.M. Van de Sanden, W.M.M. Kessels, Deposition of TiN and Hf O 2 in a commercial 200 mm remote plasma atomic layer deposition reactor, *Journal of Vacuum Science & Technology A: Vacuum, Surfaces, and Films.* 25 (2007) 1357–1366.
[50] H.B. Profijt, S.E. Potts, M.C.M. Van de Sanden, W.M.M. Kessels, Plasma-assisted atomic layer deposition: basics, opportunities, and challenges, *Journal of Vacuum Science & Technology A: Vacuum, Surfaces, and Films.* 29 (2011) 50801.
[51] H. Kim, Characteristics and applications of plasma enhanced-atomic layer deposition, *Thin Solid Films.* 519 (2011) 6639–6644. https://doi.org/10.1016/j.tsf.2011.01.404.
[52] D. Munoz-Rojas, J. MacManus-Driscoll, Spatial atmospheric atomic layer deposition: a new laboratory and industrial tool for low-cost photovoltaics, *Materials Horizons.* 1 (2017) 314–320.
[53] R.W. Johnson, A. Hultqvist, S.F. Bent, A brief review of atomic layer deposition: from fundamentals to applications, *Materials Today.* 17 (2014) 236–246.
[54] V. Miikkulainen, K. Väyrynen, V. Kilpi, Z. Han, M. Vehkamäki, K. Mizohata, J. Räisänen, M. Ritala, Photo-assisted ALD: process development and application perspectives, *ECS Meeting Abstracts.* MA2017–02 (2017) 1093–1093. https://doi.org/10.1149/ma2017-02/25/1093.
[55] M. Smith, R. McMahon, M. Voelskow, D. Panknin, W. Skorupa, Modelling of flash-lamp-induced crystallization of amorphous silicon thin films on glass, *Journal of Crystal Growth.* 285 (2005) 249–260.
[56] T. Henke, M. Knaut, C. Hossbach, M. Geidel, L. Rebohle, M. Albert, W. Skorupa, J.W. Bartha, Flash-enhanced atomic layer deposition: basics, opportunities, review, and principal studies on the flash-enhanced growth of thin films, *ECS Journal of Solid State Science and Technology.* 4 (2015) P277.

5 Characterization Techniques of Smart Thin Films

5.1 INTRODUCTION

Characterization techniques play a vital role in the field of materials science and engineering by providing valuable insights into the properties and behavior of materials. These techniques encompass a wide range of methods used to examine and analyze various aspects of a material, such as its structure, composition, morphology, and properties. The characterization of materials is essential for understanding their performance, improving their properties, and guiding the development of new materials with tailored functionalities. In particular, the characterization of thin films, which are thin layers of materials deposited on substrates, is of significant interest due to their unique properties and wide-ranging applications in fields such as electronics, optics, energy, and coatings.

Thin film characterization techniques enable researchers to investigate the thickness, composition, crystal structure, surface morphology, optical and electrical properties, magnetic behavior, and other important parameters of thin films. By utilizing these techniques, scientists and engineers can gain a comprehensive understanding of the structure-property relationships of thin films, ultimately leading to the optimization of their performance and the design of novel applications. This article provides an overview of commonly employed thin film characterization techniques and discussing their principles. It highlights techniques such as spectroscopy, microscopy, diffraction, magnetic and electrical measurements, and surface profiling. Each technique offers unique capabilities and insights into different aspects of thin film behavior, and their combined use provides a comprehensive characterization toolkit.

Understanding and employing appropriate thin film characterization techniques are crucial for researchers and engineers working in the field of thin film materials. These techniques enable the evaluation, comparison, and optimization of thin film properties, supporting the advancement of various technologies and facilitating the development of innovative devices and materials.

5.2 CLASSIFICATION OF CHARACTERIZATION TECHNIQUES

There are two main categories of techniques used for characterizing materials: destructive and non-destructive methods.

DOI: 10.1201/9781003331940-5

1. **Destructive Characterization:** These techniques involve modifying or damaging the material under study to gather information about its properties. Examples of such techniques are differential scanning calorimetry, sample preparation for electron microscopy, thermal analysis, and mechanical testing.
2. **Non-destructive Characterization:** These techniques are employed to gather valuable information about a material's properties, composition, and structure without causing any damage or alteration. These techniques also allow for detailed analysis while preserving the integrity of the material. Some examples of non-destructive characterization techniques include: scanning electron microscopy (SEM) and transmission electron microscopy, atomic force microscopy (AFM), X-ray diffraction (XRD), Raman spectroscopy, x-ray photoelectron spectroscopy (XPS), Fourier transform infrared (FTIR) spectroscopy, x-ray absorption near-edge spectroscopy (XANES), current-voltage (*I-V*) and capacitive-voltage (C-V), superconducting quantum interference device (SQUID), vibrating sample magnetometer (VSM), electron spin resonance (ESR), ultraviolet visible (UV-Vis) spectroscopy and near-infrared (NIR) spectroscopy and photoluminescence (PL).

Additionally, characterization techniques can also be classified based on the type of information they provide, such as:

1. **Structural Characterization:** Techniques used to determine the structure of a material, including XRD, TEM, and SEM.
2. **Composition Characterization:** Techniques used to determine the chemical composition of a material, including XRF, energy-dispersive X-ray spectroscopy (EDS), and inductively coupled plasma-mass spectrometry.
3. **Physical Property Characterization:** Techniques used to determine physical properties such as electrical conductivity, optical, and magnetic properties, including *I-V* and C-V, UV-Vis, PL, SQUID, VSM, and ESR.
4. **Surface Characterization:** Techniques used to determine the surface characteristics of a material, including AFM, scanning probe microscopy (SPM), and surface energy analysis.

Each characterization technique has its own strengths and limitations, and the choice of which technique to use depends on the specific information required and the properties of the material being studied. In this chapter, non-destructive characterization techniques were discussed.

5.3 STRUCTURAL AND SURFACE CHARACTERIZATION

5.3.1 Scanning Electron Spectroscopy

SEM spectroscopy is a method used to analyze the microstructure and morphology of nanomaterials through imaging. It involves scanning the surface of a material using a beam of electrons, which emits a sequence of radiations that are utilized to generate images. Some of the radiation emitted in SEM spectroscopy includes the following: (i) Characteristic X-rays are generated when there is a disparity in energy between two electrons. (ii) Secondary electrons, with energies below 50 eV, are expelled from the outer orbital of the sample's atoms because of multiple inelastic scattering effects. (iii) Electrons above 50 eV energy that undergo elastic scattering are reflected out of the specimen's contact volume, causing them to scatter backward. The SEM technique is utilized to generate images and analyze the morphological arrangement of particles in a sample. It allows for easy observation of various samples such as flakes, nanotubes, nanowires, and more. The imaging process in SEM relies on the interaction between the sample and electron beams [1]. The key components of an SEM system consist of an electron source, a medium through which electrons travel, an electron detector, a sample chamber, and a computer display. A comprehensive technical description of these components can be found in the available literature [2,3]. The system comprises several components, including an electron gun, condenser lens, objective lens, deflection coil, and detectors, as illustrated in Figure 5.1. The electron beam is generated by either a tungsten filament or a field emission gun. High voltage is applied to accelerate the electron beams, enabling them to scan the surface of the sample. As electrons are generated at the top of the column, they travel down the column and pass through various configurations of lenses and apertures.

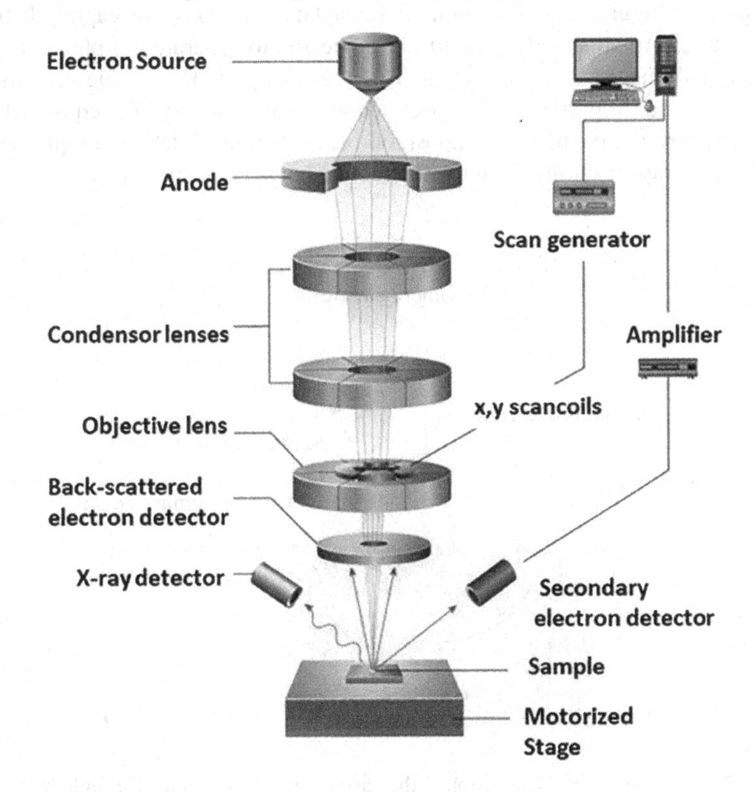

FIGURE 5.1 Schematic working principle of SEM [4].

This process culminates in the formation of a collimated beam of electrons, which is directed toward the sample surface. To control the interaction between the sample and the electron beam, scan coils are positioned above the objective lens (not shown in the diagram but located within the scanning coil). These coils scan the surface of the sample, while the electron-sample interaction generates signals that are captured by a detector. This allows for precise control and measurement of the interaction between the sample and the electron beam [3].

Scanning a sample with high-energy electron beams results in the generation of images. The interaction between the electrons and the sample produces three distinct signals: secondary electrons, back-scattered electrons, and characteristic X-rays. These signals carry valuable information about the sample's properties, including its surface topography, composition, and elemental characteristics. Detectors capture the signals generated during the electron-sample interaction, and these signals are then displayed on a computer monitor. The penetration depth of the electron beams into the sample, typically a few microns, is determined by the accelerating voltage and sample density. The quality of the resulting image relies on factors such as the electron spot size and the volume of the sample that interacts with the electron beam [3].

The interaction between electrons and the specimen yields various additional signals, such as the emission of characteristic X-rays, cathodoluminescence, and Auger electrons (see Figure 5.2). The emitted secondary electrons are captured, transformed into electrical signals, and further processed to generate morphological and compositional images of the specimen. These signals provide valuable insights into the structure and properties of the specimen under analysis. SEM is equipped with EDS to analyze the elemental composition of a specimen. EDS techniques utilize the emitted X-rays resulting from the interaction between primary electrons and the

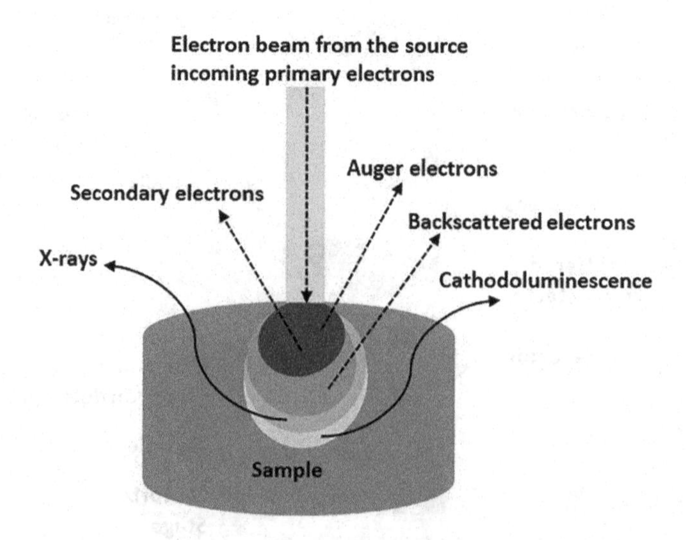

FIGURE 5.2 Schematic representation of the various signals generated through the interaction between electrons and the specimen.

specimen. These X-rays provide valuable information about the characteristic elements present in the specimen and their respective compositions [5].

5.3.2 Transmission Electron Microscopy

TEM serves as a valuable complementary technique to XRD when assessing the material's crystallography. It is commonly employed to analyze the microstructure of materials, utilizing a high-energy electron beam with an energy of at least 200 keV. TEM provides detailed insights into the internal structure and arrangement of atoms within a material. In TEM, a focused beam of electrons is directed on a specimen, and the electrons transmitted through the specimen are utilized to form an image. This image can be magnified and displayed on a fluorescent screen, recorded on photographic film, or detected by a sensor such as a charge-coupled device camera. By capturing and interpreting the transmitted electrons, TEM enables the visualization and analysis of the internal structure and fine details of the specimen at high resolution. Electrons in TEM are produced either through thermionic emission or field emission. They are then accelerated by an electrical field and focused onto the sample using electrical and magnetic fields. When the electron beam interacts with a crystalline material, diffraction plays a significant role, whereas absorption is less prominent. However, the intensity of the transmitted beam can still be influenced by the volume and density of the material it passes through. For TEM analysis, the sample is typically prepared as an ultra-thin slice (less than 100 nm) to ensure proper transmission of the electron beam. The sample allows electrons to pass through, resulting in the formation of a diffraction pattern and image at the objective lens's back focus plane and image plane. By considering the back focus plane as the objective plane for the intermediate lens and projector lens, we can project the diffraction pattern onto a screen. This utilization of diffraction characterizes the functioning of a TEM, often referred to as operating in diffraction mode [6,7].

When we designate the image plane of the objective lens as the objective plane for the intermediate lens and projector lens, we can generate an image on a screen. This configuration is known as the image mode. Typically, TEM techniques offer a resolution of around 0.3 nm. In advanced diffraction contrast TEM instruments, it is possible to examine crystal structures using high-resolution transmission electron microscopy (HRTEM), which is also referred to as phase contrast imaging. In HRTEM, images result from variations in the phase of electron waves scattered by a thin specimen, enabling detailed investigation of crystal structures. Interpreting the contrast in HRTEM images can be challenging due to the significant aberrations caused by the imaging lenses in the microscope. However, despite these challenges, HRTEM has achieved remarkable resolutions as fine as 0.5 Å [6,7]. To illustrate the operational principle of TEM, refer to Figure 5.3.

5.3.3 Atomic Force Microscopy

AFM is an imaging technique renowned for its high-resolution capabilities in examining the surface properties of various materials, particularly thin films. AFM operates by employing a cantilever probe that scans across the sample surface and detects

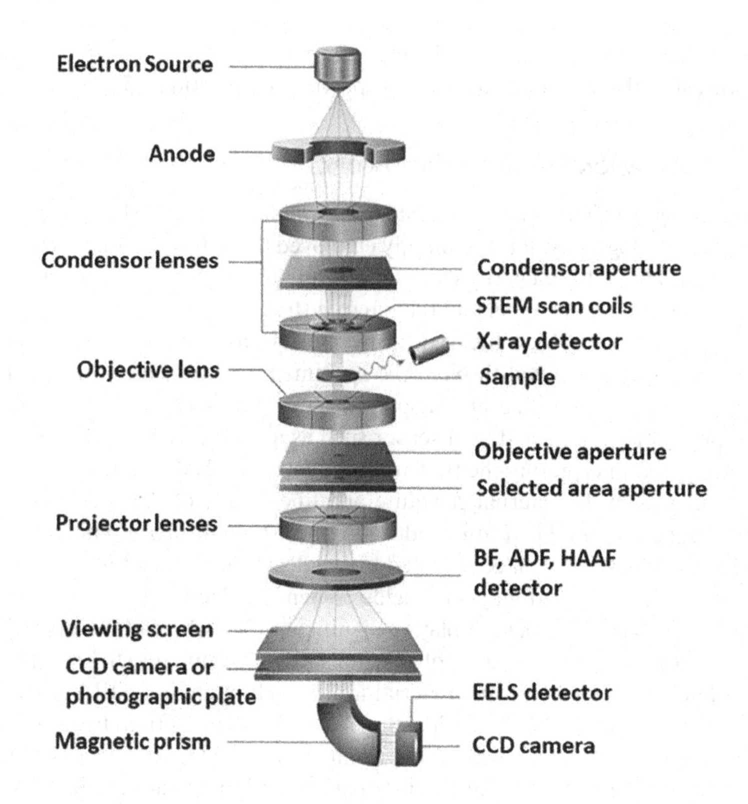

FIGURE 5.3 Schematic of TEM system [4].

the force exerted between the probe and the sample. By measuring this force, AFM generates detailed images revealing the topography, roughness, and mechanical characteristics of the sample [8].

AFM can be operated in two distinct modes: contact mode and non-contact mode, each suited to different imaging needs. In contact mode, the probe maintains continuous contact with the sample surface while being scanned. The deflection of the probe caused by its interaction with the sample is precisely recorded during the scanning process. Non-contact mode in AFM involves maintaining a constant separation between the probe and the sample surface. This mode relies on laser interferometry [9] to monitor the interaction between the probe and the sample. Figure 5.4 illustrates the setup and mechanism employed in non-contact mode.

The high-resolution imaging capabilities of AFM enable the generation of nanometer-scale images, rendering it a valuable tool for investigating materials with intricate structures, such as thin films. AFM is employed to explore diverse material properties, including adhesion, conductivity, surface roughness, and elasticity. In addition to its imaging capabilities, AFM is extensively utilized for studying the morphology of materials, which involves analyzing the shape, size, and distribution of individual particles on a surface. Furthermore, AFM offers the remarkable ability to manipulate materials at the nanoscale. The ability to

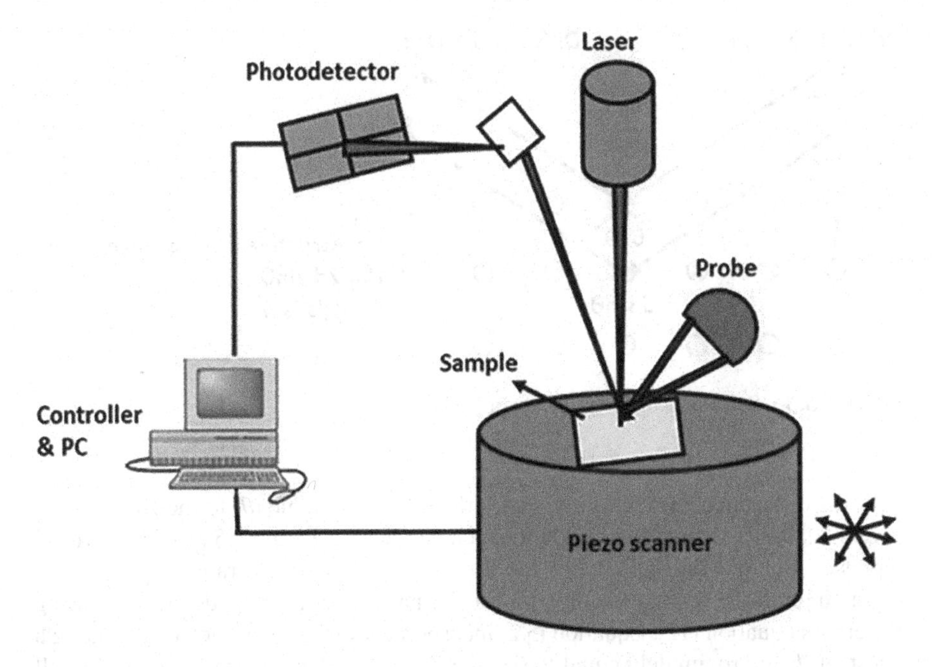

FIGURE 5.4 Schematic of AFM technique.

control the position and exert force through the cantilever probe empowers AFM to manipulate and assemble materials at the nanoscale. This capability renders AFM an indispensable tool in diverse fields such as electronics, materials science, and nanotechnology [10].

5.3.4 X-Ray Diffraction

XRD is a powerful and effective technique used to study the phase characteristics of crystalline materials. Researchers rely on XRD to extract valuable data about the lattice parameters, crystal structure, crystal orientation, and size of these materials. Due to its speed and dependability, XRD has become a widely employed method for investigating unknown crystalline substances and acquiring a deeper understanding of their structural properties. XRD operates based on constructive interference between monochromatic X-rays and a crystalline sample. To generate the monochromatic X-rays, energetic electrons bombard a cathode ray tube, and the resulting radiation is filtered and concentrated on the material of interest. This process forms the foundation of the XRD technique. The interaction between incident rays and the material leads to constructive interference, resulting in diffracted rays, according to Bragg's Law [11]. This law governs the relationship between the incident and diffracted angles in XRD and is given as:

$$n\lambda = 2d\sin\theta \tag{5.1}$$

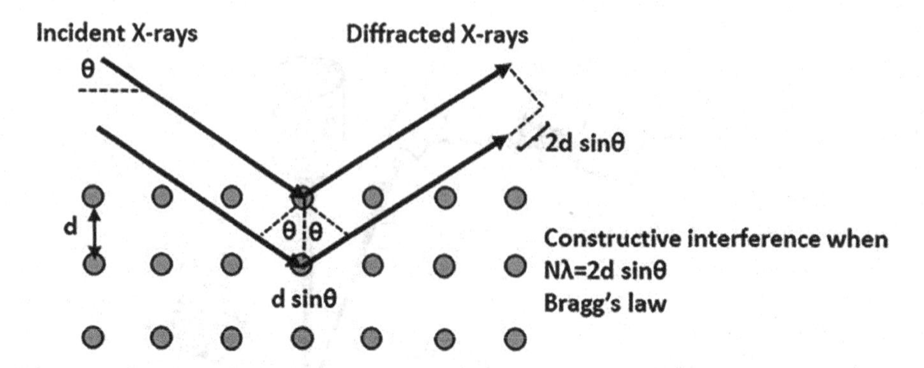

FIGURE 5.5 Schematic of Bragg's Law in XRD.

In X-ray diffraction, Bragg's Law relates the diffracted angle (θ) to the X-ray wavelength (λ), lattice distance (d), and reflection number (n). Figure 5.5 provides a visual representation of Bragg's Law in XRD through a schematic diagram.

In addition, the crystallite size (D) of the nanomaterials was determined using Scherrer's equation [12]. Equation (5.2) incorporates various parameters, including a constant (k) approximately equal to 0.9, the X-ray wavelength (λ) of 1.54 Å, the full width at half maximum (β), and the Bragg's angle (θ). By plugging in these values, the crystallite size can be calculated:

$$D = k\lambda / \beta \cos\theta \tag{5.2}$$

Crystals exhibit a regular arrangement of atoms in space, forming crystal lattices. X-ray diffraction (XRD) has been widely employed for structural characterization purposes, as it is a non-destructive technique. This technique can determine the crystal structure of various materials, including minerals, metals, organic substances, inorganic compounds, alloys, and polymers. XRD patterns exhibit a distinct series of peaks, where the intensity of each peak is plotted on the y-axis, while the diffraction angle (2θ) is plotted on the x-axis. These peaks, referred to as reflections, serve as distinctive signatures of the crystal structure inherent to the material being analyzed. The locations of these peaks in the XRD pattern are determined by the crystal structure of the material, while the intensities of the peaks are influenced by several factors, including slit width, number of grains present, incident intensity, and atomic structure factors [13]. X-rays are a form of high-energy electromagnetic radiation characterized by wavelengths ranging from 0.1 to 100 Å. Their discovery can be attributed to Wilhelm Conrad Röntgen, who first identified X-rays in 1895 and was subsequently honored with the Nobel Prize in Physics in 1901 for this ground-breaking achievement. As a result, X-ray diffraction (XRD) has become an indispensable analytical technique in various research fields, including solid-state chemistry and material science. In a typical XRD configuration, X-rays are produced within a high-vacuum tube [13].

5.4 RAMAN SPECTROSCOPY

Raman spectroscopy plays a crucial role in the characterization of the electronic properties of materials [14,15]. This technique is invaluable in assessing variations in element concentrations within a sample. The examination of Raman spectra and the observed Raman shifts provide information about changes in concentration, which can be indicative of the quality of the analyzed sample. A high-quality sample often exhibits prominent peaks in the Raman spectra, while the presence of minimal background noise further enhances the overall clarity of the results. These characteristics of the Raman spectra contribute to the assessment of sample quality and aid in the interpretation of the obtained data. Figure 5.6 illustrates the working principle of a Raman spectroscopy system. This technique relies on the phenomenon of light scattering when a sample is illuminated with a laser beam of high intensity [16]. Upon illuminating a sample with a laser of a specific wavelength, a significant portion of the incident light undergoes scattering. However, the scattered light with the same wavelength as the incident laser source is not typically noted, as it does not provide useful information about the sample being analyzed [16].

Rayleigh scattering refers to the scattered light that shares the same wavelength as the incident laser beam. However, the detector in a Raman spectroscopy system focuses on capturing and recording the relatively small proportion of scattered light waves that possess wavelengths different from those of the laser source. Typically, these scattered light waves account for a minuscule fraction of approximately 0.0000001%. The handful of scattered light waves that exhibit different wavelengths than the incident laser light are highly informative about the chemical structure of the sample. These specific sets of scattering are commonly

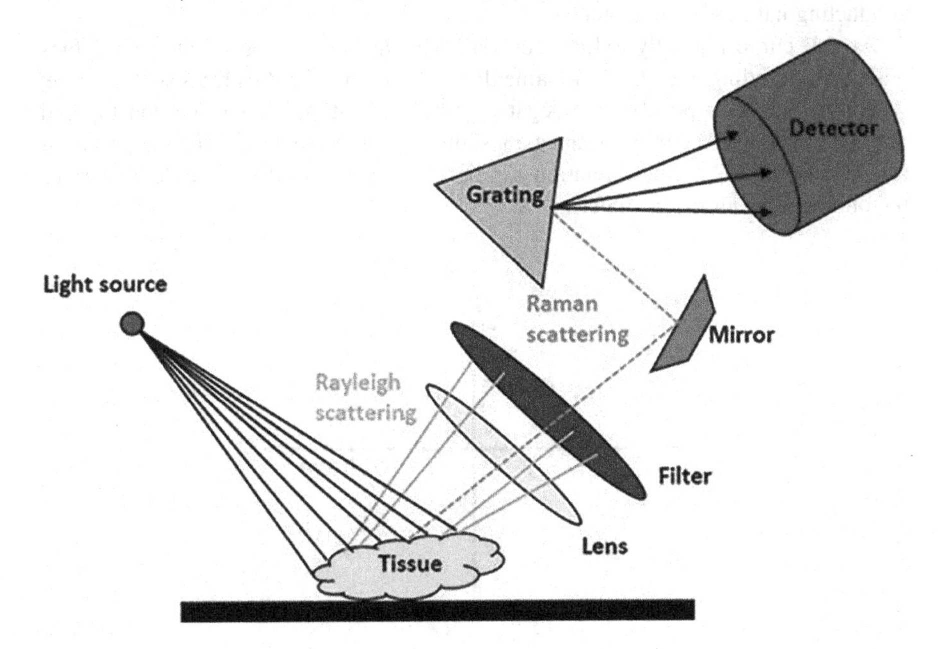

FIGURE 5.6 Schematic of Raman spectroscopy.

known as Raman scattering or Raman scatter [16]. The Raman spectrum is characterized by peaks that provide information about the intensity and wavelength positions of the scattered light. Each peak in the spectrum corresponds to specific molecular bonding vibrations present in the material under analysis [17]. Carbon materials typically exhibit distinct peaks in their Raman spectra. Two prominent peaks, known as the G band and 2D band, are commonly observed. The G band is typically located around 1,580 cm^{-1}, while the 2D band appears at approximately 2,700 cm^{-1} [18].

5.5 ELECTRICAL CHARACTERIZATION

5.5.1 *I-V* Measurement

The *I-V* technique is a crucial method used to assess the electrical properties and conductivity of materials. Conducting materials, such as metals, are characterized by a robust interaction among their electrons. This strong interaction contributes to their high conductivity and distinguishes them as effective electrical conductors. The strong interaction among electrons in conducting materials leads to the elimination of the forbidden gap, enabling the unrestricted movement of free electrons between the valence and conduction bands. This phenomenon facilitates the smooth flow of electric current or heat through the material when it is subjected to such stimuli. The electrical conductivity of a material is determined by the abundance and ease of movement of free electrons transitioning to the conduction band [19]. This fundamental characteristic governs the material's ability to conduct electricity. The *I-V* curves presented below serve as illustrations showcasing the conducting and semiconducting natures of the material's structure.

An *I-V* curve typically exhibits current flow in both forward and reverse bias with corresponding voltage in the same direction. Figure 5.7 illustrates such a curve, which intersects the positive and negative paths, indicating a linear relationship and ohmic behavior. The curve reveals a constant slope, represented by the reciprocal of resistance (1/*R*). This linear relationship between current and voltage demonstrates the ohmic behavior of the material.

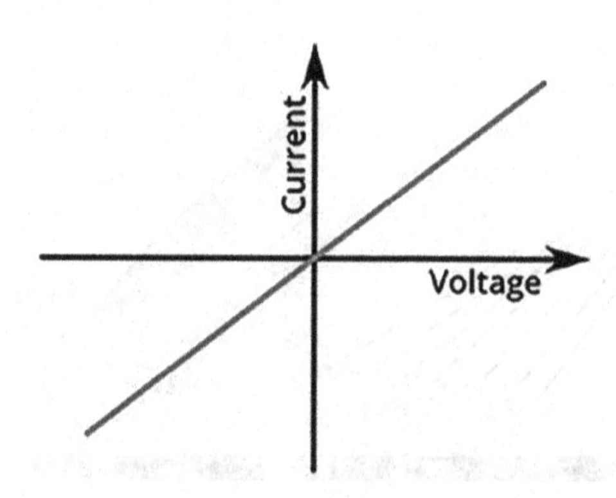

FIGURE 5.7 Schematic of a conducting *I-V* curve.

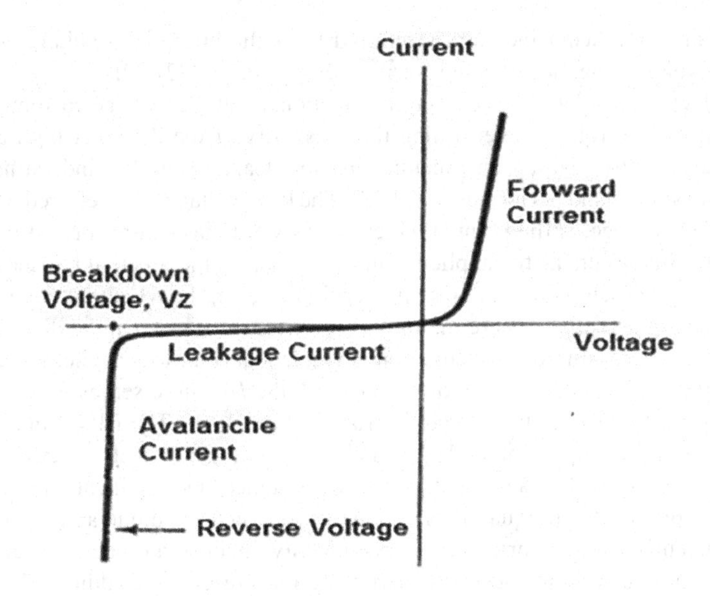

FIGURE 5.8 Schematic of a semiconducting *I-V* curve.

In contrast, Figure 5.8 displays that a semiconducting material exhibits a non-ohmic and nonlinear *I-V* curve. The current flowing through the positive path gradually increases until it reaches an internal voltage barrier limit, resulting in a distinctive nonlinear curve. This behavior highlights the unique characteristics of semiconductors, where the flow of current is significantly influenced by the voltage applied. When a diode is reverse-biased, it acts as a barrier to the flow of current. This results in a leakage current, which arises due to the inherent characteristics of the diode that prevent current from freely passing through in the reverse direction. Thus, the reverse-biased configuration effectively blocks the flow of current in a diode. Continuing this process, the negative voltage progressively surpasses the breakdown voltage threshold. Consequently, the combination of the voltage breakdown and diode voltage results in a rapid surge in reverse current. This phenomenon signifies a critical point where the diode experiences a significant increase in reverse current due to the combined effect of the applied voltage and the breakdown characteristics.

The utilization of the *I-V* technique enables the acquisition of valuable data regarding a device by observing the varying current passing through the diode as the voltage is adjusted. The plotted *I-V* characteristics offer crucial insights into the functionality of the diode. By analyzing the *I-V* curve, various details can be deduced about the diode, such as the breakdown voltage, leakage current, and knee voltage. The breakdown voltage helps in identifying the operational voltage range within which the diode functions effectively [20]. Under ideal conditions, the reverse current in a diode remains relatively low as it arises from minority carriers. However, as the reverse voltage progressively escalates, there comes a threshold where the reverse current undergoes a rapid surge. The voltage at which the reverse current experiences a sudden and significant increase is referred to as the breakdown voltage. At this particular voltage, the diode becomes irreversibly damaged [21,22].

The *I-V* curve also provides crucial information about the leakage current, which plays a significant role in determining the sensitivity of the diode. A high-quality diode is expected to exhibit an exceptionally low leakage current, indicating both superior sensitivity and performance [23,24]. The knee voltage, also referred to as the switching-on voltage, is the point at which the forward bias causes a rapid increase in current. This occurs as the applied voltage overcomes the junction barrier within the diode [23]. When plotted on a semilogarithmic scale, the *I-V* curve allows us to identify the voltage range where the impact of series resistance can be disregarded. Additionally, examining the *I-V* curve on a double logarithmic scale helps establish the conduction mechanism of the diode. Overall, the *I-V* curve serves as a valuable tool for assessing the quality and characteristics of the diode. The *I-V* technique also allows us to extract important diode parameters, including the ideality factor, saturation current, Schottky barrier height, parasitic resistance, and rectification ratio. The ideality factor specifically quantifies the degree to which the diode adheres to pure thermionic emission (TE) principles. If the ideality factor of a diode exceeds unity, it signifies non-ideal characteristics, indicating the presence of additional current mechanisms beyond pure TE.

The saturation current is caused by the diffusion of minority carriers from the neutral regions to the depletion region. The variation of the saturation current provides information about the minority carrier density. The Schottky barrier height is the potential needed for an electron to move from a semiconductor bulk to a metal. This parameter confirms the existence of the metal-semiconductor configuration [16], and it can be used to study the properties of charge carriers because of the induced defects.

The parasitic resistance in terms of the shunt resistance and series resistance provides information about the diode resistance [25]. An increase in the resistivity of the diode would indicate a recombination or compensation of carriers, resulting in the properties of the induced defects. The rectification ratio provides information on the rectifying behavior of a diode, and its variation can be used to study the quality of the diode. In an ideal situation, TE theory describes the diode current [26,27] as

$$I = I_s \left[\exp\left(\frac{qV}{\eta kT} - 1 \right) \right] \tag{5.3}$$

where $\eta = 1$ and the effect of series resistance (R_s) is negligible, the forward *I-V* curve is expected to be linear throughout the voltage region. However, this is not the case because the linear trend disappears in high-voltage regions due to η always being greater than unity and R_s is always present in high-voltage regions. To account for these non-ideal situations ($\eta > 1$ and presence of R_s), Eq. (5.3) is modified into

$$I = I_s \left[\exp\left(\frac{q(V - IR_s)}{\eta kT} - 1 \right) \right] \tag{5.4}$$

where

$$I_s = AA^{**}T^2 \exp\left(\frac{-q\phi_B}{kT} \right). \tag{5.5}$$

In the above equations, A (=2.83 × 10^{-3}cm^2) is the diode active area, A^{**} (=112 and 32 A cm^{-2} K^{-2}) is the effective Richardson constant for n and p-type Si, respectively, T is the temperature in kelvin, k is the Boltzmann constant, q is the electronic charge, and ϕ_B is the Schottky barrier height. In extracting the diode parameters, Rhoderick assumed that for highly doped semiconductors, the I-V characteristics at $V > 3kT/q$ can be represented in a simplified form and given [26] as

$$I = I_s \exp\left(\frac{qV}{\eta kT} \right). \tag{5.6}$$

Solving further by linearizing will give

$$\ln(I) = \frac{qV}{\eta kT} + \ln I_s \tag{5.7}$$

indicating that a plot of $\ln(I) - V$ will give a linear trend at the low voltage region where the effect of R_s is negligible. The diode parameters are extracted from the linear region using Rhoderick method [28]. The ideality factor, η, is obtained from the slope of the linear region as

$$\eta = \frac{q}{kT} \frac{dV}{d\ln(I)} \tag{5.8}$$

and the saturation current (I_s) is obtained from the y-axis intercept. The Schottky barrier height (ϕ_B) is evaluated [27,29] as

$$\phi_B = \frac{kT}{q} \ln\left(\frac{AA^{**}T^2}{I_s} \right) \tag{5.9}$$

The accuracy of the parameters derived using Rhoderick's method cannot be guaranteed because the parameters are extracted from the linear region where the effects of R_s are negligible. However, Cheung's method [30] is employed to account for the nonlinear region where the effect of R_s is noticeable and is given [24] as

$$\frac{dV}{d\ln(I)} = \left(\frac{\eta kT}{q} \right) + IR_s \tag{5.10}$$

$$H(I) = V - \left(\frac{\eta kT}{q}\right)\ln\left(\frac{I}{AA^{*}T^{2}}\right) \quad (5.11)$$

and $H(I)$ is given as

$$H(I) = \eta\phi_B + IR_s \quad (5.12)$$

The value of R_s can be obtained from the slope of $dV/d\ln(I) - I$ plot from Eq. (5.10) and η can be calculated from the y-axis intercept. The value of η determined from Eq. (5.10) was inserted in Eq. (5.12) to obtain the plot of $H(I) - I$ and the y-axis intercept is used to calculate ϕ_B. The slope $H(I)$–I plot can be used for the second determination of R_s.

5.5.2 C-V Measurement

The *C-V* technique is another useful tool to investigate the electrical properties of a semiconducting device, and it measures the capacitance at the junction of a diode. The space charge region (*SCR*) acts like a capacitor. By varying the applied voltage to the junction, the *SCR* width also varies. The dependence of the *SCR* width upon the applied voltage provides information on the diode's characteristics, such as doping profile, doping concentration, full depletion voltage (FDV), image force barrier lowering, and Schottky barrier height.

The doping profile provides information on the uniformity of the doping concentration and is determined by the linear regions identified in the C^{-2}-V plot. The doping profile can be used to study the properties of the induced defects, as a single linear region would indicate a uniform doping profile in the material, while multiple linear regions would indicate a non-uniform doping profile.

The doping density is the density of charge carriers responsible for the conductivity in the diode. An increase or decrease in the doping density would provide information on the properties of the induced defects. An increase in the doping density would indicate that the induced defects generate carriers, while a decrease would suggest that they are responsible for recombination or compensation of charge carriers. The FDV is the voltage needed to withdraw all charge carriers from the SCR. A variation in FDV would provide information on the properties of the induced defects. A high FDV would indicate that a high density of carriers is available in the SCR, while a low FDV would indicate that the density of carriers available in the SCR is low. FDV also provides information about the operating voltage of a diode.

The image force barrier lowering provides information about the image charges that accumulate on the metal electrode as charge carriers approach the metal-semiconductor interface. The potential associated with these image charges reduces the Schottky barrier height. Though the value of image force barrier lowering is small in comparison to the barrier height itself, it is still of interest because it is voltage-dependent and causes the reverse current to be voltage-dependent. This parameter has a direct impact on the Schottky barrier height and the reverse current; hence, it is useful in the study of the properties of the induced defects.

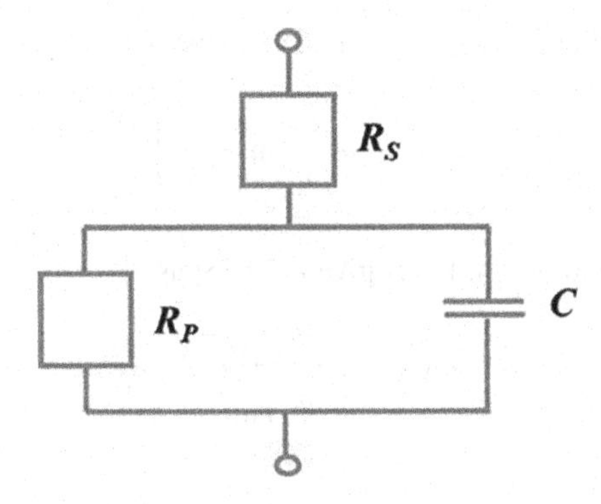

FIGURE 5.9 Equivalent circuit representing a reverse-biased detector.

The electric field is formed because of the static charges formed at the boundary as charge carriers move from the semiconductor to the metal. The maximum electric field is at the metal-semiconductor interface [31]. In this electric field, which opposes the movement of majority carriers, the SCR is formed. The SCR width can be used to study the properties of the induced defects. The variation of the width of the SCR indicates recombination or generation of charge carriers available in the SCR because of the induced defects.

During the capacitance measurement, for example, of a Si diode, the diode is biased with a direct current voltage. An inductance, capacitance, and resistance (LCR) meter adds an alternating current to the bias voltage. An equivalent circuit representing a reverse-biased diode is shown in Figure 5.9. The diodes can be described by a capacitor C in parallel with a parallel resistor R_p that accounts for the leakage current [32]. In addition, there is a series resistor, R_s, which represents the resistance of the undepleted detector bulk.

The capacitance, C, of a Schottky diode is a function of the doping concentration, $N_{D(A)}$ in the semiconductor material and is given [24,27] as

$$C^{-2} = \frac{2(V_{bi} + V)}{q\varepsilon_s\varepsilon_o A^2 N_{D(A)}} \tag{5.13}$$

where V_{bi} is the built-in potential, ε_s is the dielectric constant of semiconductor material, and ε_o is the permittivity in a vacuum.

Ideally, the plot of $C^{-2} - V$ should result in a linear graph, provided the doping profile is uniform. The V_{bi} is calculated from the voltage-axis intercept, while $N_{D(A)}$ is determined from the slope of the linear region of plot $C^{-2} - V$. However, in most cases, the doping profile is not uniform, resulting in two or more linear regions in $C^{-2} - V$ plot. In this case, the parameters are calculated from the different linear regions observed. The relationship between the slope (m) and the doping concentration ($N_{D(A)}$) is given [33] as

$$N_{D(A)} = \frac{2}{mqA^2\varepsilon_s\varepsilon_0}. \tag{5.14}$$

The Schottky barrier height (ϕ_B) is given [28,34] as

$$\phi_B = V_{bi} + \frac{kT}{q}\ln\left(\frac{N_C}{N_D}\right) \tag{5.15}$$

and the depletion region width, W_D is given [29,35] as

$$W_D = \sqrt{\frac{2\varepsilon_o\varepsilon_s V_d}{qN_{D(A)}}}. \tag{5.16}$$

5.6 ELECTRONIC CHARACTERIZATION

5.6.1 Fourier Transform Infrared Spectroscopy

FTIR spectroscopy is a highly adaptable method utilized for material analysis. The distinctiveness of infrared spectra makes them akin to blueprints that allow for the identification and differentiation of various materials [36]. The absorption peaks linked to atomic bond vibrations are specific to the material's constituents. The infrared spectrum of a compound is determined by its unique combination of atoms, making it impossible for two compounds to have identical spectra due to their distinct atomic compositions. By cross-checking the infrared spectrum obtained through FTIR spectroscopy with a database, it becomes possible to identify a material due to its distinct spectral characteristics [36]. FTIR spectroscopy relies on the interaction between infrared radiation and a molecule undergoing a dipole change, leading to infrared absorption. This absorption takes place at specific wavelengths where infrared photons have sufficient energy, surpassing an energy threshold [36]. The presence of threshold energy facilitates the transition of energy from lower to higher states. This energy transition allows the material to move from its ground state to a more excited state with higher energy levels [36]. Analyzing the spectrum produced during this transition provides valuable structural information that is directly associated with the sample being examined. The appearance of absorption peaks in the spectrum is a direct consequence of changes in dipole moments and energy-level transitions, giving rise to valuable structural information [36]. The number of recorded peaks primarily depends on the level of vibrational freedom exhibited by the molecules present in the sample [17,36]. Figure 5.10 illustrates the FTIR diagram.

5.6.2 X-Ray Photoelectron Spectroscopy

The application of XPS technique is pivotal in investigating the elemental composition and comprehending the chemical and electronic properties of a material. By utilizing photoelectron spectroscopy, the technique effectively utilizes high-energy radiation to extract core electrons from a sample, enabling detailed analysis of its properties [37,38]. Figure 5.11 illustrates the process of core electron ejection from a sample, providing valuable insights.

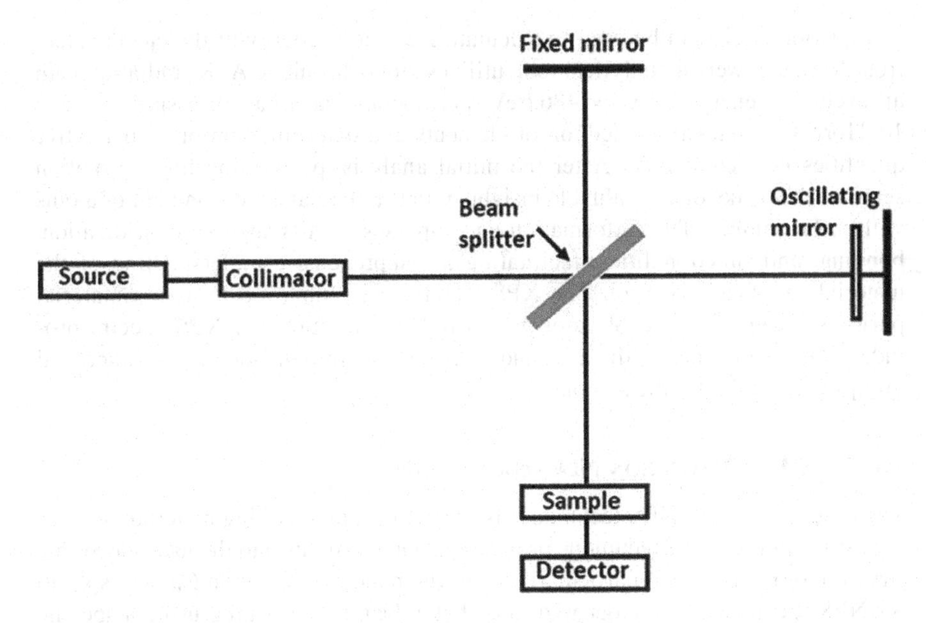

FIGURE 5.10 Schematics of FTIR technique.

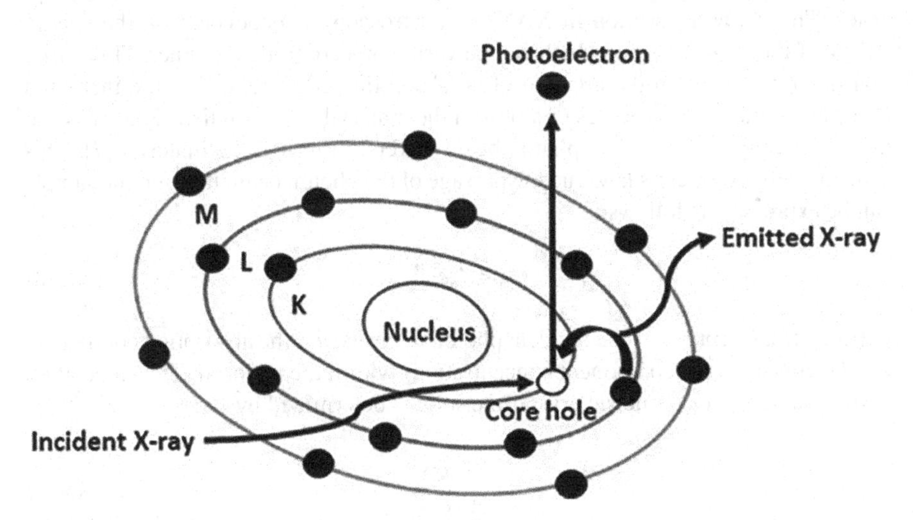

FIGURE 5.11 Schematic principle of XPS.

The ejection of core electrons is achieved by imparting kinetic energy to them, which is quantified as follows:

$$KE = h\upsilon - E_b - \varphi \tag{5.17}$$

The kinetic energy (KE) of the ejected electron can be calculated using the equation involving Planck's constant (h), the frequency (υ) of the incident radiation, the binding energy (E_b), and the work function (φ).

Equation (5.17) can be used to calculate the kinetic energy of the ejected electron. XPS, a powerful analytical tool, utilizes monochromatic Al K_α radiation with an excitation energy of $h\nu = 1486.6\,\text{eV}$ and operates at a base pressure of 1.2×10^{-9} Torr. It enables the detection of elements and determination of their relative quantities in each sample. After the initial analysis, performing high-resolution scans of the peaks offers valuable insights into the state and environment of atoms within the sample. This information encompasses details such as hybridization, bonding, and functionalities, facilitating a comprehensive understanding of the material's surface structure. The XPS technique is employed to assess material purity, serving as a valuable tool in this regard. Additionally, XPS spectra provide evidence of functionalization and offer insights into the identity of functional groups present within the sample.

5.6.3 X-Ray Absorption Near-Edge Spectroscopy

The utilization of XANES techniques is crucial in characterizing materials such as carbon nanotubes. This technique provides valuable insights into the local electronic structure of the material, particularly regarding bond hybridization (sp^2 and sp^3). In XANES spectroscopy, the decay of the photon beam is measured using a specific energy source, offering detailed information about the material's electronic properties. The X-ray absorption in XANES spectroscopy is dependent on the energy source of the photon beam, which can be varied in a controlled manner. This variation in energy leads to the creation of an absorption edge, reflecting the increased X-ray absorption by molecules or atoms in the material being studied. The measurement of energy occurs as the photon beam traverses through the material [7]. This principle relies on Beer's law, and the passage of the photon beam through the sample can be expressed as follows:

$$I = I_o e^{-D} \tag{5.18}$$

In Eq. (5.18), I_o represents the incident photon intensity, μ is the absorption coefficient, and D represents the component concentrations within the material. The calculation of the μ value for a symmetric crystalline solid is determined by:

$$= \frac{1}{V} \sum_{i=1}^{n} {}_i \tag{5.19}$$

In Eq. (5.19), V denotes the volume of the unit cell, σ_i represents the absorption cross-section, which is associated with the unit cell, and the number of elements (n). The value of σ_i typically falls within the range of magnitude order of

$$1\,\text{Mbarn} = 10^{-18}\,\text{cm}^2 \text{ [7].}$$

5.7 MAGNETIC CHARACTERIZATION

5.7.1 Superconducting Quantum Interference Device

The SQUID is a valuable tool for measuring the magnetic behavior of materials. It is capable of detecting even extreme changes in magnetic fluxes. By utilizing a SQUID magnetometer, highly sensitive measurements of magnetic moment can be obtained, with a sensitivity level of less than 0.1 memu (electromagnetic unit). In commercial systems, the applied magnetic field is generated by a superconducting magnet, while the sample is located at a variable environmental temperature. This technique enables the measurement of magnetic properties across a wide temperature range, providing insights into the field cooling and zero-field cooling characteristics of the material. SQUID magnetometers are widely utilized for precise measurements in scenarios where low temperatures down to 1.9 K and high magnetic fields up to 7 Tesla are necessary. The term "SQUID" derives from the transformer/amplifier system utilized to detect the signal obtained from a set of second-order gradiometer coils [39], as depicted in Figure 5.12.

The sample is moved gradually (at a speed of mm/s) through the set of detector coils by means of a transport mechanism, while the amplified signal is noted. Once

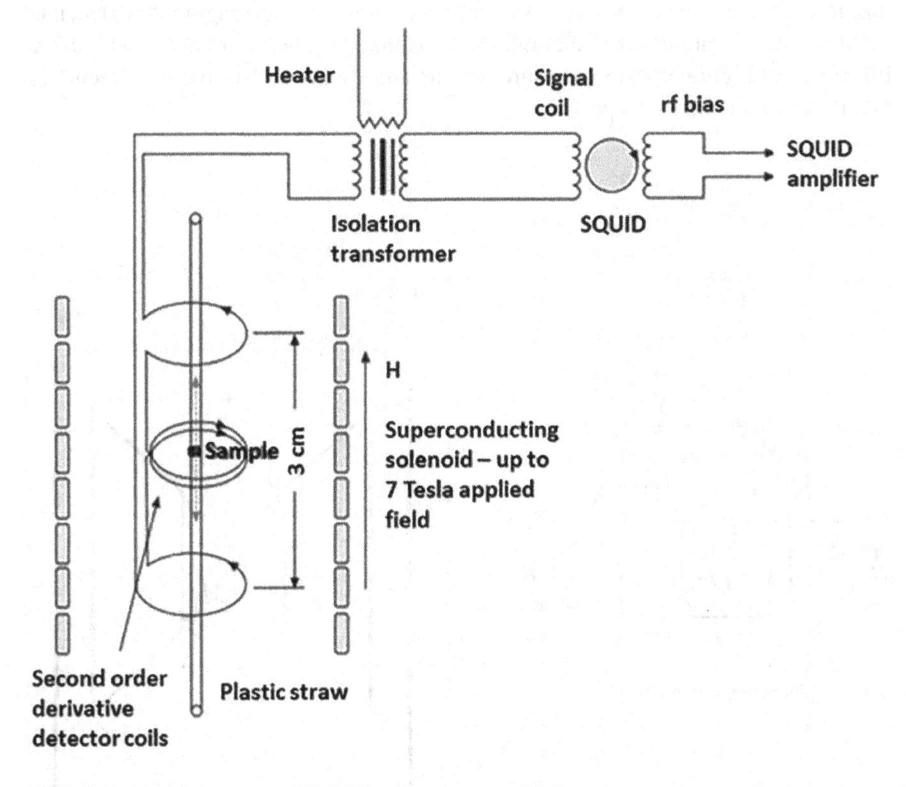

FIGURE 5.12 Schematic of SQUID magnetometer [39].

the measurements are obtained, they are graphically represented. Subsequently, these data points are analyzed using a dipole model and calibrated against a reliable reference standard, such as Ni or Pd. The precise calibration process ensures an accurate determination of the absolute magnetic moment of the sample, providing reliable and trustworthy results. The investigation of magnetic properties in a material is made possible by manipulating the applied magnetic field (such as through a hysteresis loop) or adjusting the temperature. However, the creation of a reliable instrument for this purpose demands substantial engineering expertise. It involves the critical task of maintaining the superconducting magnet at extremely low temperatures using liquid helium while also effectively isolating it from the sample space at variable temperatures. To minimize interference from the superconducting magnet, careful arrangement of the SQUID amplifier and transformer is necessary. This ensures that the signals are accurately detected without being affected by the magnetic field generated by the magnet [39].

5.7.2 Vibrating Sample Magnetometer

VSM is a highly precise measurement technique used to determine the magnetic moment of a sample. It operates based on Faraday's law, which states that a change in flux through a coil generates an electromagnetic force. By leveraging this principle, VSM enables accurate determination of the magnetic properties of a sample [40]. As illustrated in Figure 5.13, the measurement arrangement involves the movement of a magnetic sample near two pickup coils.

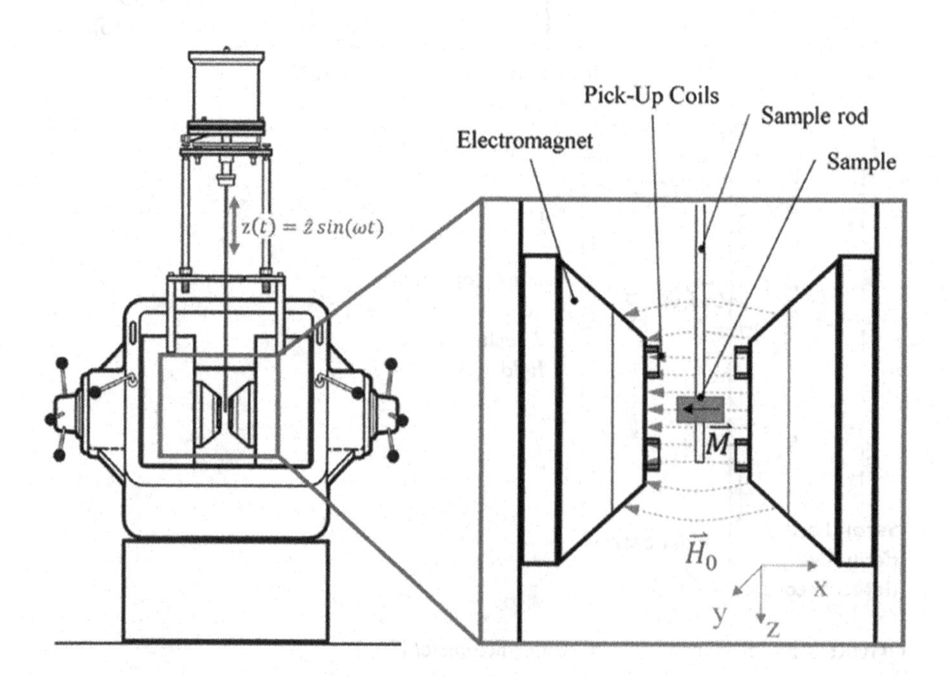

FIGURE 5.13 Schematic of VSM technique [41].

VSM systems are widely used to measure the magnetic properties of materials, considering factors such as magnetic field, temperature, and time. These versatile systems find extensive application in diverse fields, including research, process control, production testing, and quality control. VSMs can accommodate a wide range of sample types, including powders, liquids, solids, single crystals, and thin films [41]. The oscillator generates a sinusoidal signal that is converted into vertical vibrations by the transducer assembly. These vibrations are then transferred to the sample, which is securely attached to the sample rod. The sample undergoes controlled vibration with a specific amplitude and frequency. The static sample is precisely positioned between the two pole pieces of an electromagnet, which produces a magnetic field $\bar{H0}$ with exceptional homogeneity. Mounted on the poles of the electromagnet are stationary pickup coils, positioned in such a way that their symmetry center aligns perfectly with the magnetic center of the sample. As a result, the vertical movement of the magnetized sample leads to a change in magnetic flux, which in turn induces a voltage *Uind* in the coils. When the sample is subjected to a homogeneous field $\bar{H0}$ along the x-axis, it becomes magnetized in the same direction as the field, thereby acquiring a magnetic moment $\bar{m}$. Next, the sample is subjected to periodic movement in relation to the pickup coils. The measurement setup exhibits high sensitivity, capable of detecting even extremely low magnetic moments. Modern VSMs have the capability to detect magnetic moments in the range of micro-electromagnetic units (μemu), corresponding to approximately 10^{-9} g of iron [40].

Achieving precise measurements requires careful calibration of the VSM and considering factors such as magnetic shielding, temperature, and the applied magnetic field. Furthermore, selecting the appropriate measurement frequency is crucial, as different magnetic materials exhibit distinct resonant frequencies. By considering these factors, accurate and reliable magnetic property measurements can be obtained using a VSM.

5.8 ELECTRON SPIN RESONANCE

ESR focuses on the phenomenon of magnetically induced splitting of electronic spin states, allowing for phase analysis of materials. During ESR experiments, the sample is placed within a robust static magnetic field and subjected to a low-amplitude, high-frequency field orthogonal to it. Typically, microwaves of specific frequencies

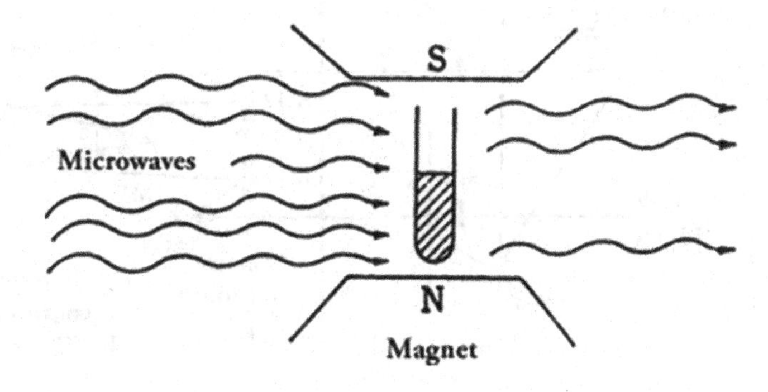

FIGURE 5.14 Schematic diagram of ESR process.

are employed in ESR, and the sample absorbs energy when the radiation frequency matches the energy difference between the two electron states present in the sample. ESR is a valuable technique employed for the investigation of materials containing unpaired electrons. It relies on the absorption or emission of photon energy by these unpaired electrons when exposed to a magnetic field. By introducing the material into an ESR system, the interplay between microwaves and a static magnetic field becomes crucial for observing the behavior of the unpaired electrons. This analysis of electron behavior yields valuable insights into the material's properties and condition. Figure 5.14 shows the schematic diagram of ESR process.

5.9 OPTICAL CHARACTERIZATION

5.9.1 UV-Vis Spectroscopy

Optical spectroscopy is a scientific method that centers on the investigation of the interactions between various wavelengths of electromagnetic radiation and unfamiliar substances, solutions, or solid materials. When these materials are exposed to electromagnetic radiation of a particular wavelength, the resulting light waves can be either reflected, absorbed, or transmitted, thereby offering significant details regarding the composition and properties of the sample. Through the analysis of these interactions between light and matter, optical spectroscopy empowers researchers to acquire valuable knowledge about the attributes of the substances under study.

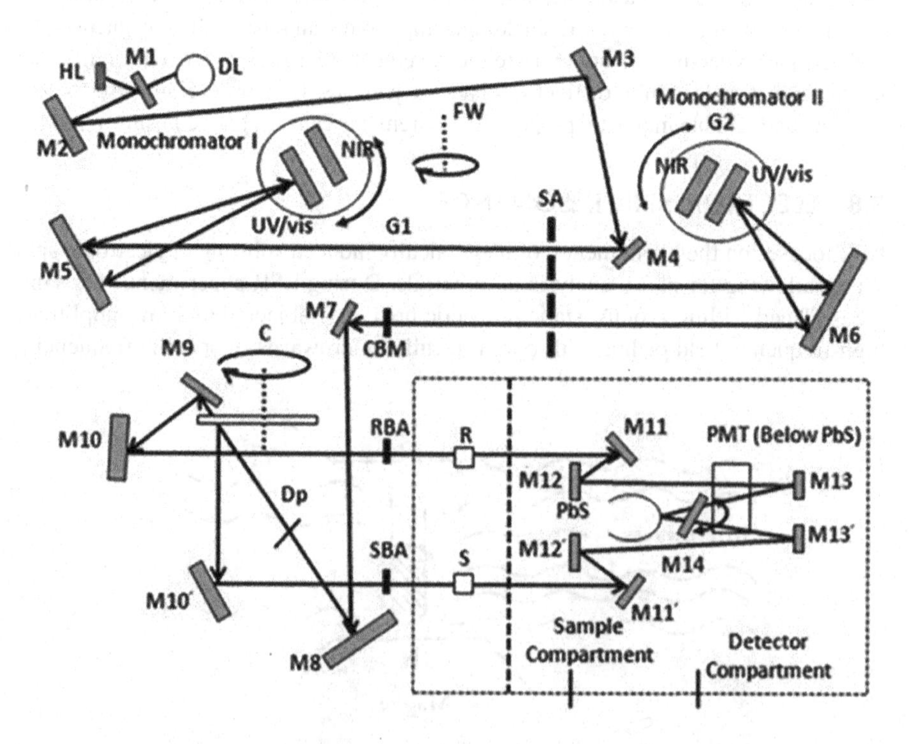

FIGURE 5.15 Schematic of UV-Vis-NIR spectrophotometer optical layout [44].

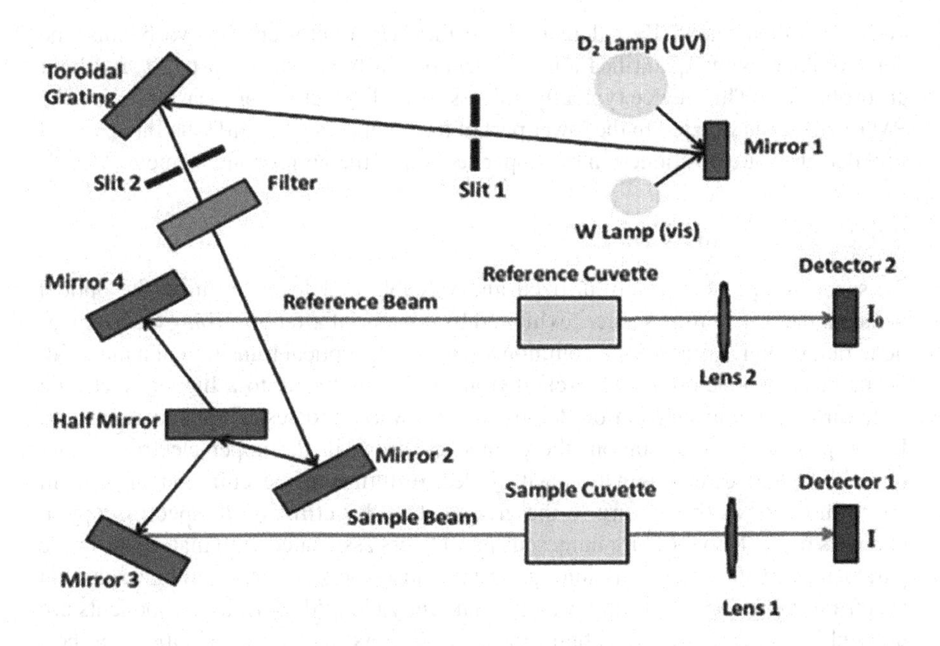

FIGURE 5.16 Schematic of UV-Vis double beam instrument [44].

Optical spectroscopy is employed to measure the reflectance, absorbance, or transmittance properties of materials. Among the various optical spectroscopic techniques, UV-Vis absorption spectroscopy [42] is particularly valuable for studying light absorption. This method utilizes an instrument equipped with a lamp compartment, double holographic grating monochromator, dual sampling compartment, and three detector modules, enabling the examination of a wide spectral range spanning from 175 to 3,300 nm (Figure 5.15).

In most cases, the lamp compartment in optical spectroscopy setups contains a deuterium lamp and a tungsten-halogen lamp. This combination allows for a broad range of wavelengths spanning UV, visible, and NIR regions, achieved through the utilization of a source-doubling mirror. Within the double monochromator, a beam splitting system is responsible for the precise selection of a single wavelength from the wide spectrum of primary electromagnetic radiation. During this procedure, one of the beams is directed toward the dual sampling compartment, while the other beam serves as a reference and is directed toward the detector module. The beam splitting mechanism employed in the Perkin Elmer Lambda 950 UV-Vis-NIR absorption spectrometer combines filters, lenses, mirrors, diffraction gratings, and slits to achieve its functionality. A visual representation of the beam splitting process can be observed in Figure 5.16, demonstrating a typical setup. Inside the dual sampling compartment, the sample is carefully positioned within an integrated sphere. This sphere plays a crucial role in efficiently gathering the widely dispersed light generated when the incident beam interacts with the sample [43,44]. The light collected from the sample and reference beams (I and I_0) is reflected through a series of mirrors in the optics and directed toward the relevant detector housed within the detector compartment. By calculating the ratio (I/I_0) between the two beams, the detector enables the quantification of reflection (%*R*), transmission (%*T*), and light absorption (*A*). This device typically utilizes several detectors for measurement. The PMT detector is situated in the lower part of the compartment, while the InGaAs and PbS detectors are positioned in the upper section of the optics compartment [43,44].

5.9.2 Photoluminescence

PL spectroscopy is a widely utilized and valuable tool for examining the optical behavior, such as luminescence, exhibited by a material after absorbing electromagnetic radiation. Luminescence commonly denotes the optical transition of a material, occurring when electrons are excited from the ground state to a higher electronic state through the absorption of electromagnetic waves possessing adequate energy. In the process of deexcitation, the excited electrons in the upper electronic state undergo radiative recombination with holes, resulting in the emission of light in the form of PL as they return to the ground state. By utilizing PL spectroscopy, it becomes possible to measure numerous parameters associated with materials. These parameters include the emission spectrum, anisotropy or polarization, excitation spectrum, luminescence lifetimes, and quantum yield [45]. Various components are utilized in the instrument to obtain these parameters, including excitation sources, excitation and emission monochromators, a sample compartment, and a detector. One of the most popular systems in PL spectroscopy is the Fluorolog-3 spectrofluorometer (depicted in Figure 5.17), known for its widespread usage and effectiveness.

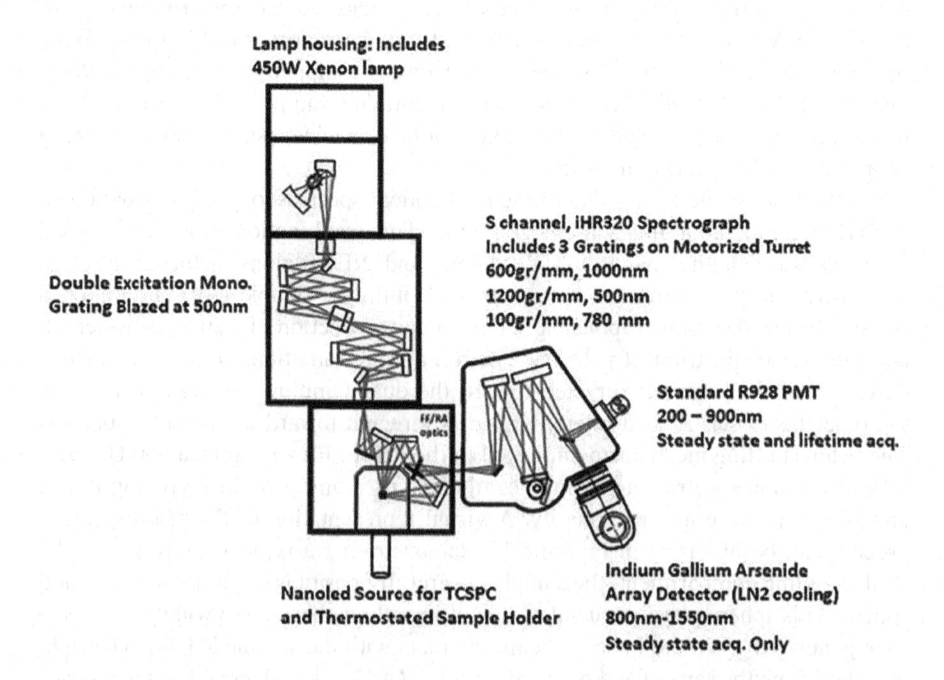

FIGURE 5.17 Schematic PL spectrofluorometer setup [48].

Its distinctive modular configuration includes essential components such as a 450 W xenon lamp, a few pulsed laser diodes, and a selection of detectors. The array of detectors used in this instrument includes the Hamamatsu R928P photomultiplier, specifically designed for the UV-visible range; the Hamamatsu R2658P photomultiplier, optimized for the UV-Vis-NIR range; and a specialized PbS solid-state detector dedicated to the NIR-IR range [46].

In this system, the light sources consist of a high-power broadband 450 W Xenon lamp and a tunable laser that spans the excitation and emission spectra, spanning from UV to NIR wavelengths. Within the instrument, the excitation source is directed through diffraction gratings and focusing optics contained within the monochromator. It is then focused on the material for excitation. Afterwards, the emitted light from the material is gathered using focusing optics and directed toward the detector compartment. To avoid any interference with the transmitted light from the source, the detector compartment is strategically positioned at a 90° angle. Lastly, the detector captures the emission spectrum, which is then presented as the intensity of luminescence plotted against the wavelength of the emitted light at RT [47]. Moreover, the spectrofluorometer is equipped with the Janis cryostat [46], enabling measurements to be conducted at low temperatures. This capability allows for the examination of luminescence properties under controlled temperature conditions.

5.10 SUMMARY

The techniques mentioned in this chapter represent only a subset of the wide range of methods employed in thin film characterization. The selection of a particular technique depends on the specific properties under investigation, the characteristics of the film itself, and the equipment accessible. Typically, a combination of multiple techniques is utilized to achieve a comprehensive understanding of the thin film's properties. Thin film characterization plays a vital role in the advancement of various technological applications. Various techniques are available to determine the physical, chemical, and optical properties of thin films. Comprehensive knowledge of these properties is crucial for enhancing the performance and reliability of these applications.

REFERENCES

[1] A. Bogner, P.-H. Jouneau, G. Thollet, D. Basset, C. Gauthier, A history of scanning electron microscopy developments: towards "wet-STEM" imaging, *Micron*. 38 (2007) 390–401.

[2] A. Ul-Hamid, *A Beginners' Guide to Scanning Electron Microscopy*, Springer, 2018.

[3] P.J. Goodhew, J. Humphreys, R. Beanland, *Electron Microscopy and Analysis*, CRC Press, 2000.

[4] B.J. Inkson, Scanning electron microscopy (SEM) and transmission electron microscopy (TEM) for materials characterization, in: G. Hübschen, I. Altpeter, R. Tschuncky, H-G. Herrmann, eds., *Materials Characterization Using Nondestructive Evaluation (NDE) Methods*, Elsevier, 2016: pp. 17–43.

[5] H. Yao, K. Kimura, Field emission scanning electron microscopy for structural characterization of 3D gold nanoparticle superlattices, *Modern Research and Educational Topics in Microscopy*. 2 (2007) 568–576.

[6] C. Kisielowski, B. Freitag, M. Bischoff, H. Van Lin, S. Lazar, G. Knippels, P. Tiemeijer, M. van der Stam, S. von Harrach, M. Stekelenburg, Detection of single atoms and buried defects in three dimensions by aberration-corrected electron microscope with 0.5-Å information limit, *Microscopy and Microanalysis.* 14 (2008) 469–477.

[7] D. Shindo, H. Kenji, *High-Resolution Electron Microscopy for Materials Science*, Springer Science & Business Media, 2012.

[8] O. Sahin, S. Magonov, C. Su, C.F. Quate, O. Solgaard, An atomic force microscope tip designed to measure time-varying nanomechanical forces, *Nature Nanotechnology.* 2 (2007) 507–514.

[9] V. Karoutsos, Scanning probe microscopy: instrumentation and applications on thin films and magnetic multilayers, *Journal of Nanoscience and Nanotechnology.* 9 (2009) 6783–6798.

[10] G.A. Mansoori, *Principles of Nanotechnology: Molecular-Based Study of Condensed Matter in Small Systems*, World Scientific, 2005.

[11] P.M. Chaikin, T.C. Lubensky, T.A. Witten, *Principles of Condensed Matter Physics*, Cambridge University Press, Cambridge, 1995.

[12] L.A.A. Rodriguez, D.N. Travessa, Core/shell structure of TiO2-coated MWCNTs for thermal protection for high-temperature processing of metal matrix composites, *Advances in Materials Science and Engineering.* 2018 (2018) 7026141. https://doi.org/10.1155/2018/7026141.

[13] G. Temesgen, Thin film deposition and characterization techniques, *Journal of 3D Printing and Applications.* 1 (2022) 1–24.

[14] A.C. Ferrari, D.M. Basko, Raman spectroscopy as a versatile tool for studying the properties of graphene, *Nature Nanotechnology.* 8 (2013) 235–246. https://doi.org/10.1038/nnano.2013.46.

[15] M.S. Dresselhaus, G. Dresselhaus, R. Saito, A. Jorio, Raman spectroscopy of carbon nanotubes, *Physics Reports.* 409 (2005) 47–99. https://doi.org/10.1016/j.physrep.2004.10.006.

[16] I.R. Lewis, H. Edwards, *Handbook of Raman Spectroscopy: From the Research Laboratory to the Process Line*, CRC Press, 2001.

[17] S.Y. Lin, C.W. Dence, *Methods in Lignin Chemistry*, Springer Science & Business Media, 2012.

[18] J.R. Dennison, M. Holtz, G. Swain, Raman spectroscopy of carbon materials, *Spectroscopy.* 11 (1996) 38–45.

[19] S.K. Sharma, D.S. Verma, L.U. Khan, S. Kumar, S.B. Khan, *Handbook of Materials Characterization*, Springer, 2018.

[20] V.S. Waman, M.M. Kamble, M.R. Pramod, S.P. Gore, A.M. Funde, R.R. Hawaldar, D.P. Amalnerkar, V.G. Sathe, S.W. Gosavi, S.R. Jadkar, Influence of the deposition parameters on the microstructure and opto-electrical properties of hydrogenated nanocrystalline silicon films by HW-CVD, *Journal of Non-Crystalline Solids.* 357 (2011) 3616–3622.

[21] R.L. Merlino, Understanding Langmuir probe current-voltage characteristics, *American Journal of Physics.* 75 (2007) 1078–1085.

[22] B.K. Jones, J. Santana, M. McPherson, Ohmic I-V characteristics in semi-insulating semiconductor diodes, *Solid State Communications.* 105 (1998) 547–549.

[23] J.O. Bodunrin, S.J. Moloi, Current-voltage characteristics of 4 MeV proton-irradiated silicon diodes at room temperature, *Silicon.* 14 (2022) 10237–10244.

[24] D.K. Schroder, *Semiconductor Material and Device Characterization*, John Wiley & Sons, 2015.

[25] S. Alialy, H. Tecimer, H. Uslu, Ş. Altındal, A comparative study on electrical characteristics of Au/N-Si Schottky diodes, with and without Bi-doped PVA interfacial layer in dark and under illumination at room temperature, *Journal of Nanomedicine & Nanotechology.* 4 (2013) 1000167.

[26] S.M. Sze, *Semiconductor Devices: Physics and Technology*, John Wiley & Sons, 2008.
[27] G. Lutz, *Semiconductor Radiation Detectors*, Springer, 2007.
[28] E.H. Rhoderick, E.H. Rhoderick, *Metal-Semiconductor Contacts*, Clarendon Press, Oxford, 1978.
[29] S.M. Sze, *Physics of Semiconductor Devices*, John Wiley, New York, 1981: pp. 96–98.
[30] S.K. Cheung, N.W. Cheung, Extraction of Schottky diode parameters from forward current-voltage characteristics, *Applied Physics Letters.* 49 (1986) 85–87.
[31] M.A. Mastro, Power MOSFETs and diodes, in: S. Pearton, F. Ren, M. Mastro, eds., *Gallium Oxide*, Elsevier, 2019: pp. 401–418.
[32] S.J. Moloi, M. McPherson, Current-voltage behaviour of Schottky diodes fabricated on p-type silicon for radiation hard detectors, *Physica B: Condensed Matter.* 404 (2009) 2251–2258.
[33] M. Msimanga, M. McPherson, Diffusion characteristics of gold in silicon and electrical properties of silicon diodes used for developing radiation-hard detectors, *Materials Science and Engineering: B.* 127 (2006) 47–54.
[34] C. Morosanu, C. Cesile, S. Korepanov, P. Fiorini, C. Bacci, F. Meddi, F. Evangelisti, A. Mittiga, a-Si: H based particle detectors with low depletion voltage, *Journal of Non-Crystalline Solids.* 164 (1993) 801–804.
[35] I.S. Yahia, H.Y. Zahran, F.H. Alamri, M.A. Manthrammel, S. AlFaify, A.M. Ali, Microelectronic properties of the organic Schottky diode with pyronin-Y: admittance spectroscopy, and negative capacitance, *Physica B: Condensed Matter.* 543 (2018) 46–53.
[36] R. Salzer, P.R. Griffiths, J.A. de Haseth, Fourier transform infrared spectrometry (2nd edn.), *Analytical and Bioanalytical Chemistry.* 391 (2008) 2379–2380.
[37] J.F. Watts, X-ray photoelectron spectroscopy, *Surface Science Techniques.* 45 (1994) 5.
[38] H. Liu, H. Zhang, H. Yang, Photocatalytic removal of nitric oxide by multi-walled carbon nanotubes-supported TiO2, *Cuihua Xuebao/Chinese Journal of Catalysis.* 35 (2014) 66–77. https://doi.org/10.1016/S1872-2067(12)60705-0.
[39] T. Thomson, Magnetic properties of metallic thin films, in: K. Barmak, K. Coffey, eds., *Metallic Films for Electronic, Optical and Magnetic Applications*, Elsevier, 2014: pp. 454–546.
[40] Christian-Albrechts-Universitat zu Kiel, Vibrating Sample Magnetometry, Basic Laboratory Materials Science and Engineering, https://www.tf.uni-kiel.de/servicezentrum/neutral/praktika/anleitungen/m106, (accessed 01 May 2023).
[41] Lake Shore Cryotronics, Magnetic Media Measurements with a VSM, www.lakeshore.com, (accessed 01 May 2023).
[42] C.A. De Caro, C. Haller, *UV/VIS Spectrophotometry-Fundamentals and Applications*, Mettler-Toledo International, 2015: pp. 4–14.
[43] C. Tams, N. Enjalbert. The Use of UV/Vis/NIR Spectroscopy in the Development of Photovoltaic Cells, Application Note.
[44] Experimental Techniques, https://shodhganga.inflibnet.ac.in/bitstream/10603/42071/14/8_chapter-2.pdf, (accessed 10 May 2022).
[45] Y. Povrozin, B. Barbieri, Fluorescence spectroscopy, in: M. Kutz, ed., *Handbook of Measurement in Science and Engineering*, John Wiley & Sons, 3 (2016), pp. 2475–2498.
[46] Photoluminescence Measurements, www.eagleregpot.eu/EfI/index.php?pid=52&l=en, (accessed 20 May 2023).
[47] Fluorescence Spectroscopy, https://fluorescence.ttk.pte.hu/devices.php, (accessed 07 June 2022).
[48] An Introduction to Fluorescence Spectroscopy, https://www.azooptics.com/Article.aspx?ArticleID=1327, (accessed 20 May 2023).

6 Optoelectronics and Spintronics in Smart Thin Films for Solar Cells

6.1 INTRODUCTION TO SOLAR CELLS

Solar cells, also known as photovoltaic (PV) cells, are devices that convert sunlight directly into electricity. They are a key technology for harnessing the abundant and renewable energy of the sun and are increasingly used as a clean and sustainable source of power for homes, businesses, and other applications [1]. The basic principle of solar cells is the PV effect, in which certain materials can convert light energy into electrical energy [2]. When sunlight strikes a solar cell, it excites electrons in the material, creating a flow of electric current. This current can be harnessed, used to power electrical devices, or stored in batteries for later use. Solar cells come in a variety of shapes and sizes and can be made from a range of known materials, including silicon, cadmium telluride, and copper indium gallium selenide. They are typically arranged in panels or arrays, which can be installed on rooftops, in fields, or even in space [3]. One of the major advantages of solar cells is their ability to generate electricity without producing any greenhouse gas emissions or other harmful pollutants [1]. They are also highly reliable and require minimal maintenance, making them an attractive option for remote or off-grid locations [1].

Despite their many benefits, however, solar cells still face some challenges. The cost of manufacturing and installing solar panels can be relatively high, and their efficiency in converting sunlight into electricity is still relatively low. However, ongoing research and development efforts are working to improve the efficiency and affordability of solar cells, and many experts predict that they will play an increasingly important role in meeting the world's energy needs in the years to come. The two emerging technologies in the field of solar cells are optoelectronics and spintronics.

Optoelectronics is a branch of electronics that deals with the interaction between light and electronic devices. It is concerned with the study of materials that can convert light into electrical energy and vice versa. In the context of solar cells, optoelectronics can be used to improve their efficiency and reduce their cost. One approach to improving the efficiency of solar cells using optoelectronics is by incorporating materials with a high absorption coefficient. These materials can absorb a greater amount of light, resulting in higher current and power output. Another approach is to use thin-film solar cells, which are less expensive to produce than conventional silicon solar cells.

Spintronics, on the other hand, is a field of electronics that deals with the interaction between the electron's spin and its magnetic moment. In the context of solar

DOI: 10.1201/9781003331940-6

cells, spintronics can be used to improve their efficiency by controlling the electron's spin [4]. One approach to improving the efficiency of solar cells using spintronics is by using materials with a high spin polarization. These materials can convert the spin of the electrons into electrical current, resulting in higher power output [4]. Another approach is by using spin-selective contacts, which can selectively allow electrons with a specific spin to pass through, resulting in higher efficiency [5].

6.2 DEVICE PRINCIPLE

The basic device structure of a solar cell consists of a thin layer of semiconductor material sandwiched between two electrical contacts. The semiconductor layer is typically made of a material such as silicon, which is a good absorber of sunlight. When sunlight strikes the semiconductor material, it excites electrons, creating electron-hole pairs. The electrical contacts on either side of the semiconductor layer are typically made of metals such as silver or aluminum. When the electron-hole pairs are generated in the semiconductor material, they are separated by an electric field created by a junction within the device. This electric field causes the electrons to flow toward one of the contacts, while the holes flow toward the other contact. This flow of electrons and holes creates a flow of electrical current, which can be harnessed for use in electrical devices. To maximize the efficiency of a solar cell, it is important to optimize several key parameters, such as the material properties of the semiconductor, the thickness of the semiconductor layer, and the design of the electrical contacts. For example, the semiconductor material should have a bandgap that is well-matched to the energy of sunlight to absorb the maximum amount of light energy. Additionally, the electrical contacts should be designed to minimize losses due to resistance and reflectance.

6.3 DIFFERENT TYPES OF SOLAR CELLS

There are several different types of solar cells, each with its own unique set of advantages and disadvantages. Some of the most common types of solar cells include thin-film solar cells, dye-sensitized solar cells (DSSCs), organic solar cells, perovskite solar cells, and quantum solar cells.

6.3.1 SILICON SOLAR CELLS

Silicon solar cells are a type of solar cell that uses crystalline silicon to absorb sunlight and convert it into electrical energy. They are the most commonly used type of solar cell in commercial PV systems. Silicon solar cells work by absorbing photons from sunlight, which knock electrons in the silicon atoms to a higher energy level, creating a flow of electrons that can be collected as an electric current [6]. Silicon solar cells are typically made from a thin layer of silicon that is doped with impurities to create a p-n junction, which helps to separate the electrons and holes created by the absorbed photon [7]. One of the main advantages of silicon solar cells is their high efficiency. Silicon solar cells can achieve efficiencies of over 20% [8], which means that they produce a high amount of energy per unit area of the cell. Another advantage of silicon solar cells is their long lifespan. Silicon solar cells can last for over 25 years with proper maintenance, making them a reliable and cost-effective source of renewable energy [9]. The main challenge of silicon solar cells is their high cost. The manufacturing process for silicon solar cells is complex and involves a high degree of precision, which can make them more expensive compared to other types of solar cells. A further challenge is the rigid nature of silicon solar cells. Silicon solar cells are typically made from rigid, crystalline materials, which makes them less flexible and limits their use in certain applications [10].

6.3.2 Thin-Film Solar Cells

Thin-film solar cells are a type of solar cell that uses thin layers of semiconductor materials to absorb sunlight and convert it into electrical energy. They are typically made from materials such as cadmium telluride, copper indium gallium selenide, or amorphous silicon [11]. Thin-film solar cells are different from traditional silicon solar cells in that they use much less semiconductor material, making them thinner and more flexible [12]. This allows for a wider range of applications, such as building-integrated PVs and portable electronic devices [13]. One of the main advantages of thin-film solar cells is their lower cost compared to traditional silicon solar cells. Thin-film solar cells use much less material, reducing the cost of manufacturing and making them a more cost-effective option for large-scale solar energy projects [14]. Thin-film solar cells can be produced on flexible substrates, such as plastic or metal, which makes them suitable for a wider range of applications, including building-integrated PVs and portable electronic devices [13]. However, one of the main challenges of thin-film solar cells is their lower efficiency compared to traditional silicon solar cells. Thin-film solar cells typically have lower efficiency, which means that they produce less energy per unit area of the cell [11]. The stability of thin-film solar cells is another challenge. Thin-film solar cells can be sensitive to moisture and other environmental factors, which can degrade the performance of the cells over time [15].

6.3.3 Dye-Sensitized Solar Cells

DSSCs are a type of solar cell that uses a thin layer of dye molecules to absorb sunlight and convert it into electrical energy. They are also known as Grätzel cells, after the Swiss scientist Michael Grätzel, who developed the technology in the 1990s [16,17]. DSSCs are made up of several layers, including a transparent conductive oxide layer, a porous layer of titanium dioxide nanoparticles coated with a dye, an electrolyte solution, and a counter electrode. When sunlight hits the dye-coated titanium dioxide layer, it excites the electrons in the dye molecules, which are then transferred to the titanium dioxide layer, generating a flow of electrons and creating an electric current. The current is then collected by the electrodes and used as electrical power [18].

The advantage of DSSCs is their potential for low-cost production. DSSCs can be made using simple and low-cost manufacturing techniques [19], such as screen printing or spray coating, which can reduce the cost and complexity of production. The main advantage of DSSCs is their ability to work in low-light conditions, such as on cloudy days or indoors. This makes them suitable for a wide range of applications, such as powering portable electronic devices, indoor lighting, and building-integrated PVs [18].

However, DSSCs have a main problem of lower efficiency compared to traditional silicon solar cells. The efficiency of DSSCs is typically between 5% and 10%, while silicon solar cells can achieve efficiencies of over 20%. Researchers are working to improve the efficiency of DSSCs by developing new materials, improving the structure of the cells, and optimizing the manufacturing process. Another challenge is the stability of DSSCs. The dye molecules used in DSSCs can degrade over time due to exposure to air, light, and moisture [18].

6.3.4 Organic Solar Cells

Organic solar cells, also known as organic photovoltaics (OPV), are a type of solar cell that uses organic materials, such as polymers and small molecules, to absorb sunlight and convert it into electrical energy. They are a relatively new and emerging technology in the field of PVs. Organic solar cells are made by sandwiching a thin layer of organic material between two electrodes [20]. When sunlight hits the organic material, it absorbs the energy and generates a flow of electrons, which creates an electric current. The current is then collected by the electrodes and used as electrical power.

The main advantage of organic solar cells is their potential to be made using low-cost, flexible, and lightweight materials. This makes them suitable for a wide range of applications, such as powering portable electronic devices, wearable technology, and building-integrated PVs. Organic solar cells can also be manufactured using simple, solution-based techniques such as spin-coating or inkjet printing. This reduces the cost and complexity of production, making organic solar cells a more accessible technology than traditional silicon solar cells.

Nevertheless, the challenges of organic solar cells are their lower efficiency compared to traditional silicon solar cells. The efficiency of organic solar cells is typically between 5% and 10%, while silicon solar cells can achieve efficiencies of over 20%. Another challenge is the stability of organic solar cells. Organic materials can be susceptible to degradation over time due to exposure to air, light, and moisture [21].

6.3.5 Perovskite Solar Cells

Perovskite solar cells are a type of solar cell that uses a material called perovskite to absorb sunlight and convert it into electrical energy. Perovskite is a mineral that has a unique crystalline structure and is made up of a combination of elements such as lead, iodine, and methylammonium. Perovskite solar cells are relatively new technology but have gained a lot of attention in recent years due to their high efficiency and potential for low-cost production [22]. The efficiency of perovskite solar cells has increased rapidly in the last decade, with some cells achieving efficiencies of over 20% [23].

The major advantage of perovskite solar cells is their potential for low-cost production. Perovskite materials can be produced using simple, low-cost manufacturing processes such as solution processing, which involves depositing the perovskite material onto a substrate using a liquid solution. This is a much simpler and less expensive process than the manufacturing of traditional silicon solar cells. Another advantage of perovskite solar cells is their ability to absorb a broad range of light wavelengths, including both visible and infrared light [24]. This means that they can potentially capture more energy from sunlight than traditional silicon solar cells.

Despite their advantages, perovskite solar cells also face several challenges. The stability of the perovskite material is one of the main challenges [25]. Perovskite materials can degrade over time due to exposure to moisture, heat, and light, which can limit the lifespan of the solar cells [26]. Another challenge is the toxicity of some of the materials used in perovskite solar cells, such as lead [27].

6.3.6 Quantum Dot Solar Cells

Quantum dot solar cells are a promising new technology in the field of PVs, which harnesses the energy of sunlight to generate electricity. These cells use tiny semiconductor particles, called quantum dots, to absorb sunlight and convert it into electrical energy [28]. A quantum dot is a nanoscale particle that is only a few nanometers in size. They are typically made from materials such as cadmium selenide or lead sulfide, which have unique electronic and optical properties that make them suitable for use in solar cells. In a typical quantum dot solar cell, a layer of quantum dots is deposited onto a substrate, which is usually made of glass or plastic. When sunlight hits the quantum dots, they absorb the energy and become excited, releasing electrons that can be harnessed to produce electricity.

The main advantage of quantum dot solar cells is their ability to absorb a broader range of light than traditional solar cells. Because quantum dots can be tuned to absorb specific wavelengths of light, they can be used to create solar cells that are optimized for different parts of the solar spectrum. This means that quantum dot solar cells can potentially capture more energy from sunlight than traditional solar cells. Another advantage of quantum dot solar cells is their potential to be made using low-cost, scalable manufacturing processes. Because quantum dots can be deposited onto substrates using simple solution-based techniques, it may be possible to produce large quantities of quantum dot solar cells at low cost.

However, quantum dot solar cells are still in the early stages of development and face several challenges. One of the biggest challenges is improving their efficiency, which currently lags traditional silicon solar cells.

In general, researchers are constantly working to improve the efficiency, durability, and cost-effectiveness of materials used in solar cell applications. Some of the key challenges they are trying to address include:

1. **Efficiency:** One of the main challenges in solar cell technology is improving the efficiency of solar cells. Researchers are working on developing materials that can absorb more light, convert a greater percentage of the absorbed light into electrical energy, and reduce energy loss due to heat.
2. **Durability:** Solar cells are often exposed to harsh environmental conditions, which can cause degradation and reduce their lifespan. Researchers

are developing materials that can withstand exposure to sunlight, temperature fluctuations, and moisture to improve the durability of solar cells.

3. **Cost-effectiveness:** The cost of solar cells remains a major barrier to their widespread adoption. Researchers are developing materials and manufacturing processes that can reduce the cost of producing solar cells while maintaining or improving their performance.
4. **Sustainability:** Many current solar cell materials use rare and expensive elements that may not be sustainable in the long term. Researchers are exploring the use of more sustainable and abundant materials, such as perovskite, that can be produced using environmentally friendly methods.
5. **Flexibility:** Traditional solar cells are rigid and inflexible, limiting their use in certain applications. Researchers are developing flexible and lightweight solar cell materials that can be integrated into a wider range of products and structures.
6. **Stability:** They are working on improving the stability and encapsulation techniques to protect the cells from environmental factors.
7. **Toxicity:** They are also exploring the use of alternative materials that are less toxic to humans and the environment.

6.4 OPTOELECTRONICS IN SMART THIN FILMS FOR SOLAR CELLS

6.4.1 Chalcogenide in Solar Cells

Chalcogenide thin films have gained significant attention in the field of solar cells due to their unique optoelectronic properties. In recent years, there have been advancements in the synthesis techniques and commercial development of chalcogenide solar cells. Chalcogenide thin films have been predicted as a potential material for efficient solar cell applications [29].

A recent study focused on the synthesis and characterization of Cu_2FeSnS_4 (CFTS) nanoparticles, an environmentally benign compound with potential applications in PVs [30]. The nanoparticles were synthesized using a low-cost and simple reaction method and then spin-coated onto a fluorine-doped tin oxide (FTO) substrate to form a thin film [30]. Various characterization techniques were employed to assess the properties of the synthesized CFTS nanoparticles. X-ray diffraction (XRD) and field-emission scanning electron microscopy (FE-SEM) confirmed the good crystalline nature and slight agglomeration of the nanoparticles, respectively. The UV-Vis absorption spectrum indicated that the nanoparticles exhibit wide absorption in the visible region, making them suitable for PV applications. The optical energy bandgap was calculated using Tauc's plot and found to be 1.32 eV. The performance of the fabricated thin film was measured employing the current-voltage (*I-V*) characteristic under simulated illumination of AM 1.5. The solar power conversion efficiency of the fabricated solar cell with the structure FTO/ZnO/CdS/CFTS was reported to be 1.32%.

More recently, antimony sulfide selenide thin-film solar cells were fabricated through thermal evaporation [31]. The $Sb_2S_xSe_{3-x}$ sources used for the evaporation were obtained via an in situ reaction of $SbCl_3$, Se, and Sb_2S_3 with a desired mole ratio. Figure 6.1a–d displays the results and outcomes of this research. A bandgap

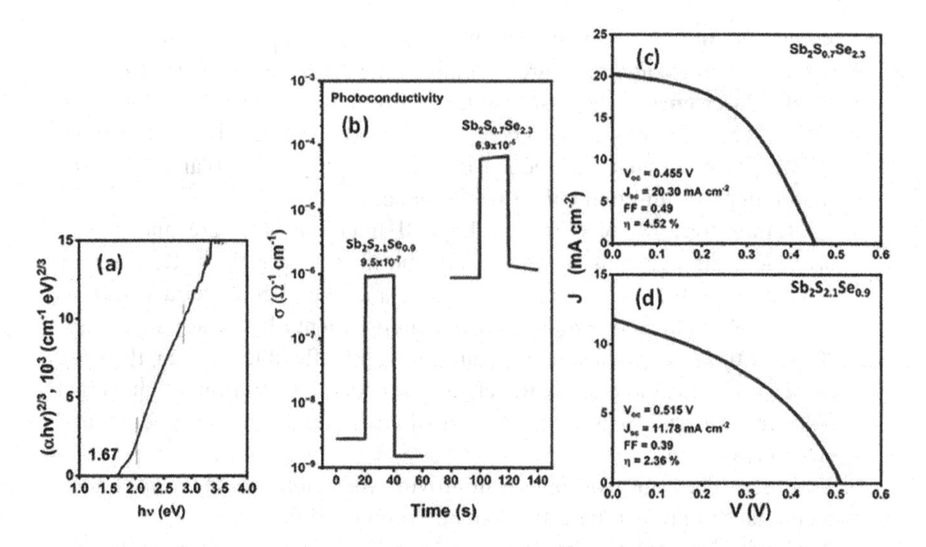

FIGURE 6.1 (a) Direct band gap of $Sb_2S_{2.1}Se_{0.9}$ thin film. (b) Photoconductivity response of $Sb_2S_{2.1}Se_{0.9}$ and $Sb_2S_{0.7}Se_{2.3}$ thin films. (c) and (d) *J-V* characteristics of $Sb_2S_{0.7}Se_{2.3}$ and $Sb_2S_{2.1}Se_{0.9}$ thin-film solar cell structures, respectively [31].

and photoconductivity of 1.67 eV and 10^{-6} $\Omega^{-1}cm^{-1}$ were observed for $Sb_2S_{2.1}Se_{0.9}$ thin-film (175 nm) solar cell, as presented in Figure 6.1a and b. The solar cell structure consisted of SnO_2:F(FTO)/CdS (100 nm)/$Sb_2S_{2.1}Se_{0.9}$/C–Ag and exhibited an open circuit voltage (V_{OC}), short-circuit current density (J_{SC}), and energy conversion efficiency (η) of 0.514 V, 11.78 mA cm^{-2}, and 2.36%, respectively, as depicted in Figure 6.1d. Another composition of the $Sb_2S_xSe_{3-x}$ thin films was prepared using a powder mixture with increased Se content was produced by heating a mixture of $SbCl_3$, Se, and Sb_2S_3 in an oven at 450°C. This powder mixture was then used as a source for thermal evaporation. A bandgap and photoconductivity of 1.29 eV and 7×10^{-5} $\Omega^{-1}cm^{-1}$ (see Figure 6.1b) were also confirmed for the $Sb_2S_{0.7}Se_{2.3}$ (190 nm) thin film. The corresponding solar cell structure, FTO/CdS(100 nm)/$Sb_2S_{0.7}Se_{2.3}$/C–Ag, demonstrated a V_{OC}, J_{SC} and η of 0.455 V, 20.3 mA cm^{-2}, and 4.52%, respectively, as presented in Figure 6.1c. This research highlights the versatility of vacuum thermal evaporation for fabricating $Sb_2S_xSe_{3-x}$ thin-film solar cells with varying compositions. By adjusting the mole ratio of the precursor materials, it is possible to achieve thin films with different bandgaps and photoconductivity.

In a study of antimony sulfide (Sb_2Se_3) thin films, which were prepared using vapor transport deposition (VTD), solar cell devices with a superstrate structure of ITO/CdS/Sb_2Se_3/Au were fabricated [32]. The authors developed an effective Sb_2Se_3 post-treatment method to optimize the performance of Sb_2Se_3 thin film without modifying the main structure or phase of Sb_2Se_3. By optimizing the time of the post-treatment, a device efficiency of 4.02% was achieved, accompanied by improvements in J_{SC} and *FF*. Detailed comparative studies were conducted to analyze the electrical properties, carrier transport, and carrier recombination in cells with and without Sb_2Se_3 post-treatment. This analysis included current density versus voltage (*J-V*) measurements under both light and dark conditions, energy-dispersive X-ray spectroscopy, deep-level transient spectroscopy, and V_{OC} measurements at various temperatures and light intensity levels. The results demonstrated that the Sb_2Se_3 post-treated cells outperformed the cells without post-treatment. These cells exhibited reduced trap-assisted recombination, longer carrier lifetimes, parallel current pathways, and a favorable distribution of elements with S-rich features. These characteristics contribute to the improved performance of the Sb_2Se_3-post-treated cells.

The development of high V_{OC} in Sb_2Se_3 thin-film solar cells presents a significant challenge in the advancement of earth-abundant PV devices. Traditionally, CdS selective layers have been employed as the standard electron contact in Sb_2Se_3 [33]. However, long-term scalability concerns related to cadmium toxicity and environmental impact have prompted researchers to explore alternative materials. A ZnO-based buffer layer combined with a polymer-film-modified top interface was proposed as a replacement for CdS in Sb_2Se_3 PV devices [34]. The introduction of a branched polyethylenimine layer at the ZnO and transparent electrode (TE) interface resulted in enhanced performance of Sb_2Se_3 solar cells. Notably, this modification led to a significant increase in V_{OC}, rising from 243 to 344 mV, and achieved a maximum efficiency of 2.4%. The utilization of conjugated polyelectrolyte thin films in chalcogenide PVs demonstrates the potential for improving device performance. This research offers promising strategies for enhancing the performance of earth-abundant PV devices by exploring alternative materials and interface modifications.

In the context of a confined-space selenium-assisted tellurization (c-SeTe) post-treatment strategy, c-SeTe was used in Sb_2Se_3 to fabricate solar cell devices [35]. Material characterizations confirm that most of the tellurium is concentrated at the back of the film, with minimal presence in the bulk. Further investigations into the physical properties reveal the role of c-SeTe in enhancing device performance. The alloying of Se and Te at the back of the film helps to mitigate the back-contact barrier, leading to improved extraction efficiency. Moreover, the co-doping of Se and Te in the bulk of the film contributes to the passivation of interface and bulk defects, thereby enhancing the quality of the CdS/Sb_2Se_3 heterojunction and improving the quantum yield of long-wavelength photons. As a result of the developed c-SeTe post-treatment strategy, a champion power conversion efficiency of 4.95% has been achieved, which is 0.5% higher than the control sample. This robust treatment method holds promise for advancing the rapid development of antimony chalcogenide solar cells.

In another study, Cu_2ZnSnS_4 (CZTS) thin films were deposited using the vacuum spray pyrolysis technique [36], and their structural, elemental, optical, and electrical properties were investigated. The deposition process involved varying substrate temperatures from 325°C to 400°C. Structural analysis revealed that the deposited CZTS thin films exhibited a crystalline nature, with the formation of the CZTS kësterite phase observed at 375°C. Additionally, X-ray photoelectron spectroscopy (XPS) confirmed the elemental composition of the films. The optical properties of the CZTS films were characterized using UV-visible spectroscopy with an observed 1.5 eV bandgap and $10^5 cm^{-1}$ absorption coefficient for the as-deposited films, as seen in Figure 6.2a and b. These values are close to the ideal range for an absorber layer in solar cells, suggesting the potential of CZTS for efficient light absorption. The electrical properties of the CZTS thin films were evaluated through Hall effect measurements, which confirmed their p-type nature (see Figure 6.2c). Furthermore,

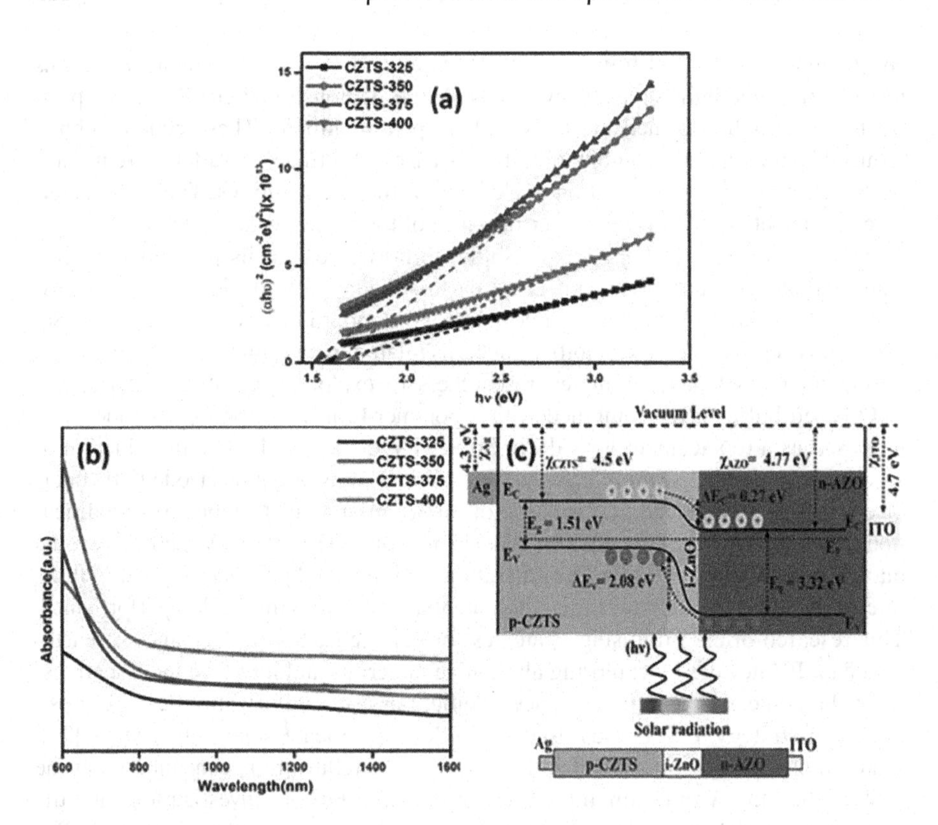

FIGURE 6.2 CZTS (a) Tauc plot for bandgap, (b) absorption spectra, and (c) band structure of the solar cell device [36].

CZTS simple heterojunction was fabricated as the absorber layer and Al:ZnO as the window layer. The estimated solar cell parameters were as follows: V_{OC}, I_{SC}, fill factor (*FF*), and η, exhibiting values of 0.85 V, 16.68 mA, 36.54%, and 2.87%, respectively. However, the calculated efficiency is very low, but the authors concluded that the CZTS thin films can be useful for the absorption layer in solar cells due to the observed high absorption coefficient.

The increasing relevance of multijunction solar cells as single-junction cells approach theoretical limits has sparked interest in low-cost, wider bandgap light harvesters for tandem applications with silicon in thin-film PV research. While Cu-based chalcogenide compounds have shown success as absorbers, their performance for bandgaps above 1.5 eV is still limited. Additionally, transparent back contacts pose challenges for this class of materials. To address these issues, the authors of this work have focused on fabricating wide-bandgap $CuGaSe_2$ absorbers using a combination of metallic sputtering and reactive thermal annealing on transparent fluorine-doped tin oxide-coated glass substrates [37]. The annealing temperature is carefully optimized to achieve desirable material and device properties. A key improvement strategy involves introducing an ultrathin Mo interlayer at the $CuGaSe_2$/back interface, which enhances contact ohmicity and significantly improves various figures of merit. The incorporation of the Mo interlayer leads to a record conversion efficiency of 5.4%, the highest reported for this class of absorber on a transparent back contact. Fundamental material characterization of the as-grown $CuGaSe_2$ films demonstrates better homogeneity in Cu distribution throughout the absorber's thickness when using the Mo interlayer, resulting in enhanced crystalline quality. However, the sub-bandgap transparency of the final device still has room for improvement, and the authors propose utilizing transfer matrix-based optical modeling to explore pathways for enhancing optical transmission, such as employing more specular interfaces.

CdS films are commonly used as a buffer layer in chalcogenide solar cells due to their excellent performance. However, the commercial application of CdS films is limited due to the environmental concerns associated with the production of wastewater containing cadmium. In this study, the researchers have developed an efficient method for preparing CdS films using the spin-coating technique [38], which offers advantages in terms of simplicity and potential for large-scale processing. The spin solution used in this technique is anhydrous, and the oxygen content in the films can be controlled by adjusting the Cd/S content ratio in the solution. The researchers characterized the crystal structure, morphology, and optical and electrical properties of the CdS films with different Cd/S ratios. They also investigated the effect of oxygen content on the performance of the solar cell devices. The findings of the study indicate that the introduction of an appropriate amount of oxygen improves carrier transport efficiency by reducing interface recombination. Additionally, the spin-coated CdS films demonstrate good coverage of the substrate for thicknesses below 50 nm. Through optimization of the Cd/S content ratio, the authors achieved an enhanced efficiency of 5.76% for the Sb_2Se_3 solar cells.

Barium zirconium sulfide ($BaZrS_3$) is a chalcogenide perovskite material that is gaining attention for its potential application in thin-film PV devices. Unlike lead halide perovskites, $BaZrS_3$ is stable and does not contain toxic elements. The numerical investigations were conducted using a Solar Cell Capacitance Simulator software-1D to analyze PV devices incorporating barium zirconium sulfide (BZS) as a photoabsorber with a back-contact layer of both crystalline and amorphous p+-type silicon. The authors examined titanium (Ti) alloyed BZS, which has an electron-energy bandgap close to the optimum for a single-junction PV device. The study involved a systematic exploration by varying the thickness, doping density, and defect density in the BZS layer. Among the two phases of silicon (crystalline and amorphous), the amorphous phase showed higher photoconversion efficiency due to favorable energy band alignment with BZS. The study predicted the best photoconversion efficiency values to be 19.7% for BZS films and 30% for $Ba(Zr_{0.95}Ti_{0.05})S_3$ films with amorphous Si as the back-contact material.

These findings indicate the potential of utilizing $BaZrS_3$ and Ti-alloyed BZS in thin-film PV devices, particularly when combined with an amorphous silicon back-contact layer. The numerical simulations provide insights into the influence of various parameters on the performance of these devices and highlight the promising photoconversion efficiencies that can be achieved with these configurations.

6.4.2 Carbon Nanotube in Solar Cells

Carbon nanotubes (CNTs) have been extensively studied for their optoelectronic properties in solar cells. Incorporating CNTs into the electrodes and active layers of solar cells has shown the potential to improve their efficiency and stability [39]. By optimizing the optoelectronic properties of CNTs, solar cells have the potential to compete with commercialized silicon solar cells.

In recent years, significant progress has been made in the development of CNT/Si junction solar cells, which have shown promising device performance. Wei et al. first reported a CNT/Si solar cell in 2007 with an η of 1.31% [40]. Since then, researchers have made notable advancements in improving the PV performance of CNT/Si solar cells [41]. Tune et al. conducted investigations on metallic, semiconducting, and mixed CNT/Si heterojunction solar cells and found that the metallic CNT/Si junction exhibited better performance [42]. Di et al. reported a CNT/Si heterojunction with well-aligned CNTs, achieving a J_{SC}, V_{OC}, FF, and η of 33.4 mA/cm^2, 540 mV, 58%, and 10.5%, respectively [43]. Cui et al. utilized single-walled CNTs (SWCNTs) and achieved an η of 10.8% with J_{SC} of 29.7 mA/cm^2, V_{OC} of 535 mV, and FF of 68% [44]. However, SWCNTs tend to agglomerate easily due to strong van der Waals forces, resulting in reduced transmission of light. To address this issue, researchers have explored the use of isolated SWCNTs in fabricating heterojunction devices. Hu et al. successfully fabricated an SWCNT/Si heterojunction using isolated SWCNTs synthesized through floating catalyst chemical vapor deposition (FCCVD). This approach yielded the highest reported J_{SC}, V_{OC}, FF, and η of 31.8 mA/cm^2, 580 mV, 63.7%, and 11.8%, respectively [45]. Despite these advancements, there are some drawbacks associated with CNT/Si heterojunctions. One significant limitation is the small device area, typically less than 0.1 cm^2. Additionally, the use of FCCVD for device fabrication adds complexity, and heterojunctions formed by random CNTs may result in sparse conformal coverage, leading to non-homogeneous depletion regions and potential effects on device performance. Addressing these challenges will be crucial for further enhancing the efficiency and scalability of CNT/Si heterojunction solar cells.

Researchers have explored the incorporation of CNTs into organic polymers such as PEDOT:PSS and PANI, which enables better conformal coverage [46] to address the above challenges. Yu et al. reported the use of PANI-SWCNT/Si solar cells with an η of 7.4%. The performance of the device was found to be influenced by the PANI loading, with better performance observed at lower PANI concentrations [46]. Numerous studies have also investigated the use of multi-walled carbon nanotubes (MWCNTs) and single-walled carbon nanotubes (SWCNTs) in solar cells, focusing on charge separation and electron transport between conjugated polymers such as P3OT and nanotubes [47–49]. Khatri et al. utilized green-tea-modified MWCNTs to enhance solar cell properties. The CNTs were spin-coated onto a Si substrate followed by PEDOT:PSS, resulting in an efficiency of 10.93% and a V_{OC} of 540 mV for a cell area of 5 mm × 5 mm [50]. Qingxia et al. transferred CVD-grown CNTs to a silicon substrate and spin-coated PEDOT:PSS layers with varying thicknesses. The best-performing cell, with a nitric acid-grown SiO_2 passivation layer (3 mm × 3 mm), achieved an efficiency of 10.2% and a V_{OC} of 548 mV [51]. In these approaches, the fabrication process involves two steps where the CNTs are initially transferred or coated onto the Si substrate, followed by the spin-coating of PEDOT:PSS. This results in a bilayer structure rather than a composite film. These studies demonstrate the potential of incorporating CNTs into organic polymers to enhance the performance of CNT/Si solar cells.

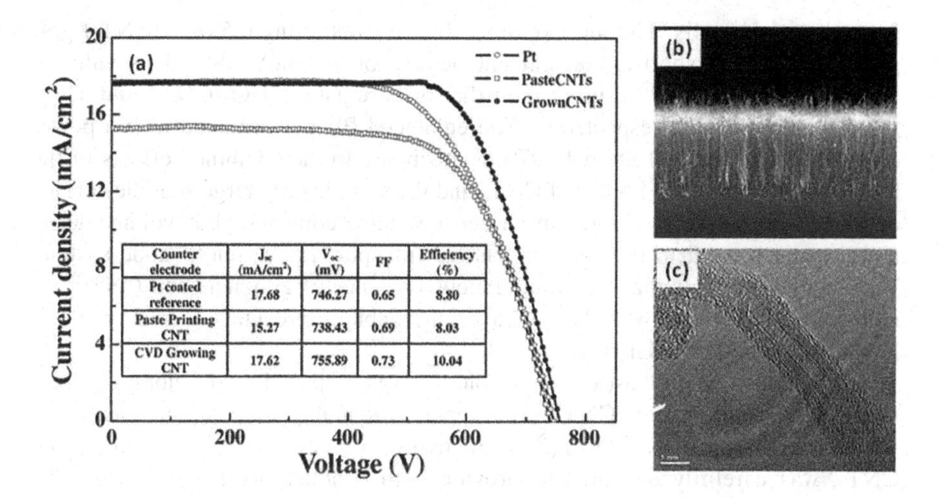

FIGURE 6.3 (a) DSSCs *I–V* characteristics Pt, paste printing, and CVD CNTs counter electrodes. (b) SEM image of CVD CNTs, and (c) HR-TEM CVD CNT counter electrodes [52].

CNTs have also been used to functionalize other materials, apart from Si and polymers, to improve the performance of solar cells. The authors of this study investigated the performance of DSSCs with counter electrodes coated with CNTs using two methods: screen printing and direct-growing [52]. They successfully applied screen-printed and thermally CVD-grown multi-wall CNTs on a FTO glass substrate as counter electrodes. They also utilized reference electrodes such as Pt and randomly dispersed CNTs, which achieved efficiencies of 8.80% and 8.03%, respectively. However, the DSSC with a direct-grown CNT counter electrode achieved an efficiency of 10.04% due to the remarkable enhancement of the *FF* to 0.73 (see Figure 6.3). Their study observed that while all three PV parameters were affected by an aligned CNT counter electrode, the variation in FF had a significant positive impact on the overall conversion efficiency. By employing a thermal CVD approach to synthesize the CNT electrode, the electron conduction path between the electrolyte and the FTO glass was significantly improved. The CNT counter electrode exhibited a large specific surface area of $490\,m^2\,g^{-1}$, which contributed to the faster redox reaction on the counter electrodes, leading to higher FF and overall efficiency in DSSCs. The large surface area and high electron conductivity of CNTs were believed to be responsible for the enhanced DSSC cell efficiency. The findings of this study suggest that CNT-based counter electrodes could provide a new avenue for producing non-platinum-based counter electrodes in DSSC devices. This research highlights the potential of CNTs in improving the performance and efficiency of DSSCs, offering an alternative to platinum-based counter electrodes.

The utilization of CNTs and graphene in PVs, particularly in the form of TiO_2-CNT or TiO_2-graphene nanocomposite photoanodes, has garnered significant interest. In this investigation, a hybrid TiO_2-CNT-graphene mesoporous thin film is fabricated using a mixed-solvent and in situ hydrothermal synthesis, along with a two-step calcination process [53]. This film is then employed as a photoanode in perovskite solar cells. The most optimal device, containing 0.5 wt% (CNT + graphene), exhibits a short-circuit current density of 24.8 mA cm^{-2}. This value is 43.4%, 39.3%, and 19.2% higher than that of bare TiO_2-, TiO_2-CNT-, and TiO_2-graphene-based cells, respectively. The enhanced PV performance with a power conversion efficiency of up to 13.97% is attributed to the combined effects of the high electron extraction ability of CNTs and the significantly large specific surface area of graphene sheets. Photoluminescence spectroscopy and photovoltage decay measurements confirm the superior charge transport in the photoanode system. This study highlights that the simultaneous presence of graphene and CNTs in a nanocomposite system with TiO_2 yields remarkable improvements in charge collection in perovskite solar cells.

In a similar study, a cost-effective solution was proposed by developing a free-standing hybrid graphene/ CNT film to replace the noble metal counter electrodes in perovskite solar cells (PSCs) [54]. The hybrid film, consisting of graphene and CNTs, was carefully designed to provide high conductivity and stability. The hybrid graphene/CNT film demonstrated excellent properties, such as high conductivity and stability, and could be easily transferred to various substrates using a simple rolling process. PSCs incorporating the hybrid film as counter electrodes exhibited a notable power conversion efficiency (PCE) of 15.36%. Furthermore, these devices exhibited remarkable stability, retaining 86% of their initial PCE after being stored for 500 hours in a high-moisture atmosphere with 50% relative humidity. The enhanced stability of PCEs in the devices can be attributed to the effective moisture blocking provided by the multilayered graphene/CNT structure present in the hybrid film. The thin, flexible, and easily synthesized free-standing hybrid graphene/CNT film with its high conductivity shows great potential for enabling the low-cost production of highly stable PSCs. This innovation by the authors could address the limitations of costly noble metal counter electrodes and highlight the feasibility of achieving stable and cost-effective PSCs by utilizing hybrid graphene/CNT films.

Thin-film organic-inorganic halide PSCs have also gained significant attention due to their high power conversion efficiency and advantageous properties such as solution processability and low fabrication cost. However, the existing halide perovskite solar cell technology has been limited by structural and material constraints. A solution has been proposed by replacing conventional metal-oxide photoelectrodes with CNT-doped materials [55]. A mixed-cation PSC was fabricated in ambient air but exhibited instability when the perovskite (PVK) was deposited on top of zinc oxide (ZnO) due to the presence of hydroxyl groups on the ZnO surface [55]. CNTs were incorporated into the ZnO film by Mohammed et al. [55] to mitigate this issue. The addition of CNTs improved charge extraction from the ZnO film and reduced recombination rates in the PSCs. This enhancement was attributed to the higher conductivity and lower trap states achieved with CNT-doped ZnO. The optimized PSC with the CNT additive demonstrated a PCE of up to 15%. Furthermore, the surface modification approach resulted in hysteresis-free and stable PSCs, exhibiting minimal decomposition even after approximately 2,000 hours of storage in a moist environment.

6.4.3 Graphene in Solar Cells

Graphene, a two-dimensional carbon allotrope, has exceptional optoelectronic properties that make it an attractive material for solar cell applications. In the context of solar cells, graphene and CNTs have been explored as electrode materials [56]. These solar cells utilize the unique optoelectronic properties of graphene and CNT-based electrodes to enhance their performance. Such are discussed in this chapter.

In this particular study, CZTS (Copper Zinc Tin Sulfide) samples were fabricated on both Mo and graphene/Mo-coated glass substrates using a quaternary target [57]. RF magnetron sputtering was employed to deposit the CZTS thin films, which were subsequently annealed using rapid thermal processing in a sulfur atmosphere at temperatures of 500°C, 525°C, and 550°C. This process aimed to obtain glass/Mo/CZTS and glass/Mo/graphene/CZTS (g-CZTS) structures. The optical band gap values of the samples ranged from 1.41 to 1.44 eV, depending on the Sn content. Solar cells fabricated using the more promising absorber layers demonstrated that the incorporation of graphene in the CZTS cell structure led to an enhancement in conversion efficiency from 2.40% to 3.52%. The integration of graphene into the solar cell structure resulted in improvements in various cell parameters, such as J_{SC}, V_{OC}, and *FF*. The J_{SC}, V_{OC}, and *FF* increased from 25.8 to 29.9 mA cm^{-2}, 0.252 to 0.295 V, and 0.37 to 0.40, respectively, with the inclusion of graphene. There are several explanations for the efficiency enhancement of CZTS. For instance, the study by Zhou et al. utilized carbon layers as an interlayer in CZTS thin films deposited over Mo-coated glass substrates. During the sulfurization process, carbon atoms settled on the inner wall of the void structures formed in the CZTS, improving the conductivity between the Mo-CZTS structures and consequently enhancing cell efficiency [58]. Another study by Vishwakarma et al. reported that the use of graphene flakes as an interlayer reduced the formation of MoS_2 at the Mo-CZTS interface and prevented the formation of detrimental structures such as pores. The presence of graphene flakes reduced defects at the interface due to their high thermal conductivity and mechanical stability, thereby improving cell performance [59]. The high mobility and carrier density of graphene film also contributed to the enhancement of cell efficiency. These superior properties allowed for faster electron collection by the back contact and reduced recombination processes within the cell structure [60]. Furthermore, the use of graphene in the cell structure improved the crystalline quality of the CZTS compound. This improvement, characterized by larger crystallite sizes and fewer grain boundaries, led to a decrease in interface recombination and voltage loss, resulting in an increased V_{OC} value [61]. This improvement can be attributed to the enhanced crystalline quality of the absorber layer facilitated by graphene.

In a similar study, a novel combination of TE of Li-doped graphene oxide (LGO), electron-transporting layer (ETL), and Mg-Ga-co-doped ZnO (MGZO) was utilized in $Cu_2ZnSn(S,Se)_4$ (CZTSSe) thin-film solar cells (TFSCs) for the first time [62]. The use of MGZO as the TE offered a wider optical spectrum with high transmittance compared to conventional Al-doped ZnO (AZO), enabling additional photon harvesting. Furthermore, MGZO exhibited low electrical resistance, leading to an increased electron collection rate. These exceptional optoelectronic properties resulted in significant improvements in the I_{SC} density and FF of the TFSCs. To prevent plasma-induced damage to the chemically bath-deposited cadmium sulfide (CdS) buffer, a solution-processable alternative LGO ETL was employed. This allowed for the maintenance of high-quality junctions using a thin CdS buffer layer of approximately 30 nm. Through interfacial engineering with LGO, the V_{OC} of the CZTSSe TFSCs was enhanced from 466 to 502 mV. Additionally, the tunable work function achieved through Li doping created a more favorable band offset in the interfaces of CdS/LGO/MGZO, thereby improving electron collection. The combination of MGZO/LGO/TE/ETL demonstrated a remarkable power conversion efficiency of 10.67%, surpassing that of conventional AZO/intrinsic ZnO (8.33%) configurations.

In DSSCs, the semiconductor photoanode plays a crucial role in collecting photoexcited electrons from the sensitizer. However, electron-hole recombination often hampers the photocoversion efficiency of the photoanode. To address this issue, Kandasamy et al. [63] have explored the modification of TiO_2 nanotubes with amine-functionalized GO. In their study, the nanocomposites of ethylenediamine functionalized GO/NH_2 and titania nanotubes (TiO_2 NTs) were prepared through a hydrothermal method. These nanocomposites were further incorporated with 3 wt% of silver nanoparticles (Ag). The properties of the nanocomposites were analyzed using various techniques, including diffuse reflectance spectroscopy, Raman spectroscopy, XRD, TEM, and XPS. The results revealed that the amine functionalization of GO led to the formation of C-N bonds, causing the GO surface to acquire a defective structure and disorder. The incorporation of silver nanoparticles into the amine-functionalized GO-modified TiO_2 nanotubes (TiO_2 NTs-GO/NH_2/Ag) resulted in a higher dye adsorption capacity and improved light-gathering capabilities. DSSCs fabricated with the TiO_2 NTs-GO/NH2/Ag photoanode exhibited a remarkable photocoversion efficiency of 8.18%. This efficiency can be attributed to efficient charge transport at the interfaces, high dye loading, and broad absorption by the sensitizer-adsorbed TiO_2 NTs-GO/NH_2/Ag thin film [64]. These factors contribute to significant light-to-electrical energy conversion efficiency in the TiO_2 NTs-GO/NH_2/Ag photoanode-based DSSCs.

In their research, they investigated the influence of a graphene layer and zinc oxide nanorods (NRs) on the performance of ZnO thin-film-based heterojunction solar cells [65]. Initially, a thin film of ZnO was deposited on p-type silicon using radio frequency sputtering, serving as both an anti-reflective layer and an n-type semiconductor for the heterojunction solar cell. Next, ZnO NRs were synthesized through a cost-effective chemical bath deposition technique. To explore the optical properties of ZnO NRs, they incorporated reduced graphene oxide (rGO) into the colloidal suspension of ZnO NRs, resulting in the formation of rGO:ZnO NRs. These optimized nanostructures were then applied to the surface of the ZnO/Si sample using a drop-cast method. The resultant outcomes are depicted in Figure 6.4. Through morphological examination, it was observed that the ZnO NRs exhibited various orientations and the composite was not destroyed (see Figure 6.4a), which effectively increased the trapping of incident photons. XRD analysis demonstrated improved structural properties of ZnO NRs when rGO flakes were introduced. The reduction of graphene oxide (GO) in the rGO:ZnO NRs sample was confirmed using Raman spectroscopy, as indicated by a higher I_D/I_G ratio, suggesting a chemical reduction process. Optical investigations revealed a higher light absorption and

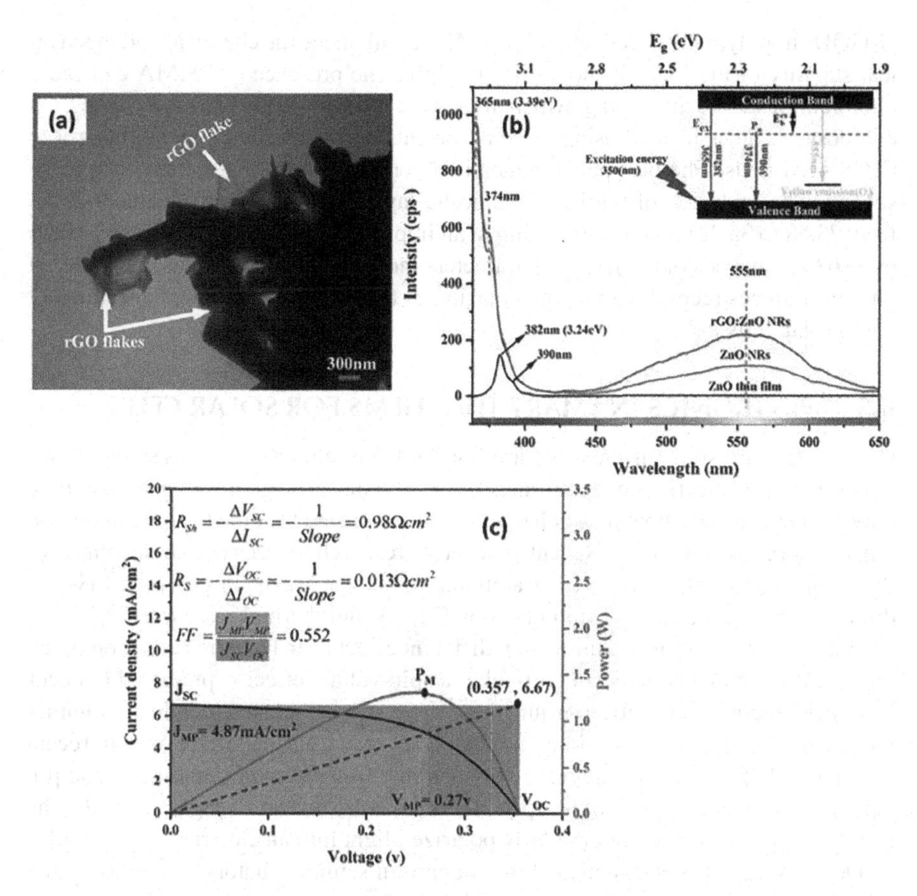

FIGURE 6.4 (a) TEM image of rGO:ZnO, (b) PL of rGO:ZnO and ZnO NRs and ZnO thin film, and (c) *I-V* characteristic used for the determination of rGO:ZnO NRs solar cell efficiency [65].

photoluminescence intensity in the samples containing rGO, as seen in Figure 6.4b, indicating that graphene played a role in inhibiting electron-hole recombination. To assess the optoelectrical properties, the researchers fabricated three different solar cells based on ZnO thin film, ZnO NRs, and rGO:ZnO NRs. The results showed that the solar cell incorporating rGO:ZnO NRs exhibited enhanced efficiency and a *FF* of 14.12% and 0.552 (see Figure 6.4c), respectively, compared to the other solar cells. This suggests that the addition of rGO can improve the PV characteristics of solar cells utilizing ZnO nanorods.

The use of Cu(In, Ga)Se_2 (CIGS) as an absorbing material in thin film-based PV technologies is gaining significant attention in the PV community due to its higher absorption capability and adjustable band gap. However, CIGS solar cells often exhibit lower current values in the short wavelength region, leading to a reduction in J_{SC} density and overall conversion efficiency. To address this issue, N-doped graphene quantum dots (N-GQDs) are incorporated as luminescent photon downshifters (LDS) in the solar cell structure by Khan et al. [66]. Additionally, embedding the

N-GQDs in poly(methylmethacrylate) (PMMA) enhances the chemical and mechanical stability of the LDS. It should be noted that the presence of PMMA can cause photoluminescence quenching, which affects the efficacy of the LDS layer. The LDS composite layer is formed using various solvents and is evaluated by applying it to CIGS solar cells. The best performance is observed when using the chlorobenzene solvent. The inclusion of the LDS composite layer leads to an enhancement in J_{SC} from 33.58 to 35.32 mA cm^{-2}, resulting in an improvement in efficiency from 13.08% to 14.62%. Analysis of the PV cell parameters indicates that the increased number of photons reduces recombination and contributes to the performance enhancement of CIGS solar cells [67].

6.5 SPINTRONICS IN SMART THIN FILMS FOR SOLAR CELLS

Generating pure spin currents that are both efficient and energy-conserving is one of the primary objectives of spintronics research today. The goal is to generate these pure spin currents without using charge currents that lead to Joule heating, mainly to reduce energy consumption. Recent spintronics research focuses on novel techniques that avoid charge currents in the traditional sense, such as spin pumping [68–70], direct or non-local electrical spin injection [71,72], and thermal injection [73].

When a p-n junction is illuminated, different effects can be anticipated compared to a uniform semiconductor. In particular, a spin-voltaic effect is predicted to occur in a magnetic/non-magnetic p-n junction where one of the semiconductors (either the p-type or n-type) has a spin-split band. This effect has been proposed in recent research [74,75]. Experiments have demonstrated this effect in a non-magnetic p-n junction consisting of n-GaAlAs/p-GaInAs/p-GaAs under a magnetic field. The result was the conversion of circularly polarized light into an electrical signal [76].

Decades ago, it was established that in certain semiconductors, electrons with a specific spin orientation can be generated through a phenomenon called optical orientation [77]. When circularly polarized light is used, the carriers receive non-zero angular momentum, resulting in a higher concentration of electrons with a particular spin direction and a lower concentration with an opposite spin direction. This leads to carrier spin polarization. Unfortunately, this method relies on circularly polarized light and materials with appropriate electronic properties. Although III–V semiconductors, such as gallium arsenide, can be successfully used, a technologically crucial semiconductor like Si is unsuitable for this process [77].

However, another option to generate spin flows is using thermally induced mechanisms. This involves utilizing the energy from light to create a thermal gradient that propels a spin current. Previous research has demonstrated that when heat flows through a ferromagnetic tunnel contact on a semiconductor, a spin current is generated [73]. This led to the suggestion that light could induce spin polarization in any semiconductor without requiring circularly polarized light. Recently, this hypothesis was validated for Ge with CoFe/MgO tunnel contacts. The CoFe electrode, which was ferromagnetic, was exposed to a laser, leading to a heat flow and subsequently a spin current across the contact [78].

A solar cell that generates spin has been proposed by Endres et al. [79], consisting of a p-n junction rather than thermally induced mechanisms. The p-side of the junction

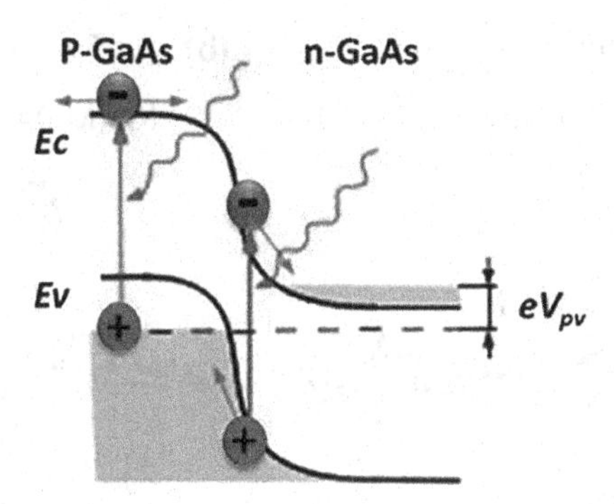

FIGURE 6.5 Band diagram of GaAs p-n junction [79].

is a ferromagnetic semiconductor, allowing efficient optical spin injection even with unpolarized light. The diagram in Figure 6.5 displays the GaAs p-n junction.

In their research, a solar cell of the ferromagnetic semiconductor Ga,Mn:As on the p-side was employed. GaAs was on the n-side, and a depletion zone that was narrow was observed, which allowed tunneling across the bandgap owing to the high doping with an observed Esaki diode current-voltage characteristic [80]. The band-bending region in a p-n junction is primarily restricted to the n-GaAs due to the highly p-doped Ga,Mn:As. Consequently, the resulting photocurrent from illuminating this junction will predominantly consist of photoexcited electrons from the n-GaAs side, with only a small fraction of spin-polarized electrons created in the Ga,Mn:As. As a result, the generated photocurrent exhibits minimal spin polarization. When the n-GaAs accumulate charge, it generates a photovoltage that prompts electrons to tunnel into the Ga,Mn:As through a narrow barrier. Due to varying tunneling probabilities for spin-up and spin-down electrons, spins amass in the n-GaAs. This leads to a light-induced spin extraction, shown in Figure 6.6a, which counters the spin accumulation induced by the photocurrent.

In junctions that are not heavily doped with n-type impurities and where the reverse direction of tunneling is suppressed, a different operational mode can be utilized, as depicted in Figure 6.6b. When a negative voltage is applied to the p-side of the junction (reverse bias), the depletion width is increased, and tunneling is suppressed. As a result of the spin-dependent density of states in the valence band [81], photoexcited electrons on the p-side become spin-polarized and move toward the conduction band of n-GaAs by drifting in the electric field of the junction. In 2001, Žutić et al. predicted the occurrence of the spin photodiode effect, which leads to spin accumulation [82]. However, the orientation of the spin is opposite to that of the spin solar cell effect, as shown in Figure 6.6a. Thus, the effect of light-induced spin injection takes place.

6.6 APPLICATIONS OF SOLAR CELLS

Solar cells have a wide range of applications, from powering small electronics to providing electricity for entire buildings and communities. Some of the most common applications of solar cells include:

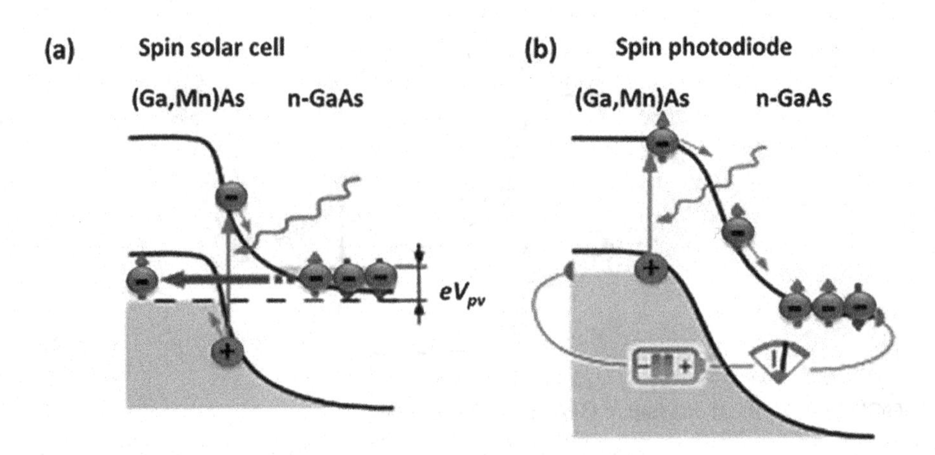

FIGURE 6.6 Band diagram of (a) p-GaMn:As and n-GaAs p-n junction for spin solar cell. (b) p-GaMn:As and n-GaAs p-n junction for spin photodiode [79].

1. **Residential Solar Power:** Solar panels can be installed on rooftops to provide electricity for individual homes. This can help homeowners reduce their dependence on the grid and lower their electricity bills.
2. **Commercial and Industrial Solar Power:** Solar panels can be installed on the roofs or grounds of commercial and industrial buildings to provide electricity for the businesses that occupy them. This can help businesses reduce their energy costs and their carbon footprint.
3. **Off-Grid Power:** Solar panels can be used to provide electricity for locations that are not connected to the grid, such as remote cabins, campsites, and research stations.
3. **Transportation:** Solar cells can be used to power small vehicles, such as golf carts, as well as larger vehicles, such as buses and trains.
4. **Consumer Electronics:** Solar cells can be used to power small electronics such as calculators, watches, and smartphones. They can also be used to recharge batteries for these devices.
5. **Emergency Power:** Solar panels can be used to provide electricity in the event of a power outage or natural disaster. They can be used to power lights, refrigerators, and other essential appliances.
6. **Space Exploration:** Solar cells are used to provide power for spacecraft and satellites. They are designed to be highly efficient and lightweight to maximize their power output while minimizing their weight.

6.7 CONCLUSION, CHALLENGES, AND PERSPECTIVES

Chalcogenide, CNTs, and graphene have shown potential for enhancing the efficiencies of different types of solar cells due to their unique properties. Despite their unique properties, the efficiencies of solar cells are still low. These materials still hold significant promise for next-generation solar energy technologies. Overcoming challenges related to low efficiency, stability, scalability, toxicity, and cost-effectiveness is crucial for their future development and commercialization. Continued research and innovation in materials synthesis, device design, and manufacturing processes are expected to drive the advancement and realization of efficient and sustainable solar cells based on chalcogenide, CNT, and graphene thin-film materials.

Spintronic solar cells are still new, and GaAs and Ga,Mn:As are also explored in this chapter because they open new vistas for solar cells. A key aspect in this domain is the generation of spin currents using light, which opens new possibilities for spin-based solar cell devices. While the correlation between spintronics and optics has been established for III–V semiconductors, recent breakthroughs have extended this connection to a wider range of mainstream semiconductors, enabling the utilization of light with any polarization to generate spin currents. These findings indicate a promising future for spin optics in group IV semiconductors, which include silicon and germanium. Considering the significance of these developments, researchers and engineers should closely follow the advancements in this field.

REFERENCES

[1] O.A. Al-Shahri et al., Solar photovoltaic energy optimization methods, challenges and issues: a comprehensive review, *Journal of Cleaner Production*. 284 (2021) 125465.

[2] S. Sharma, K.K. Jain, A. Sharma, Solar cells: in research and applications-a review, *Materials Sciences and Applications*. 6 (12) (2015) 1145.

[3] L. Mok, S. Hyysalo, Designing for energy transition through value sensitive design, *Design Studies*. 54 (2018) 162–183.

[4] J. Puebla, J. Kim, K. Kondou, Y. Otani, Spintronic devices for energy-efficient data storage and energy harvesting, *Communications Materials*. 1 (1) (2020) 24.

[5] D. Giovanni et al., Tunable room-temperature spin-selective optical Stark effect in solution-processed layered halide perovskites, *Science Advances*. 2 (6) (2016) e1600477.

[6] M.A. Green, *Solar Cells: Operating Principles, Technology, and System Applications*, Prentice Hall, 1982.

[7] J.G. Yañuk, F.M. Cabrerizo, F.G. Dellatorre, M.F. Cerdá, Photosensitizing role of R-phycoerythrin red protein and β-carboline alkaloids in dye sensitized solar cell. Electrochemical and spectroscopic characterization, *Energy Reports*. 6 (2020) 25–36.

[8] M. Green, E. Dunlop, J. Hohl-Ebinger, M. Yoshita, N. Kopidakis, X. Hao, Solar cell efficiency tables (version 57), *Progress in Photovoltaics: Research and Applications*. 29 (1) (2021) 3–15.

[9] V.M. Fthenakis, H.C. Kim, Photovoltaics: life-cycle analyses, *Solar Energy*. 85 (8) (2011) 1609–1628.

[10] R.M. Swanson, A vision for crystalline silicon photovoltaics, *Progress in Photovoltaics: Research and Applications*. 14 (5) (2006) 443–453.

[11] M. Imamzai, M. Aghaei, Y.H.M. Thayoob, M. Forouzanfar, A review on comparison between traditional silicon solar cells and thin-film CdTe solar cells, in: *Proceedings of National Graduate Conference (Nat-Grad)*, 2012: pp. 1–5.

[12] U. Gangopadhyay, S. Jana, S. Das, State of art of solar photovoltaic technology, in: *Conference Papers in Science*, 2013, vol. 2013.

[13] S. Mishra, S. Ghosh, T. Singh, Progress in materials development for flexible perovskite solar cells and future prospects, *ChemSusChem*. 14 (2) (2021) 512–538.

[14] M.A. Green, Y. Hishikawa, E.D. Dunlop, D.H. Levi, J. Hohl-Ebinger, M. Yoshita, A.W. Ho-Baillie, Solar cell efficiency tables (version 53), *Progress in Photovoltaics*. 27 (1) (2019) 3–12.

[15] J. Ramanujam et al., Flexible CIGS, CdTe and a-Si: H based thin film solar cells: a review, *Progress in Materials Science.* 110 (2020) 100619.
[16] S. Sivaraj et al., A comprehensive review on current performance, challenges and progress in thin-film solar cells, *Energies.* 15 (22) (2022) 8688.
[17] B. O'regan, M. Grätzel, A low-cost, high-efficiency solar cell based on dye-sensitized colloidal TiO2 films, *Nature* 353 (6346) (1991) 737–740.
[18] K.K.S. Lau, M. Soroush, Chapter 1 – Overview of dye-sensitized solar cells, in: M. Soroush and K. K. S. Lau, eds., *Dye-Sensitized Solar Cells*, Elsevier, 2019: pp. 1–49.
[19] E.T. Efaz et al., A review of primary technologies of thin-film solar cells, *Engineering Research Express.* 3 (3) (2021) 32001.
[20] S. Schubert, M. Hermenau, J. Meiss, L. Müller-Meskamp, K. Leo, Oxide sandwiched metal thin-film electrodes for long-term stable organic solar cells, *Advanced Functional Materials.* 22 (23) (2012) 4993–4999.
[21] G.A. Chamberlain, Organic solar cells: a review, *Solar Cells.* 8 (1) (1983) 47–83.
[22] M.A. Green, A. Ho-Baillie, H.J. Snaith, The emergence of perovskite solar cells, *Nature Photonics.* 8 (7) (2014) 506–514.
[23] T. Todorov, O. Gunawan, S. Guha, A road towards 25% efficiency and beyond: perovskite tandem solar cells, *Molecular Systems Design & Engineering.* 1 (4) (2016) 370–376.
[24] H. Li et al., Near-infrared and ultraviolet to visible photon conversion for full spectrum response perovskite solar cells, *Nano Energy.* 50 (2018) 699–709.
[25] M. Saliba et al., Cesium-containing triple cation perovskite solar cells: improved stability, reproducibility and high efficiency, *Energy & Environmental Science.* 9 (6) (2016) 1989–1997.
[26] A.J. Pearson et al., Oxygen degradation in mesoporous Al2O3/CH3NH3PbI3-xClx perovskite solar cells: kinetics and mechanisms, *Advanced Energy Materials.* 6 (13) (2016) 1600014.
[27] W. Ke, M.G. Kanatzidis, Prospects for low-toxicity lead-free perovskite solar cells, *Nature Communications.* 10 (1) (2019) 965.
[28] C.R. Kagan, E. Lifshitz, E.H. Sargent, D.V Talapin, Building devices from colloidal quantum dots, *Science.* 353 (6302) (2016) aac5523.
[29] M. Suhail, H. Abbas, M.B. Khan, Z.H. Khan, Chalcogenide perovskites for photovoltaic applications: a review, *Journal of Nanoparticle Research.* 24 (7) (2022) 142.
[30] R. Deepika, P. Meena, Colloidal chemical synthesis of quaternary semiconductor Cu2FeSnS4 (CFTS) nanoparticles: absorber materials for thin-film photovoltaic applications, *Journal of Materials Science: Materials in Electronics.* 34 (1) (2023) 16.
[31] A.L.E. Santana, P.K. Nair, Antimony sulfide selenide thin film solar cells prepared from thermal evaporation sources produced via chemical reactions, *Materials Science in Semiconductor Processing.* 160 (2023) 107450.
[32] R. Wang et al., Optimisation of Sb2S3 thin-film solar cells via Sb2Se3 post-treatment, *Journal of Power Sources.* 556 (2023) 232451.
[33] Y. Zeng et al., Comparative study of TiO2 and CdS as the electron transport layer for Sb2S3 solar cells, *Solar RRL.* 6 (10) (2022) 2200435.
[34] D. Rovira et al., Polymeric interlayer in CdS-free electron-selective contact for Sb2Se3 thin-film solar cells, *International Journal of Molecular Sciences.* 24 (4) (2023) 3088.
[35] F. Xiao et al., Confined-space selenium-assisted tellurization posttreatment strategy for efficient full-inorganic Sb2S3 thin-film solar cells, *Energy Technology.* 11 (4) (2023) 2201315.
[36] P. Aabel, A. Anupama, M.C.S. Kumar, Preparation and characterization of CZTS thin films by vacuum-assisted spray pyrolysis and fabrication of Cd-free heterojunction solar cells, *Semiconductor Science and Technology.* 38 (4) (2023) 45010.

[37] A. Thomere et al., 2-step process for 5.4% CuGaSe2 solar cell using fluorine doped tin oxide transparent back contacts, *Progress in Photovoltaics: Research and Applications.* 31 (2022) 524–535.
[38] J. Cheng et al., High-efficiency Sb2Se3 thin-film solar cells based on Cd (S, O) buffer layers prepared via spin-coating, *Materials Chemistry and Physics.* 303 (2023) 127794.
[39] E. Muchuweni, E.T. Mombeshora, B.S. Martincigh, V.O. Nyamori, Recent applications of carbon nanotubes in organic solar cells, *Frontiers in Chemistry.* 9 (2022) 733552.
[40] J. Wei et al., Double-walled carbon nanotube solar cells, *Nano Letters.* 7 (8) (2007) 2317–2321.
[41] E. Shi et al., TiO2-coated carbon nanotube-silicon solar cells with efficiency of 15%, *Scientific Reports.* 2 (1) (2012) 884.
[42] D.D. Tune, B.S. Flavel, J.S. Quinton, A.V Ellis, J.G. Shapter, Single-walled carbon nanotube/polyaniline/n-silicon solar cells: fabrication, characterization, and performance measurements, *ChemSusChem.* 6 (2) (2013) 320–327.
[43] J. Di, Z. Yong, X. Zheng, B. Sun, Q. Li, Aligned carbon nanotubes for high-efficiency Schottky solar cells, *Small.* 9 (8) (2013) 1367–1372.
[44] K. Cui et al., Air-stable high-efficiency solar cells with dry-transferred single-walled carbon nanotube films, *Journal of Materials Chemistry A.* 2 (29) (2014) 11311–11318.
[45] X.-G. Hu, P.-X. Hou, C. Liu, F. Zhang, G. Liu, H.-M. Cheng, Small-bundle single-wall carbon nanotubes for high-efficiency silicon heterojunction solar cells, *Nano Energy.* 50 (2018) 521–527.
[46] L. Yu, D. Tune, C. Shearer, T. Grace, J. Shapter, Heterojunction solar cells based on silicon and composite films of polyaniline and carbon nanotubes, *IEEE Journal of Photovoltaics.* 6 (3) (2016) 688–695.
[47] I. Khatri, Z. Tang, Q. Liu, R. Ishikawa, K. Ueno, H. Shirai, Green-tea modified multiwalled carbon nanotubes for efficient poly (3, 4-ethylenedioxythiophene): poly (stylenesulfonate)/n-silicon hybrid solar cell, *Applied Physics Letters.* 102 (6) (2013) 63508.
[48] K.M. Kurias, M. Jasna, M.R.R. Menon, A. Antony, M.K. Jayaraj, Fabrication of CNT-PEDOT: PSS/Si heterojunction carrier selective solar cell, *AIP Conference Proceedings.* 2082 (1) (2019) 50008.
[49] S.P. Somani, P.R. Somani, M. Umeno, E. Flahaut, Improving photovoltaic response of poly (3-hexylthiophene)/n-Si heterojunction by incorporating double walled carbon nanotubes, *Applied Physics Letters.* 89 (22) (2006) 223505.
[50] E. Kymakis, G.A.J. Amaratunga, Single-wall carbon nanotube/conjugated polymer photovoltaic devices, *Applied Physics Letters.* 80 (1) (2002) 112–114.
[51] Q. Fan et al., Novel approach to enhance efficiency of hybrid silicon-based solar cells via synergistic effects of polymer and carbon nanotube composite film, *Nano Energy.* 33 (2017) 436–444.
[52] J.G. Nam, Y.J. Park, B.S. Kim, J.S. Lee, Enhancement of the efficiency of dye-sensitized solar cell by utilizing carbon nanotube counter electrode, *Scripta Materialia.* 62 (3) (2010) 148–150.
[53] A. Amini, H. Abdizadeh, M.R. Golobostanfard, Hybrid 1D/2D carbon nanostructure-incorporated titania photoanodes for perovskite solar cells, *ACS Applied Energy Materials.* 3 (7) (2020) 6195–6204.
[54] M. Tian et al., Printable free-standing hybrid graphene/dry-spun carbon nanotube films as multifunctional electrodes for highly stable perovskite solar cells, *ACS Applied Materials & Interfaces.* 12 (49) (2020) 54806–54814.
[55] M.K.A. Mohammed et al., "Improvement of the interfacial contact between zinc oxide and a mixed cation perovskite using carbon nanotubes for ambient-air-processed perovskite solar cells, *New Journal of Chemistry.* 44 (45) (2020) 19802–19811.

[56] K.-T. Lee, D.H. Park, H.W. Baac, S. Han, Graphene-and carbon-nanotube-based transparent electrodes for semitransparent solar cells, *Materials.* 11 (9) (2018) 1503.
[57] S. Erkan, A. Yagmyrov, A. Altuntepe, R. Zan, M.A. Olgar, Integration of single layer graphene into CZTS thin film solar cells, *Journal of Alloys and Compounds.* 920 (2022) 166041.
[58] F. Zhou et al., Improvement of J sc in a Cu2ZnSnS4 Solar Cell by using a thin carbon intermediate layer at the Cu2ZnSnS4/Mo interface, *ACS Applied Materials & Interfaces.* 7 (41) (2015) 22868–22873.
[59] M. Vishwakarma, N. Thota, O. Karakulina, J. Hadermann, B.R. Mehta, Role of graphene inter layer on the formation of the MoS2-CZTS interface during growth, *AIP Conference Proceedings.* 1953 (1) (2018) 100064.
[60] A. Altuntepe et al., Hybrid transparent conductive electrode structure for solar cell application, *Renewable Energy.* 180 (2021) 178–185.
[61] Y. Wei et al., An investigation on the relationship between open circuit voltage and grain size for CZTSSe thin film solar cells fabricated by selenization of sputtered precursors, *Journal of Alloys and Compounds.* 773 (2019) 689–697.
[62] J. Kim et al., Novel Mg-and Ga-doped ZnO/Li-doped graphene oxide transparent electrode/electron-transporting layer combinations for high-performance thin-film solar cells, *Small.* 19 (2023) e2207966.
[63] M. Kandasamy, M. Selvaraj, M.M. Alam, P. Maruthamuthu, S. Murugesan, Nano-silver incorporated amine functionalized graphene oxide titania nanotube composite: a promising DSSC photoanode, *Journal of the Taiwan Institute of Chemical Engineers.* 131 (2022) 104205.
[64] L. Chen et al., Enhanced photovoltaic performance of a dye-sensitized solar cell using graphene-TiO 2 photoanode prepared by a novel in situ simultaneous reduction-hydrolysis technique, *Nanoscale.* 5 (8) (2013) 3481–3485.
[65] M.-R. Zamani-Meymian, N. Naderi, M. Zareshahi, Improved n-ZnO nanorods/p-Si heterojunction solar cells with graphene incorporation, *Ceramics International.* 48 (23) (2022) 34948–34956.
[66] F. Khan et al., Impact of solvent on the downconversion efficiency of the N-GQDs/PMMA layer: application in CIGS solar cells, *Optik.* 253 (2022) 168569.
[67] F. Khan, J.H. Kim, N-functionalized graphene quantum dots with ultrahigh quantum yield and large stokes shift: efficient downconverters for CIGS solar cells, *ACS Photonics.* 5 (11) (2018) 4637–4643.
[68] K. Uchida et al., Long-range spin Seebeck effect and acoustic spin pumping, *Nature Materials.* 10 (10) (2011) 737–741.
[69] F.D. Czeschka et al., Scaling behavior of the spin pumping effect in ferromagnet-platinum bilayers, *Physical Review Letters.* 107 (4) (2011) 46601.
[70] K. Ando et al., Electrically tunable spin injector free from the impedance mismatch problem, *Nature Materials.* 10 (9) (2011) 655–659.
[71] M. Ciorga, A. Einwanger, U. Wurstbauer, D. Schuh, W. Wegscheider, D. Weiss, Electrical spin injection and detection in lateral all-semiconductor devices, *Physical Review B.* 79 (16) (2009) 165321.
[72] X. Lou et al., Electrical detection of spin transport in lateral ferromagnet-semiconductor devices, *Nature Physics.* 3 (3) (2007) 197–202.
[73] J.-C. Le Breton, S. Sharma, H. Saito, S. Yuasa, R. Jansen, Thermal spin current from a ferromagnet to silicon by Seebeck spin tunnelling, *Nature.* 475 (7354) (2011) 82–85.
[74] I. Žutić, J. Fabian, S. Das Sarma, Spin-polarized transport in inhomogeneous magnetic semiconductors: theory of magnetic/nonmagnetic p–n junctions, *Physical Review Letters.* 88 (6) (2002) 66603.
[75] J. Fabian, I. Žutić, S. Das Sarma, Theory of spin-polarized bipolar transport in magnetic p–n junctions, *Physical Review B.* 66 (16) (2002) 165301.

[76] T. Kondo, J. Hayafuji, H. Munekata, Investigation of spin voltaic effect in ap-n heterojunction, *Japanese Journal of Applied Physics.* 45 (7L) (2006) L663.
[77] R. Jansen, Solar spin devices see the light, *Nature Materials.* 12 (9) (2013) 779–780.
[78] K.-R. Jeon et al., Thermal spin injection and accumulation in CoFe/MgO/n-type Ge contacts, *Scientific Reports.* 2 (1) (2012) 1–7.
[79] B. Endres et al., Demonstration of the spin solar cell and spin photodiode effect, *Nature Communications.* 4 (1) (2013) 2068.
[80] L. Esaki, New phenomenon in narrow germanium p–n junctions, *Physical Review.* 109 (2) (1958) 603.
[81] A.X. Gray et al., Bulk electronic structure of the dilute magnetic semiconductor Ga1–x Mn x As through hard X-ray angle-resolved photoemission, *Nature Materials.* 11 (11) (2012) 957–962.
[82] I. Žutić, J. Fabian, S. Das Sarma, Spin injection through the depletion layer: a theory of spin-polarized pn junctions and solar cells, *Physical Review B.* 64 (12) (2001) 121201.

7 Optoelectronics and Spintronics/Magnetism in Smart Thin Films for Memory Storage

7.1 INTRODUCTION TO MEMORY STORAGE

Memory storage based on optoelectronics and spintronics is an emerging field that combines the principles of optics, electronics, and spin-based technologies to develop innovative and efficient data storage solutions. While the search results provided do not specifically mention memory storage based on optoelectronics, they do provide insights into spintronics, which is a related concept.

Spintronics, also known as spin electronics, is a branch of electronics that focuses on the intrinsic spin of electrons as a fundamental property for carrying and storing information. In traditional electronics, information is encoded in the electrical charge of electrons, whereas in spintronics, the spin state of electrons is utilized to represent data. The spin of an electron can have two orientations: "spin up" (↑) or "spin down" (↓), which can be used to represent binary digits 0 and 1, respectively [1].

One potential application of spintronics in memory storage is magnetoresistive random access memory (MRAM). MRAM is a type of nonvolatile memory (NVM) that utilizes the spin of electrons to store data. It offers several advantages, such as fast read/write speeds, low power consumption, and nonvolatility, meaning it retains data even when power is removed [2]. MRAM has the potential to combine the advantages of various memory technologies, such as static (S) RAM, dynamic (D) RAM, flash, and hard disks, while overcoming their limitations. It aims to be a compact, high-speed, low-power, and universal memory solution [3].

On the other hand, optoelectronics is a field that deals with the study and application of electronic devices that interact with light. In the context of memory storage, optoelectronics can be utilized for optical storage, which involves using light to read, write, and erase data. Optical storage technologies such as optical discs (CDs, DVDs) and Blu-ray discs are examples of memory storage based on optoelectronics. These technologies use lasers to encode and retrieve data from microscopic pits on the surface of the discs [4].

DOI: 10.1201/9781003331940-7

7.2 DEVICE PRINCIPLES

Optoelectronic and spintronic memory storage are two emerging technologies that offer great potential for high-density, low-power memory storage. While both technologies operate on different principles, they share the goal of improving memory storage efficiency and capacity.

Optoelectronic memory storage is a type of memory technology that uses light to read and write data. The basic principle of optoelectronic memory is that the electrical conductivity of a semiconductor material can be changed by exposure to light. This effect is known as the photoconductivity effect. Memory storage consists of a semiconductor material, typically a thin film of silicon or a similar material, that is coated with a layer of a light-sensitive material. When light is shone on the material, the light-sensitive layer absorbs the photons and generates electron-hole pairs, which in turn alter the conductivity of the semiconductor. By controlling the amount and intensity of light, data can be written and read from the memory storage device.

Spintronic memory storage is a type of memory technology that uses the intrinsic spin of electrons to store and process data. The basic principle of spintronic memory storage is that electrons possess both charge and spin, and the spin can be used as a binary state (either up or down) to represent data. This type of memory storage consists of a nonmagnetic material, typically a thin film, that is sandwiched between two layers of ferromagnetic (FM) material. When a voltage is applied, the electrons in the FM material align their spins, creating a magnetic field. This magnetic field can be used to write data to the device. To read data, the magnetic field is detected by a sensor, which measures the resistance of the device [5].

7.3 DIFFERENT TYPES OF MEMORY STORAGE

Memory storage is an essential component in many electronic devices, and there are different types of memory storage technologies available. Here are some of the different types of memory storage, namely volatile and nonvolatile.

7.3.1 Volatile Memories

Volatile memory, in contrast to NVM, refers to computer memory that requires a constant power supply to maintain the stored information. It retains its contents while powered on, but when the power is interrupted or turned off, the stored data is quickly lost. Volatile memory is commonly used as primary storage in computer systems and has several advantages over NVM, such as faster access times. However, its main drawback is its data loss when power is removed. Volatile memory is often implemented using technologies such as dynamic random access memory (DRAM) and static random access memory (SRAM). DRAM stores data using stored capacitance, and SRAM relies on active bistable latches. These technologies enable fast data access but do not retain data without a continuous power supply [4,5]. The data stored in volatile memory remains intact only as long as the device is powered. Once the power is interrupted for any reason, the data is lost, and the memory is reset to its initial state [3]. For example, DRAM has a relatively short retention time of about

100 milliseconds, meaning it requires constant refreshing to maintain data integrity [5]. This is why RAM is used for temporary storage of data that the CPU is actively using, and data that needs to be stored permanently is typically stored in NVM, such as a hard disk drive or flash memory.

7.3.2 Nonvolatile Memories

NVM is a type of computer memory that can retain data even without a power supply. NVM encompasses various technologies, including optical memory and MRAM, that provide persistent storage even without a continuous power supply.

Optical nonvolatile memories are a relatively recent addition to light-enabled memory technology. They take advantage of the advancements in the field of phase-change technology [6]. These memories utilize the properties of specific materials that can switch between amorphous and crystalline states when exposed to light or heat. The different states represent the binary values of 0 and 1, allowing for data storage. Optical nonvolatile memories offer potential benefits such as high storage density and fast access times.

MRAM is another type of NVM that is highly regarded for its high-density storage, endurance, and fast writing speed [7]. MRAM utilizes magnetic elements to store data. It employs a combination of magnetic tunnel junctions (MTJs) and magnetoresistive effects to achieve nonvolatile storage characteristics. MRAM has applications across different categories, including low-bandwidth scenarios where it can provide efficient and persistent data storage.

These two types of NVMs, optical memory and MRAM, represent specific advancements in the field of NVM. Optical memory relies on phase-change technology and the reversible switching of materials, while MRAM utilizes magnetic elements and magnetoresistive effects [6,7]. These technologies contribute to the development of storage-class memory (SCM), embedded NVMs, and near/in-memory computing applications [8].

7.4 OPTOELECTRONICS IN SMART THIN FILMS FOR MEMORY STORAGE

Optoelectronic memory, also known as phase-change memory (PCM) and ovonic threshold switching (OTS), is a type of NVM that combines optical and electronic properties to store data. It is based on chalcogenide glass materials, which can switch between a crystalline and amorphous state when exposed to light or heat [9]. NVM applications utilize differences in electrical resistivity through the application of a SET pulse, a long-voltage pulse, to the high-resistance amorphous region. This results in heating and subsequent recrystallization. Conversely, a RESET pulse, a high-voltage pulse, is applied to the low-resistance crystalline state, causing it to become liquid. Rapid quenching of this liquid state leads to the formation of an amorphous region [9]. PCMs find two main applications in PCRAM and optical phase-change storage, such as compact disks (CDs) and digital versatile disks (DVDs), where a laser pulse or a current pulse is used for heating in optical applications and PCRAM cells, respectively. Figure 7.1 illustrates the principle of switching in PCMs.

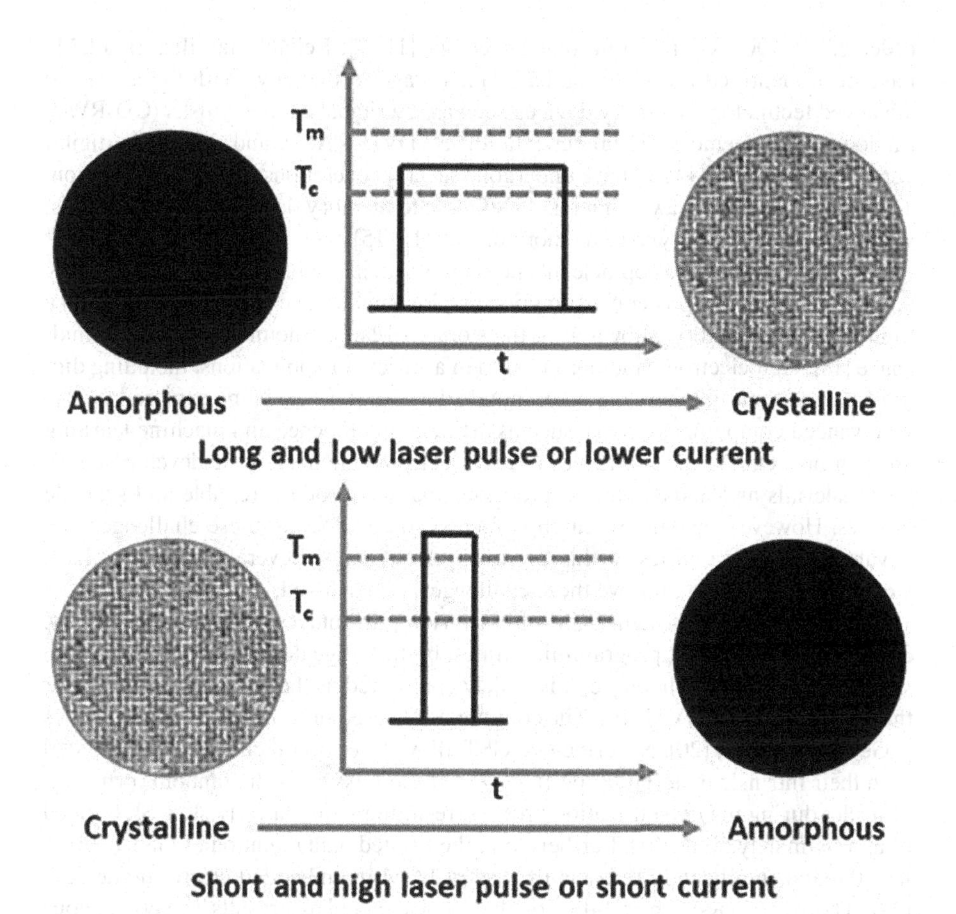

FIGURE 7.1 Schematic of the PCM process. T_c, T_m and t are the crystallization temperature, melting temperature, and time, respectively.

The quenching process is limited by power, as sufficient power is necessary to raise the material above its melting temperature (T_m). The T_m of phase-change materials (PCMs) typically falls within the range of 500–800°C [10]. Optical storage devices and PCRAM use low-power laser pulses for reflectivity and short current pulses for resistivity during the reading operation. PCMs have significant potential for technological applications, but achieving the necessary material parameters can be challenging. PCM-based storage technologies require high stability and fast crystallization of the amorphous phase, making it difficult to attain the specific parameters required for certain applications. As a result, only a few PCMs are utilized in technological applications. Although various materials can be melt-quenched to produce an amorphous state, only a few have distinguishable optical and/or electrical contrast between the amorphous and crystalline states. Consequently, a few PCMs possess unique combinations of properties that are suitable for optical data storage. The difference in atomic arrangement between crystalline and amorphous phases is indicated by optical contrast, necessitating a fast recrystallization process on the order of 10–100 ns for atomic rearrangement [11,12]. Reliable families of PCMs have been identified for use in optical data storage technology. With the advent of advanced technology, memory devices such as rewritable compact disks (CD-RWs), random access memory digital versatile disks (DVD-RAM), and rewritable digital versatile disks (DVD±RW) have undergone several generations of product evolution. This has generated renewed interest in solid-state memory devices based on PCMs, which represent a new type of memory device [13–15].

The key advantage of optoelectronic memory is its ability to store data quickly and reliably with low power consumption and high endurance [16,17]. It also offers a high level of scalability, allowing for the storage of large amounts of data in a small space [18]. Optoelectronic memory is used in a variety of applications, including digital cameras, smartphones, and other mobile devices. It is also being explored for use in advanced computing systems, such as artificial intelligence and machine learning applications. One of the challenges of optoelectronic memory is the development of new materials and manufacturing processes that can produce reliable and scalable devices. However, ongoing research is focused on addressing these challenges and advancing the capabilities of this promising technology. Several researchers have worked in this field to improve these challenges, as discussed.

Based on artificial design, PCM offers various advantages, such as nonvolatility, efficient scalability, fast programming times, high storage density, and the ability to store multilevel data. Therefore, it is widely considered as the primary contender for the upcoming era of NVM [19]. The common PCM used in commercial PCM devices is $Ge_2Sb_2Te_5$ (GST) [20]. Nevertheless, GST alloys face several challenges associated with their intrinsic material properties. One of these issues is the random formation of nuclei during the crystallization process, resulting in a relatively slow SET speed of approximately 50 ns [21]. Furthermore, the limited data retention of GST (~90°C for 10 years) significantly restricts the use of PCM in embedded electronic devices [22]. The pure long-term stability of the amorphous phase results in spontaneous structural relaxation, leading to resistance drift over time. The aforementioned issue restricts the manufacture of PCM cells with multi-bit technology, which is crucial for achieving stable, multilevel states in photonic memory devices [23]. On the other hand, Ge, Sb, and Te atoms can migrate toward the cathode and anode in opposite directions, respectively, after multiple SET to RESET cycles [23,24]. This can result in compositional segregation and significant fluctuations in resistance [23,25].

New findings have demonstrated that single-element Sb PCM may be utilized to address the issue of composition separation. The pristine Sb material possesses an A7-type structure where four elements are randomly positioned at 6C in the space group [26]. Although Sb has a hexagonal crystal structure linked by isotopes p-p, which enables it to achieve an instantaneous phase transition [26], the Sb PCM exhibits poor thermal stability and experiences rapid crystallization at RT [27]. Electronic devices typically operate at internal temperatures of 60–70°C [28], and as a result, single-element Sb is inadequate for meeting the requirements of data storage applications.

One practical approach to improving thermal stability is to reduce the thickness of the Sb thin film [29,30]. Researchers have identified that by intercalating a 4-nm-thick Sb film in a SiO_2 layer, the crystallization temperature can reach 154°C,

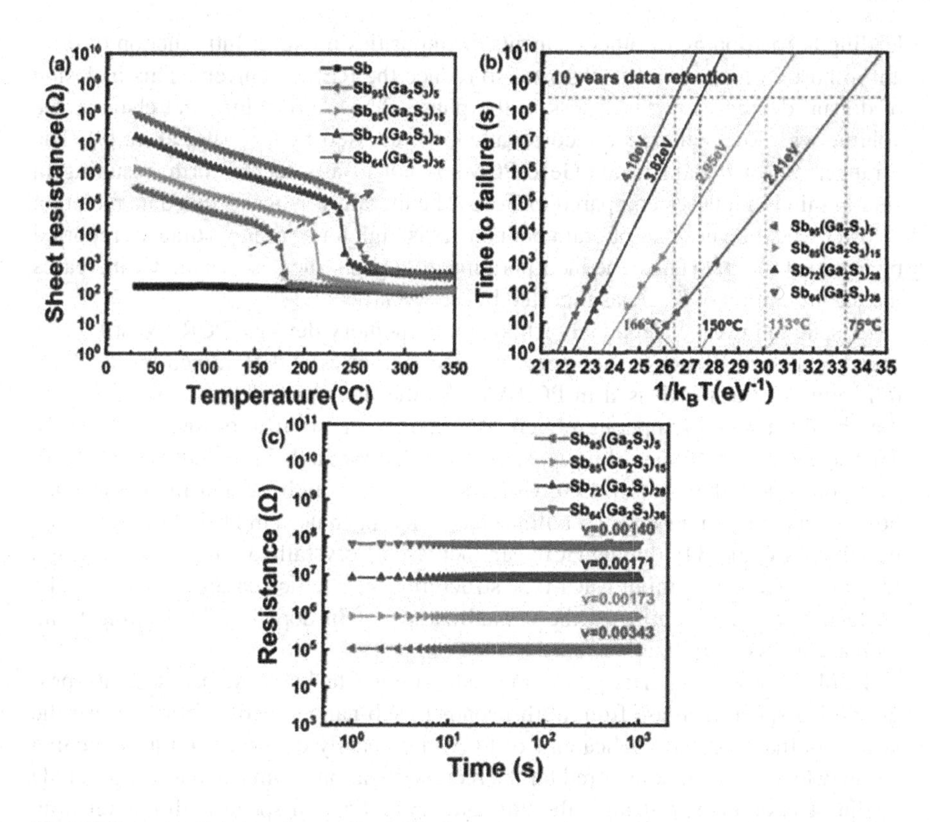

FIGURE 7.2 (a) Pristine Sb and Sb-Ga_2S_3 films resistance variation against temperature at a heating rate of 30°C min^{-1}. Ga_2S_3-Sb films (b) arrhenius fitting plots of ten-year data retention temperature, (c) resistance drift coefficient with heat treatment at 50°C [31].

and the data retention temperature over a period of 10 years can be maintained at 48°C. The activation energy of crystallization for this film was calculated to be 2.08 eV [16], resulting in a significant enhancement in the amorphous thermal stability of the Sb film. The properties of pure Sb thin films alloyed with Ga_2S_3 were investigated by Jiao et al. [31]. The formation of high binding energy of Sb-S and Sb-Ga bonds was said to improve the thermal parameters, leading to a significant enhancement of the thermal stability of the thin films (see Figure 7.2a). It was found that the Ga_2S_3-Sb films exhibit a higher crystallization temperature of ~250°C, ~166°C larger 10-year data retention temperature, and 4.10 eV higher crystallization activation energy when compared to pure Sb films (see Figure 7.2b). The enhanced thermal stability of thin films also results in a significantly lower (0.0014) resistance drift coefficient (see Figure 7.2c). By analyzing the crystallization mechanism, it was determined that the kinetic index (n) is 0.625, indicating a one-dimensional growth type in the thin films. This helps reduce nucleation randomness and decrease resistance drift.

The reliability and performance of devices are crucial for comparison with modern NVM technologies. As the density of memory devices improves, the reduction of current needed for RESET and the stability of the amorphous state must be improved to boost data retention in PCM. These issues have an impact on GeTe PCM devices, leading to the doping of other elements or impurities in GeTe. Introduction of dopant in host GeTe material is observed to reduce the RESET current. This inclusion of dopant decreases the heat loss in the phase-change procedure and changes the volume with low conductive incorporation, as reported by Czubatyj et al. [9]. The arrangement of dopant inside GeTe PCMs is not in an ordered form, resulting in small grain boundaries compared to GeTe. The literature suggests that data retention is lower when the number of grain boundaries is high for a given volume, as reported by Russo et al. [32]. Thus, the incorporation of dopants such as carbon (C) increases the data retention of GeTe, as reported by Beneventi et al. [9].

Despite being an exceptional technology in memory devices, PCRAM still faces speed-related challenges. The speed of these memories is determined by the crystallization of the PCM used in PCRAM. As the thickness of the films decreases, the phase transition temperature increases, and the crystallization speed decreases. This suggests that thinner films experience a reduction in their crystallization speed, as reported by Cheng et al. [33]. GeTe, a PCM with high Tc and high resistance between its amorphous and crystalline phases, has been used in PCRAM. According to Szkutnik et al. [34], doping GeTe can increase its crystallization speed. Although InTe and GeTe have similar NaCl-type structures, their lattice parameters differ. The nucleation process dominates the crystallization of In-doped GeTe, leading to an increase in its crystallization speed.

PCM, a promising next-generation data storage technology, has seen its programming speed increase from milliseconds to sub-nanoseconds ($\approx$500 ps) over the years. As the potential applications of PCM are heavily dependent on its switching speed, which is the time required for the recrystallization of amorphous chalcogenide media, it is crucial to identify the ultimate crystallization speed both theoretically and practically. Shen et al. [35] used a systematic analysis to discover a PCM and ab initio molecular dynamics simulations to predict that elemental Sb-based PCM has an exceptionally fast crystallization speed. Experimental cells were shown to have extremely fast crystallization speeds of only 360 ps, which are further accelerated when the device is shrunk. In fact, a record-breaking crystallization speed of only 242 ps was achieved in 60 nm-size devices. Their findings presented new opportunities for using appropriate storage materials to develop DRAM-like and even cache-like PCM.

Low-cost SCM can be achieved through the cross-point array architecture, but it requires the use of a selector, such as an OTS, in series with the memory element. This additional component increases the complexity of the bit cell [36]. However, the effectiveness of PCM for SCM applications is limited by its large RESET current [36]. Therefore, there is a need to develop a new, easy-to-fabricate, self-selecting memory cell that can overcome the limitations of PCM. Ravsher et al. observed the polarity-induced threshold voltage shift in chalcogenide-based OTS devices [37]. This effect is robust, reproducible, and semi-persistent, enabling the realization of memory functionality within the OTS selector itself. As a result, a self-rectifying cell is formed. The study conducted by Ravsher et al. demonstrated memory operation in SiGeAsSe OTS, which exhibited excellent performance characteristics including ultralow IWrite ($<$15 μA, 0.6 MA cm^{-2}), good write endurance ($>10^8$ cycles), fast switching ($\sim$10 ns), strong non-linearity ($NL^{1/2} > 10^4$), reasonable retention time ($>$1 month), and non-destructive read ($>10^6$ reads). These features, along with better manufacturability and scalability, make this concept a promising option for SCM applications. To further enhance the performance, device structure and material optimization can be explored [38]. However, to fully comprehend the physical mechanisms underlying the polarity effect and to identify fundamental trade-offs that limit device performance, additional research is necessary.

Choi et al. developed a p-1T-1P (parallel 1 transistor-1 phase-change memory cell) device by fabricating a parallel-connected IGZO-TFT (Indium Gallium Zinc Oxide Thin-Film Transistor) and line-cell GST225 PcRAM (Phase-Change Random Access Memory) device [39]. The IGZO and GST225 layers were deposited using sputtering and atomic layer deposition (ALD) techniques, respectively (see Figure 7.3a). To minimize cross-effects, a SiO_2 layer was implemented to separate the IGZO channel and active region of GST, while the ITO (Indium Tin Oxide) source and drain contacts were shared. Despite a slight decrease in the threshold voltage (Vth) of the IGZO-TFT caused by the GST ALD process involving NH_3, the device functioned smoothly and met the critical requirement of "$R_{ON} < R_{set} < R_{reset} < R_{OFF}$" for the multiply-stacked vertical TFT-PcRAM configuration. The device exhibited a switching endurance of around 10^5 cycles (see Figure 7.3b), which was comparable to current vertical NAND devices [39]. Notably, the TFT not only served as a selector but also manipulated the switching and resistance behavior of the PcRAM. The field effect from the TFT had the capability to influence the carrier distribution in the semiconducting GST layer, allowing the TFT resistance to replace the reset resistance of the PcRAM. This unique characteristic enabled unprecedented control over the properties of the PcRAM, facilitating the achievement of multiple memory states [39]. However, device failure was attributed to the reset-stuck property caused by the electromigration of the Ge/Sb and Te ions [40–42]. When projected to the ~130 nm channel-hole diameter and 25 nm layer thickness in the vertically stacked structure, the device's properties ensured a ~400-layer stacking, making it even more competitive than the current vertical NAND flash device.

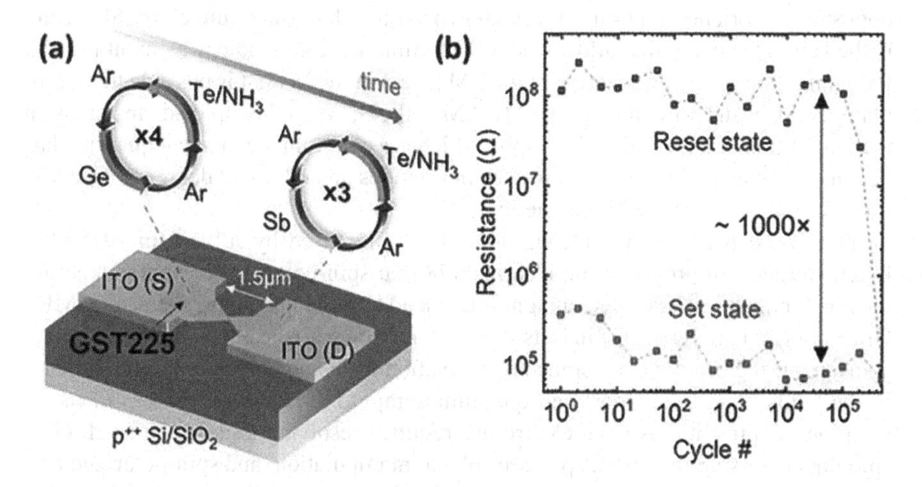

FIGURE 7.3 GST225 films (a) ALD sequences of gas injection, (b) cycling endurance results of the device [39].

7.5 SPINTRONICS/MAGNETISM IN SMART THIN FILMS FOR MEMORY STORAGE

MRAM is a type of NVM that stores data by utilizing the spin of electrons. The spin of an electron is a fundamental property that describes its intrinsic angular momentum, and it can be either "up" or "down," with the spin having the values of "+1/2" or "–1/2": an eloquent invitation to use it to encode information, in analogy to bits "0" and "1" of the binary code [43].

In an MRAM device, the magnetic state of a thin-film MTJ is used to store information. An MTJ consists of two FM layers separated by a thin, nonmagnetic layer. When a voltage is applied to the MTJ, electrons tunnel through the nonmagnetic layer from one FM layer to the other. The resistance of the MTJ depends on the relative orientation of the magnetization of the two FM layers. When the magnetization of the two layers is parallel, the resistance of the MTJ is low, and when the magnetization is antiparallel, the resistance is high [44]. To write data to an MRAM cell, a current is passed through a wire that runs above or below the MTJ. The current creates a magnetic field that can be used to switch the magnetization of one of the FM layers in the MTJ [45]. This changes the relative orientation of the two layers and modifies the resistance of the MTJ. The new resistance state can be read out by passing a small current through the MTJ and measuring the resulting voltage. There are claims that MRAM has the potential to surpass the speed of SRAM, the density of DRAM, and the nonvolatility of flash memory [45]. Furthermore, MRAM, being a nanoscale device, has low power consumption and generates less heat. MRAM represents an improved version of SRAM and DRAM, as it stores data using spin instead of electrical charges [3]. This addresses one of the drawbacks of conventional RAM, which is the loss of information during power failures.

The spin of electrons plays a crucial role in the operation of an MRAM device. When a current is passed through the MTJ, electrons with a particular spin orientation are more likely to tunnel through the non-magnetic layer than electrons with the opposite spin orientation [46]. This is known as spin-dependent tunneling (SDT) and is the key mechanism that allows the MTJ to function as a memory element [47,48]. By controlling the magnetization of the FM layers in the MTJ, it is possible to manipulate the SDT and write data to the MRAM cell [43]. With the rapid advancement in the fabrication of nanoscale MTJs, spin, which is a quantum mechanical quantity, has become a topic of significant interest. Spin provides an additional degree of freedom for electrons to interact with magnetic fields.

The first experimental evidence of SDT was reported by Julliere in 1975 [49]. Later, Berger [50] proposed the idea in 1978 that spin-polarized current can act on the local magnetization of ferromagnets and lead to giant magnetoresistance (GMR). One important property of spin is its weak interaction with the environment and other spins, resulting in a long coherence or relaxation time, which is a crucial parameter in the fields of spin transport and quantum computing. However, successful incorporation of spins into existing electronics requires resolving issues such as efficient spin injection, spin transport, spin control and manipulation, and spin-polarized current detection. Spintronics without magnetism is an attractive approach for designing semiconductor spintronic devices, as spin–orbit coupling (SOC) enables spin generation and manipulation solely by electric fields. By applying an electric field, electrons in the lattice generate a magnetic field that affects spin. The theoretical proof of spin–orbit interaction on mobile electrons dates back several decades, but practical implementation of this concept is still in its early stages.

According to research, nanostructure materials made from carbon are considered better options for spintronics compared to traditional metals and semiconductors because they have longer spin relaxation distances. This means that spin currents can travel over large distances in materials such as graphene and carbon nanotubes (CNTs) due to weaker spin orbits, and hyperfine couplings in carbon-based materials [51]. Additionally, these materials are also relatively easy to use in the production of large-scale spintronic logic circuits. Carbon-based spintronics is an exciting and promising field, particularly for creating fast memory devices, logic gates, transistors, and capacitors for computers, tablets, and handheld devices [52,53]. This is a highly attractive alternative to current technologies since the synthesis and processing of carbon-based nanostructures are comparatively inexpensive and straightforward. However, carbon materials also lack magnetic moments, which makes them diamagnetic.

7.5.1 Spintronic in CNTs

7.5.1.1 Spin Injection, Transport, and Relaxation

In their study, Tsukagoshi et al. [54] created the first two-terminal CNT spin-valve device, which demonstrated spin-dependent transport through multi-walled carbon nanotubes (MWCNTs) and exhibited a 9% magnetoresistance (MR) at 4.2 K. The MWCNTs used in the study were synthesized from graphite rods via arc discharge and had 65-nm-thick Co electrodes deposited on them to inject spin current into the MWCNTs. Later studies explored spin-dependent transport properties in single-walled carbon nanotubes (SWCNTs) or SWCNT networks, using either local or nonlocal geometry [55–57]. Jensen et al. [58] measured different two-terminal devices with various electrodes at low temperatures and observed a large MR of almost 100% and down to –150% in SWCNT devices with two Fe electrodes. However, a 10% MR was also observed in devices that only had one magnetic electrode, which is a confusing phenomenon that requires further investigation. Tombros et al. [56] conducted nonlocal spin transport measurements in SWCNTs, separating the charge current from the spin current and providing evidence of spin accumulation-induced MR in SWCNTs.

The spin injection into CNTs can be limited by the conductance mismatch problem, as discussed later in the text. In graphene spin-valve devices, a spin-dependent tunnel barrier can be introduced to improve spin injection efficiency. However, unlike in graphene, where a 1–2 nm MgO or Al_2O_3 tunneling barrier is typically required at the interface between FM electrodes and graphene, CNT-based spin-valve devices usually do not need an artificial tunneling barrier to observe the hysteretic MR. One speculation for this difference is that the small diameter of CNTs helps in forming a weak coupling between CNTs and FM electrodes, leading to the formation of a spin-dependent potential barrier and tunneling behavior of electrons, which is an effective way for spin injection [54,55,59]. Another way to improve MR is by enhancing the spin polarization of the spin injector, which can be achieved using various FM or half-metal electrodes such as Co [4], $La_{0.7}Sr_{0.3}MnO_3$ [59], Fe [58], and Ni [60], each with different spin polarizations that have been measured by tunneling experiments or point contact measurements. Hueso et al. [59] used $La_{0.7}Sr_{0.3}MnO_3$ as electrodes to inject spin current into the MWCNT and observed a large MR of up to 61% with a spin diffusion length of about 50 μm at 5 K. This large MR was attributed to the high spin polarization in their manganite electrodes, resistance of the interfacial barrier for spin injection, long spin lifetimes in CNTs, and high Fermi velocity in CNTs.

In previous studies [61–63], several theoretical spin relaxation mechanisms have been proposed for CNTs. Semenov et al. considered the hyperfine interaction with disordered nuclei spins ($I = 1/2$) of $C1_3$ isotopes in semiconducting CNTs, predicting a spin relaxation time of approximately 1 s at 4 K. However, experimental observations indicate much shorter spin relaxation times in the range of tens of nanoseconds. Borysenko et al. explored the anisotropy of the g tensor and flexural phonon modes in semiconducting CNTs and found that the spin relaxation time can reach tens of microseconds at room temperature (RT) [61]. This suggests that spin relaxation times can vary significantly depending on specific factors such as material properties and temperature. Additionally, the gate modulation properties of MR in SWCNTs and MWCNTs have also been investigated. However, specific details and results regarding these investigations were not provided in the search results [60,64,65]. Gunnarsson et al. [66] conducted a study on the gate-voltage influence on the nonlocal spin signal. They fabricated nonlocal spin-valve devices with SWCNT and observed quantum dot behavior at low temperatures. The results showed that the nonlocal voltage exhibited significant oscillations around zero with an amplitude of approximately 1 μV. Makarovski et al. [67] also observed the same phenomenon in their previous study.

7.5.1.2 Tunnel Magnetoresistance in CNTs

Since the discovery of tunnel magnetoresistance (TMR), numerous nanodevices, such as single-electron transistors and field-effect transistors (FETs), have been successfully demonstrated. However, these devices rely on the charge of the electron. On the other hand, the magnetoresistive effect known as TMR occurs in MTJ. There is a growing interest in TMR due to the potential technical advancements it could offer for the realization of nanoscale devices in more advanced spintronic applications, particularly those involving SDT. The first reproducible TMR up to 24% at RT was fabricated by Moodera et al. in 1995 [68] using $CoFe/Al_2O_3/Co$ or NiFe junctions. Since then, TMR values of up to 50% have been achieved with three-dimensional ferromagnets, making them suitable for industrial applications, as reported by Parkin et al. [69].

TMR characteristics have also been experimentally and theoretically measured in CNTs, as demonstrated by Jensen et al. [58] and Tsukagoshi et al. [54], respectively. Spintronic devices exhibiting TMR using ferromagnet-contacted SWCNTs have been successfully demonstrated by Jensen et al. However, most of the studies on CNT-TMR systems have focused on single CNTs contacted to bulk FM materials using ex-situ methods [70–72]. It is worth noting that the TMR effect is sensitive to the interface between the tunnel barrier and the electrode, and this sensitivity may be more pronounced in systems involving single CNTs. For instance, De Teresa et al.

[73] found that the type of barrier material used can even affect the sign of the TMR. Yuasa et al. [74] also investigated the effect of crystal anisotropy of the spin polarization on MTJs using single crystal iron electrodes with various crystal orientations, and they observed a clear dependence of TMR on the crystal orientation, which could be attributed to the crystal anisotropy of the electronic states in the electrodes. It is known that TMR/GMR originates from spin interactions between magnetic and nonmagnetic particles at the interface and is related to the coercivity value, as reported by Bergenti et al. [75].

CNTs are carbon-based molecular tubes known for their exceptional properties [76]. They possess high stiffness and strength, unique electronic behavior, and other remarkable properties. Due to their nanoscale size, long spin-flip scattering lengths, and ability to act as one-dimensional quantum conductors, they are considered promising for spintronic devices [77,78]. Experiments have shown that spin can be coherently transported over a distance of up to 130 nm in Co-contacted CNTs [54]. Sahoo et al. [65] investigated the gate dependence of TMR in CNTs with PdNi electrodes at a low temperature of 1.85 K. The TMR was found to oscillate between –5% and +6% in a relatively regular pattern with respect to gate voltage ($\Delta V^{TMR}{}_{g}$) ranging from 0.4 to 0.75 V. Furthermore, MWCNTs conductance was investigated at even lower temperatures of 300 mK to determine the single-electron states that could not be distinguished at the temperature at which the MR was measured. The differential conductance (dV/dI) was also measured as a function of source-drain voltage (V_{sd}) and gate voltage (V_g) within a narrow range of V_g.

The characteristic diamond-like pattern that signifies single-electron tunneling in a quantum dot is observed. The size of the diamonds varies, with the energy required for adding a single electron ranging between 0.5 and 0.75 meV, which is consistent with previous studies on quantum dots made from MWCNT with non-ferromagnetic (non-FM) leads, as reported by Buitelaar et al. [79]. These quantum dots exhibit Coulomb blockade and energy level quantization. The electron levels in the quantum dots are almost fourfold degenerate, considering the spin of the electrons, and their behavior in a magnetic field (Zeeman splitting) is consistent with a g factor of 2. However, the gate-voltage scale $\Delta V^{TM}{}_{g}$ measured at a temperature of 1.85 K is much larger than the scale $V^{e}{}_{g}$ of approximately 25 mV for adding single electrons. Instead, it corresponds to the addition of at least 16 electrons, rather than just one. A gate-voltage scale that matches the TMR signal becomes visible when the linear conductance G at low temperatures is examined over a wider range of gate voltages. The peaks in single-electron conductance are strongly modulated in amplitude, resulting in a regular beating pattern with a gate-voltage scale of ΔV_g of about 0.4 V. The conductance G and TMR of a SWCNT device was measured by Sahoo et al. [65,80]. The behavior of the quantum dot is observed at a higher temperature of 1.85 K, compared to 0.3 K in the MWCNT device, indicating that single-electron charging energy and level spacing are higher in SWCNTs compared to MWCNTs. The typical single-electron addition energy in SWCNTs is around 5 meV, which is an order of magnitude larger than that in MWCNTs. These behaviors are attributed to quantum interference in CNTs, as reported by Schäpers et al. [81]. The TMR changes sign on each conductance resonance, and the line shape of the conductance resonances is symmetric, while that of the TMR dips is asymmetric. There is a jump in the G (V_g) data at $V_g = 4.325$ V.

The amplitude of the TMR ranges from –7% to +17%, which is above what was observed in MWCNTs. This could be due to the higher charging energy in SWCNTs, as suggested by Barnas et al. [82]. Normal-SWCNTs-Ferromagnetic (N-SWCNTs-F) devices exhibit an order of magnitude lower signal, indicating that the current in the ferromagnetic-tube-ferromagnetic (F-tube-F) devices is indeed spin-polarized. Gunnarsson et al. [66] investigated the gate-voltage dependence of the nonlocal spin signal. Nonlocal spin-valve devices made from SWCNTs were fabricated, and these devices showed clear quantum dot behavior at low temperatures. They found that the nonlocal voltage oscillates around zero with a large amplitude of about 1 μV, a phenomenon also described in 2007 by Makarovski et al. [67].

7.5.1.3 MR in CNTs

The MR study of SWCNTs using ferromagnetic (FM) electrodes was conducted by Mohamed et al. [83]. The coercive force obtained from the hysteresis curve was found to be approximately 120 Oe for both temperatures. The resistance of SWCNT devices at RT was measured to be in the range of 10–200 Ω. SWCNTs with high resistance exhibited favorable FET characteristics, indicating that the channels likely consisted of metallic or semiconducting SWCNTs rather than other conductive carbon materials. In the study, the resistance remained relatively constant at around 186 Ω against the magnetic field for a channel width of approximately 500 μm at 300 K but showed a resistance peak of around 0 Oe at 4.5 K with slight variation in the direction of the sweep. When the magnetic field was swept upward, a peak appeared at approximately –110 Oe, while a peak appeared at around 110 Oe in the downward direction. Similar behavior was observed in a device with a channel width of approximately 250 μm. To rule out the possibility that the MR effect was influenced by the MR of Co nanoparticles on the electrodes, a control experiment was conducted by measuring the MR effect for one of the electrodes at the same temperatures, and no significant change in resistance was observed.

The origin of hysteretic MR in Co nanoparticles is not intrinsic to the nanoparticles themselves, as indicated previously. Despite the temperature-independent behavior of the FM component of Co, MR effects were observed at low temperatures. The resistance of a material connected to two FM electrodes for spin transport without spin scattering is high when the FM moments in the two electrodes are antiparallel and vice versa, which was originally proposed by Julliere in 1975 [49] and can explain the spin-valve effects observed in MWCNTs and SWCNTs, as reported by Sahoo et al. [65,80] and Nagabhirava et al. [60], respectively. SWCNTs are believed to grow from Co nanoparticles that are formed from a Co thin film on Mo film using alcohol catalytic chemical vapor deposition (ACCVD), as reported by Inami et al. [84]. Therefore, it is expected that the spin-dependent transport in SWCNTs is governed by the magnetic properties of the attached Co nanoparticles and strongly influenced by their size. Since each SWCNT is attached to a Co nanoparticle of varying size, it is anticipated that each SWCNT will exhibit MR peaks at different field values, corresponding to changes in the parallel and antiparallel magnetization configurations of the electrodes. However, determining the magnetic properties of individual Co nanoparticles in contact with SWCNTs remains challenging. The observed MR effects of Co nanoparticles exhibited sharp peaks at approximately ±110 Oe, which corresponded to the average coercive force measured by the SQUID. The origin of this unusual MR hysteresis remains unclear, but the results of the MR effects were found to be reproducible. A detailed analysis of the spin transport mechanism in these devices is planned for further investigation. The ratio of MR, defined as $\Delta R/R_0$, where R_0 is the resistance in the saturation region, was found to be in the range of 0.7–1.8% at 4.5 K. This value is consistent with a previous report on the MR of SWCNTs [85]. To increase $\Delta R/R_0$, it is necessary to achieve effective spin injection from the FM electrode to the SWCNTs and spin-coherent transport in the SWCNTs. Improving the quality of the Co/SWCNT interface is crucial for spin injection, while growing high-quality SWCNTs and reducing the distance between the electrodes (L) are important factors for achieving spin-coherent transport. Furthermore, the dependence of $\Delta R/R_0$ on L can be utilized to elucidate the spin diffusion length of SWCNTs, which adds an interesting aspect to the study.

7.5.1.4 Magnetism in CNTs

In Titus et al.'s (2011) study, they explored the magnetic properties of vertically aligned carbon nanotubes (VCNTs) that contained nickel nanoparticles, with the intention of determining their potential use in MTJ spintronic applications. They employed a superconducting quantum interface device (SQUID) to assess the field-dependent magnetization behavior [86]. Their findings demonstrated that the VCNTs exhibited FM behavior, with hysteresis loops observed at temperatures of both 2 and 300 K. The *M-H* hysteresis loops at both temperatures were used to obtain the saturation magnetization (M_s), coercivity (H_c), and remnant magnetization (M_r) values. The M_s values and H_c for the VCNTs with deposited nickel nanoparticles decreased from 5.3 to 4.4 emu/g and 395 to 115 Oe, respectively, as the temperature increased, which suggested characteristic FM behavior.

To gain a deeper understanding of the magnetic behavior, the researchers performed zero-field cooling (ZFC) and field cooling (FC) magnetization measurements. In the ZFC measurement, the sample was cooled from 300 to 2 K in a zero magnetic field, while in the FC measurement, the sample was cooled in a magnetic field (25 G) from 300 to 2 K, and then the magnetization was measured during the warming cycle while maintaining the field. The temperature dependence of the ZFC and FC measurements under the applied magnetic field of 25 G exhibited the main features of FM behavior [87]. The ZFC curve displayed a peak at 44 K, which corresponds to the blocking temperature (TB), indicating the transition temperature from the FM state to the superparamagnetic state. The low value of TB can be attributed to the small size of the randomly deposited nickel nanoparticles on the VCNTs [88,89].

Oke et al. [90] modified the properties of MWCNTs for potential electronic and magnetic applications by depositing silicon-oxide-nanoparticle (SiO_2-NP) heteroatoms onto the MWCNT surface. The magnetic properties were studied using SQUID, field-dependent magnetization, and electron spin resonance (ESR) techniques. The M_s of these materials are presented in Figure 7.4a-d. The M_s of SiO_2-NPs (1.5 at%) experiences a slight increase but decreases in SiO_2-NP (5.75 at%) (see Figure 7.4c and d). This decrease in M_s can be attributed to the formation of Si-C bonding in addition to Si-O bonding, resulting in a material rich in sp^3 [91,92]. The coercivity of the MWCNT deposited with 1.5 at% SiO_2 (Hc@40 K = 689 Oe) was higher than that of those decorated with 5.75 at% SiO_2 (Hc@40 K=357 Oe). As shown in Figure 7.4f and g (temperature dependence magnetization after FC and ZFC from 300 to 40 K), MWCNTs and SiO_2-NPs (1.5 at%) exhibit FM behavior, consistent with the observed *M-H* hysteresis loops in Figure 7.4b and c, respectively. From ESR, a resonant microwave absorption signal (H_r) is prominently observed at approximately 3,200 G/1,600 G for high/low field (see Figure 7.4i–l). The linewidth, along with an effective *g*-value for SiO_2 (1.5 at%), is greater than that for pure SiO_2-NPs, MWCNTs, and SiO_2-NPs (5.75 at%). This suggests that SiO_2-NPs (1.5 at%) have a higher magnetic phase in MWCNTs [93], which is comparable to values reported elsewhere for other carbon forms [94].

The M_s of nitrogen-doped multiwall nitrogenated carbon nanotubes (MW-NCNTs) that were functionalized with chlorine and oxygen plasma were studied by Ray et al. [95,96]. The results showed that at RT, nonfunctionalized MW-NCNTs had diamagnetic behavior, while chlorine- and oxygen-functionalized MW-NCNTs exhibited paramagnetic and FM behavior, respectively. The *M-H* hysteresis loop of MW-NCNTs was obtained at 300 and 5 K. Despite the presence of strong magnetic Fe particles as a catalyst in the NCNTs, the spectral features indicate a pure diamagnetic behavior. Similar observations were made by Lipert et al. [97] for Fe-based MWCNTs after postannealing at a high temperature of approximately 2,500°C, where the FM behavior changed to diamagnetic due to the complete evaporation of Fe-catalyst particles from the CNTs. They also reported diamagnetic behavior for the CNTs using nonmagnetic Fe as a catalyst [97]. According to Ray et al. [95], the diamagnetic behavior exhibited by MW-NCNTs may be attributed to the presence of nonmagnetic bonding that dominates the Fe catalyst in the structure. The authors also noted that oxygen-functionalized NCNTs display strong FM behavior in contrast to nonfunctionalized NCNTs, as evidenced by the *M-H* hysteresis loop. This finding is consistent with Del Bianco et al. [98], who observed FM behavior in the core interface of oxygen-passivated Fe nanoparticles. Ray et al. [95,96] further proposed that the observed FM behavior in oxygen-plasma-treated NCNTs may be due to oxygen passivation.

In the case of chlorine-plasma-treated NCNTs, the *M-H* loops do not resemble those of pure NCNTs or NCNTs:O but rather represent an intermediate (dia- and ferro-) phase that suggests the possibility of paramagnetic behavior. Additionally, the authors suggested that these changes in magnetic behavior may be due to the formation of different bonding between carbon/nitrogen/Fe catalysts on chlorine-/oxygen-plasma functionalization. Finally, Ray et al. [95,96] investigated the thermal evolution of the magnetization of MW-NCNTs (:Cl/O) by performing ZFC and FC procedures in an applied magnetic field of 1000 Oe between 5 and 300 K. The authors observed that the ZFC curve gradually deviates from the FC curve as the temperature decreases to 255 K for MW-NCNTs, 200 K for CNTs:Cl, and 300 K for CNTs:O when the applied magnetic field is 1,000 Oe. Upon further cooling, the MZFC curve displays a cusp centered at 45 K for MW-NCNTs and NCNTs:O, but not for NCNTs:Cl. The authors noted that this variable temperature magnetic data indicates that the Fe/NCNTs exhibit FM behavior below RT due to the presence of uncompensated surface spin states or FM Fe clusters. This FM behavior is not evident in the *M-H* curve of nonfunctionalized MW-NCNTs, which behaves completely diamagnetically. The authors attributed the FM performance of Fe/NCNTs:O to the

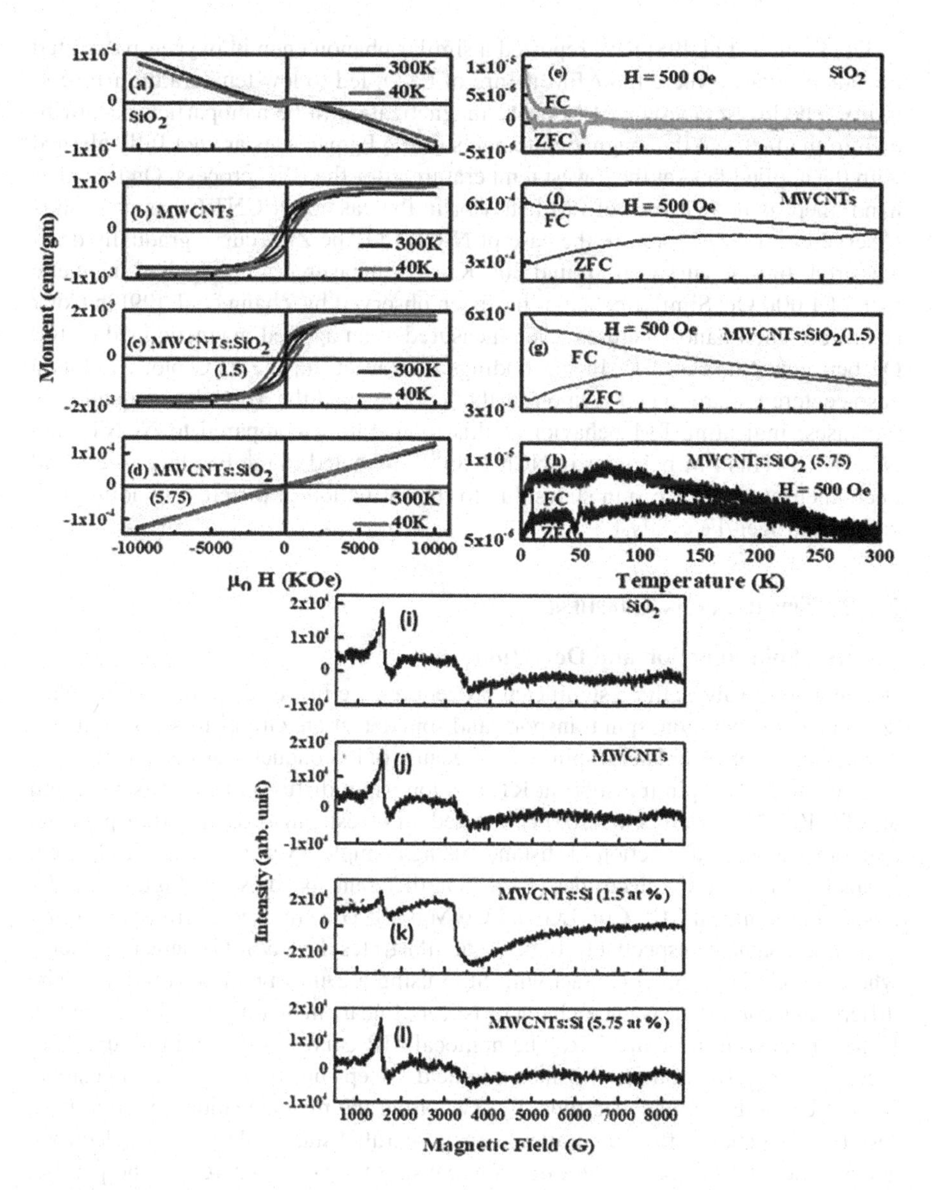

FIGURE 7.4 Magnetic hysteresis loops: (a) SiO_2 NPs, (b) MWCNTs, (c) SiO_2 NPs (1.5 at%), and (d) SiO_2 NPs (5.75 at%) at different temperatures. Temperature dependence of magnetization after FC and ZFC (e) SiO_2 NPs, (f) MWCNTs, (g) SiO_2 NPs (1.5 at%), and (h) SiO_2 NPs (5.75 at%). ESR at 300 K (i) SiO_2 NPs, (j) MWCNTs, (k) SiO_2 NPs (1.5 at%), and (l) SiO_2 NPs (5.75 at%).

presence of FM Fe clusters and the formation of different bonding with carbon/nitrogen, resulting in uncompensated surface spin states. In the case of NCNTs:Cl, the ZFC and FC curves coincide up to 200 K when measured at an applied magnetic field of 1,000 Oe, after which they split.

Del Bianco et al. [98] have reported a similar phenomenon in oxygen-passivated Fe nanoparticles, where the AFM nature of Fe_2O_3 led to low-temperature irreversibility. The low-temperature FM phase magnetization in Fe nanoparticles is attributed to the fact that the magnetic moments of the Fe particles are not fully aligned with the applied field at the lowest temperature after the ZFC process. On the other hand, there is no evidence of FM behavior in the case of N-CNT:Cl, as no cusp is observed in the ZFC plot. In the case of N-CNT:O, the ZFC curve gradually deviates from the FC curves at around 300 K when measured at an applied magnetic field of 1,000 Oe. Similar behavior has been observed by Zhang et al. [99] in CoO/CNT core-shell nanostructures when measured at an applied magnetic field of 100 Oe between 2 and 300 K. In the findings of Ray et al., the ZFC plot exhibits a cusp centered at around 45 K upon further cooling, and the MFC data sequentially increases, indicating FM behavior at this temperature compared to NCNTs and NCNTs:Cl. This FM behavior in NCNTs:O is attributed to the FM Fe clusters and uncompensated surface spin states due to the formation of different bonding with carbon/nitrogen/Fe catalyst.

7.5.2 Spintronic in Graphene

7.5.2.1 Spin Injection and Detection

In lateral spin valves, three significant procedures are involved, namely, spin-polarized current generation, spin transport, and spin detection. Graphene stands out as a promising material for lateral spin valves because of its conductivity that can be tuned by gate [100,101], spin transport at RT, and long spin diffusion lengths (several μm at RT) [102–106]. Han et al. [107] conducted an investigation on spin transport and spin precession as a function of distance using a single layer graphene (SLG) sheet contacted by seven Co electrodes (E1–E7) at different spacings (see Figure 7.5a) by employing nonlocal MR. Co/SLG and Co/MgO/SLG were used as transparent and tunneling contacts, respectively. Figure 7.5b illustrates the Co/SLG contact geometry where the Co is in direct contact with SLG using a 2 nm MgO masking layer. The differential contact resistance characteristic is linear, indicating an ohmic contact behavior, as seen in Figure 7.5c. The nonlocal MR curve, as shown in Figure 7.5d, exhibits a negative value as the magnetic field sweeps up from negative to positive. This effect is due to one electrode, E2, switching the magnetization direction first, resulting in a change from a parallel to an antiparallel state. Subsequently, electrode E3 switches the magnetization direction, causing the state to revert to the parallel state. The minor jumps observed in the nonlocal MR loop are due to the switching of electrodes E1 and E4. The arrows in Figure 7.5d indicate the magnetic directions of the four Co electrodes. The nonlocal MR loop for 1, 2, and 3 μm spacings measured at 300 K with $V_g = 0$ V is depicted in Figure 7.5e–g, respectively. The nonlocal MR value reduces from 100 to 2 mΩ as the spacing increases from 1 to 3 μm. The following equation was used to calculate the spin injection efficiency [107]:

$$\Delta R_{NL} = \frac{1}{\sigma_G} \frac{\eta^2 \lambda_G}{W} e^{-L/\lambda_G} \tag{7.1}$$

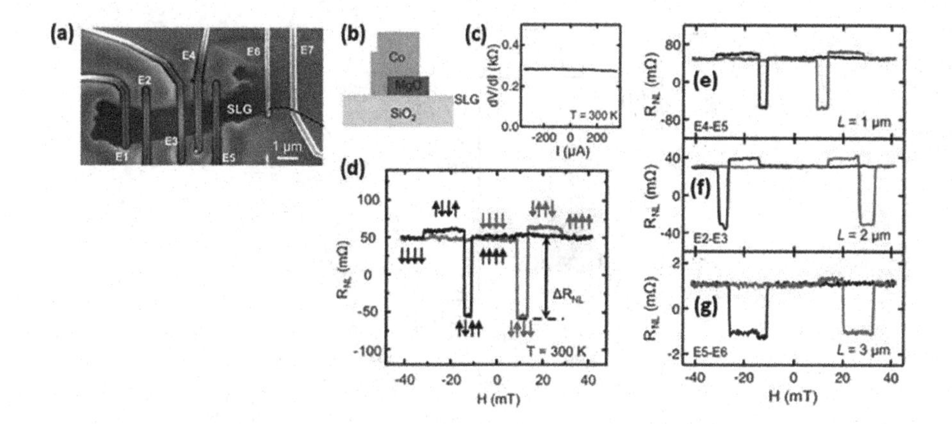

FIGURE 7.5 (a) SEM image of this device showing seven Co electrodes. SLG spin-valve device with transparent contacts Co/SLG (b) 2 nm MgO masking layer, (c) at zero gate voltage, and (d) typical nonlocal MR loop. SLG spin-valve device with seven Co electrodes with various spacings (e–g) Nonlocal MR scans for L¼ 1, 2, and 3 mm (injector, detector) (E4, E5), (E2, E3), and (E5, E6), respectively [107].

The spin injection/detection efficiency is represented by η, while the conductivity, width, and spin diffusion length of the SLG are denoted by σ_G, W, and λ_G, respectively. A low spin injection efficiency of 1% is anticipated due to the conductance mismatch between Co and SLG [108].

To address the conductance mismatch problem when injecting spin into semiconductors, tunnel barriers have been employed [109–112]. Nevertheless, the growth of uniform tunnel barriers on graphene's surface is not straightforward [107]. The growth of 1 nm MgO on highly oriented pyrolytic graphite (HOPG) tends to cluster and create "pinholes" due to the high surface diffusion and low surface energy of HOPG/graphene [107]. The same issues were also observed in studies that grew HfO_2 or Al_2O_3 on graphene [113–115]. After exploring various techniques, a method for growing atomically smooth MgO films on graphene was established through a submonolayer Ti seed layer. The MgO is grown on top of the Ti seed layer. The smoothness of the film improves with increasing Ti thickness up to 0.5 ML. To prevent a conductive path between Co and SLG, the Ti layer was oxidized to form TiO_2. This technique was then applied to SLG flakes, resulting in an RMS roughness of less than 0.2 nm for a 1 nm MgO film [107].

The SLG spin valves with tunneling contacts were produced using the Ti-seeded MgO barrier [107] (see Figure 7.6a). The *I-V* curves between the electrodes displayed highly non-linear behavior (see Figure 7.6b), and the differential contact resistance showed a sharp peak at $I_{dc}=0$ μA (see Figure 7.6c), indicating tunneling between the Co electrodes and SLG [107]. At 300 K with $V_g=0$ V, a nonlocal MR of 130 Ω was observed for a SLG spin valve. The injector and detector spacing (L) was approximately 2.1 μm, and the SLG width (W) was roughly 2.2 μm, as depicted in Figure 7.6d. Using Eq. (7.1) with experimental values of $\sigma_G=0.35$ mS and typical

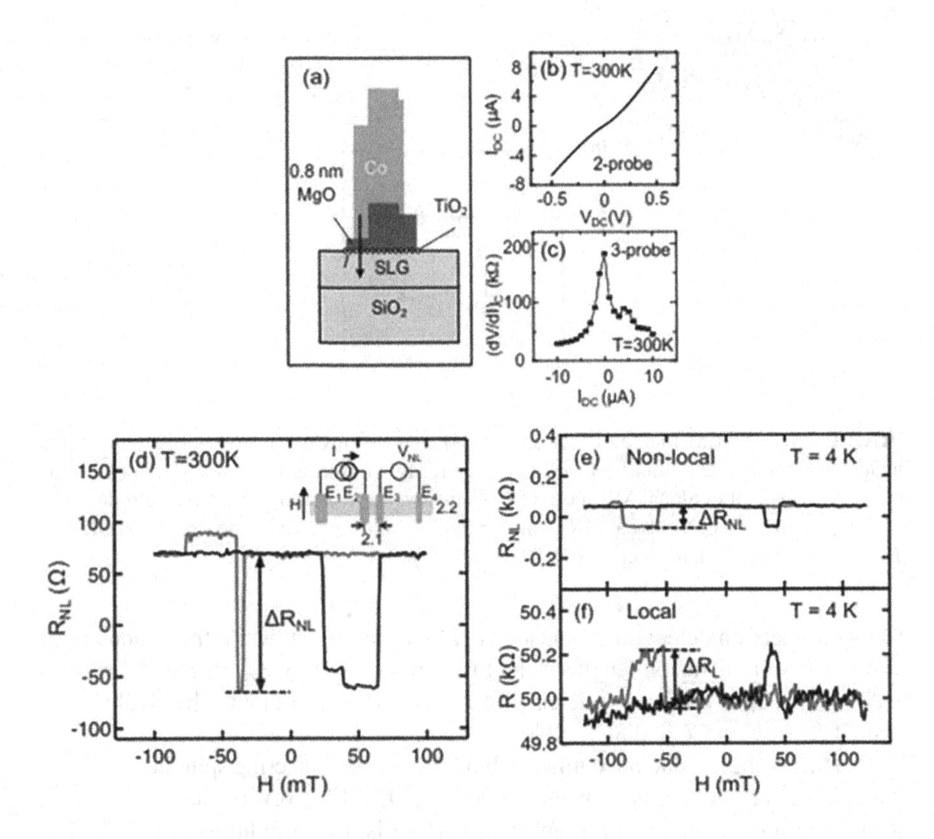

FIGURE 7.6 (a) Tunneling contacts of Co/MgO, TiO_2/SLG. (b) Typical 2-probe IV curve between Co electrodes through SLG. (c) Typical differential contact resistance $(dV/dI)C$ as a function of bias current. (d) SLG spin valves nonlocal MR scans. MR loop (e) nonlocal and (f) local [107].

experimental values of λ_G=2.5–3.0 μm, the spin injection efficiency, *P*, was calculated to be 26–30%. The successful achievement of high spin injection/detection efficiency in spin valves with tunneling contacts is attributed to the high quality of the MgO tunnel barrier, which solved the conductance mismatch issue between Co and SLG [116,117]. The tunneling spin polarization achieved by SDT from Co into a superconductor across polycrystalline Al_2O_3 barriers has been reported to be 35–42% [118–120], which is comparable to the results obtained in [107]. Figure 7.6e and f shows the nonlocal and local MR loops measured at 4 K. The nonlocal MR and local MR signals were ~100 Ω and ~200 Ω, respectively. This finding is consistent with theoretical predictions [121,122], where the local MR is expected to be approximately twice the nonlocal MR. Additionally, the nonlocal MR loop has a superior signal-to-noise ratio and is more sensitive to detecting the spin signal compared to the local MR [107].

7.5.2.2 Spin Relaxation in Graphene

SLG has attracted attention in the field of spintronics due to its predicted long spin relaxation times and diffusion length, attributed to its low intrinsic spin–orbit and hyperfine couplings [51,123–125]. However, experimental measurements have shown spin lifetimes in SLG to be significantly shorter (50–200 ps) than expected based on its intrinsic SOC [103,106,126–128]. Furthermore, a long spin lifetime has also been observed in bilayer graphene (BLG) [107].

In order to investigate spin relaxation in SLG spin valves, Han et al. [107] employed tunnel barrier devices to reduce contact-induced spin relaxation. Their study revealed that the measured spin relaxation times (τ_s) at RT (300 K) ranged between 400 and 600 ps, and there was no apparent correlation between τ_s and D (see Figure 7.7a), where D represents the carrier concentration. However, as the device was cooled to 4 K (see Figure 7.7b), τ_s and D showed a strong correlation, with both values increasing with the carrier concentration. This correlation between τ_s and D suggests a linear relationship between τ_s and the momentum scattering time (τ_p) since $D \sim \tau_p$ [126,129,130]. At low temperatures, the dominant mechanism for spin scattering in SLG is through the Elliot-Yafet mechanism, where there is a finite chance of spin-flip during a momentum scattering event. This is indicated by the strong correlation between τ_s and D at $T = 4$ K, as shown in Figure 7.7b, suggesting

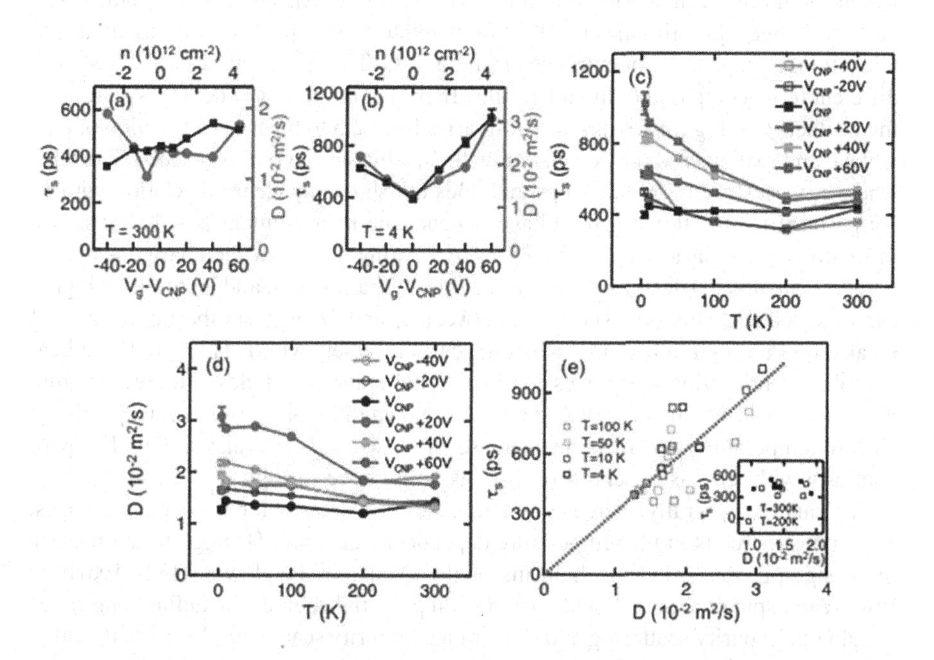

FIGURE 7.7 Spin relaxation in SLG (a, b) Spin lifetime and diffusion coefficient against gate voltage, (c) spin lifetime against temperature, (d) spin diffusion coefficient against temperature, and (e) spin lifetime against diffusion coefficient [107].

a linear relationship between τ_s and the momentum scattering time, τ_p, which has been discussed in previous studies [124]. The temperature dependencies of τ_s and D at different carrier concentrations are depicted in Figure 7.7c and d, where it can be observed that as the temperature decreases from 300 to 4 K, τ_s showed a moderate increase at higher carrier densities (e.g., from ~0.5 to ~1 ns for $V_g - V_{CNP} = +60\,V$) and little variation for lower carrier densities. The results from their study revealled that momentum scattering via the Elliot-Yafet mechanism is the main contributor to spin scattering at low temperatures [124]. This is supported by the observation that τ_s and D display a linear correlation at temperatures below or equal to 100 K [107], as shown in Figure 7.7e. However, for temperatures above 100 K, τ_s and D do not exhibit a linear relationship, indicating the presence of other sources of spin scattering. The temperature dependence of τ_s and D for different carrier concentrations is presented in Figure 7.7c and d, respectively. At higher carrier densities, τ_s exhibits a moderate increase as the temperature decreases, while D remains relatively constant [107].

BLG and SLG exhibit differences not only in thickness but also in their band structure and intrinsic SOC [107]. In contrast to the linear band structure of SLG with massless fermions, BLG has a hyperbolic band structure with massive fermions [131,132]. At 20 K, the gate-voltage dependence of τ_s and D for the BLG spin valve is shown in Figure 7.8a–c. The spin lifetimes measured at different gate voltages were 2.5, 6.2, and 3.3 ns for $V_g - V_{CNP} = -40, 0, 40\,V$. Surprisingly, longer spin lifetimes were observed in BLG, up to 6.2 ns, compared to SLG, up to 1.0 ns. Theoretically, the intrinsic SOC in BLG is predicted to be an order of magnitude larger than in SLG, which should result in shorter spin lifetimes for BLG [132]. The experimental observation of longer spin lifetimes in BLG compared to SLG supports the notion that spin relaxation in graphene is of extrinsic origin in SLG. The spin lifetimes at 20 K show a different trend compared to the RT results. In particular, the BLG device with tunneling contacts (see Figure 7.8d), τ_s at 20 K varies from 2.5 to 6.2 ns as a function of gate voltage and exhibits a weak correlation with D, which is in contrast to the RT results, where τ_s varies from 250 to 350 ps and does not show any clear correlation with D. By performing a detailed gate-voltage dependence measurement at 4 K, as shown in Figure 7.8e, τ_s varies from 1 to 2.7 ns, exhibiting a peak at the charge neutrality point. Meanwhile, D decreases near the charge neutrality point and increases at higher carrier densities. This opposite trend between τ_s and D suggests the importance of Dyakanov-Perel spin relaxation, where τ_s scales inversely with τ_p [133], and is indicative of spin relaxation via precession in internal spin–orbit fields. The temperature dependencies of τ_s and D for BLG device are shown in Figure 7.8f and g, respectively. At low temperatures, there is an increase in τ_s, while D decreases. This behavior contrasts with what is observed in SLG as seen above in Figure 7.7c and d, where both D and τ_s generally increase as temperature decreases for most gate voltages. The opposite trends in the temperature dependence of τ_s and D suggest the presence of strong spin relaxation mechanisms of the Dyakanov-Perel type [134]. Extrinsic Elliot-Yafet spin relaxation could have several potential sources, including long-range (Coulomb) impurity scattering and short-range impurity scattering [135]. On the other hand, extrinsic Dyakanov-Perel spin relaxation might occur due to the curvature of the graphene film, like random SOCs [51,136]. The shift from Elliot-Yafet dominated

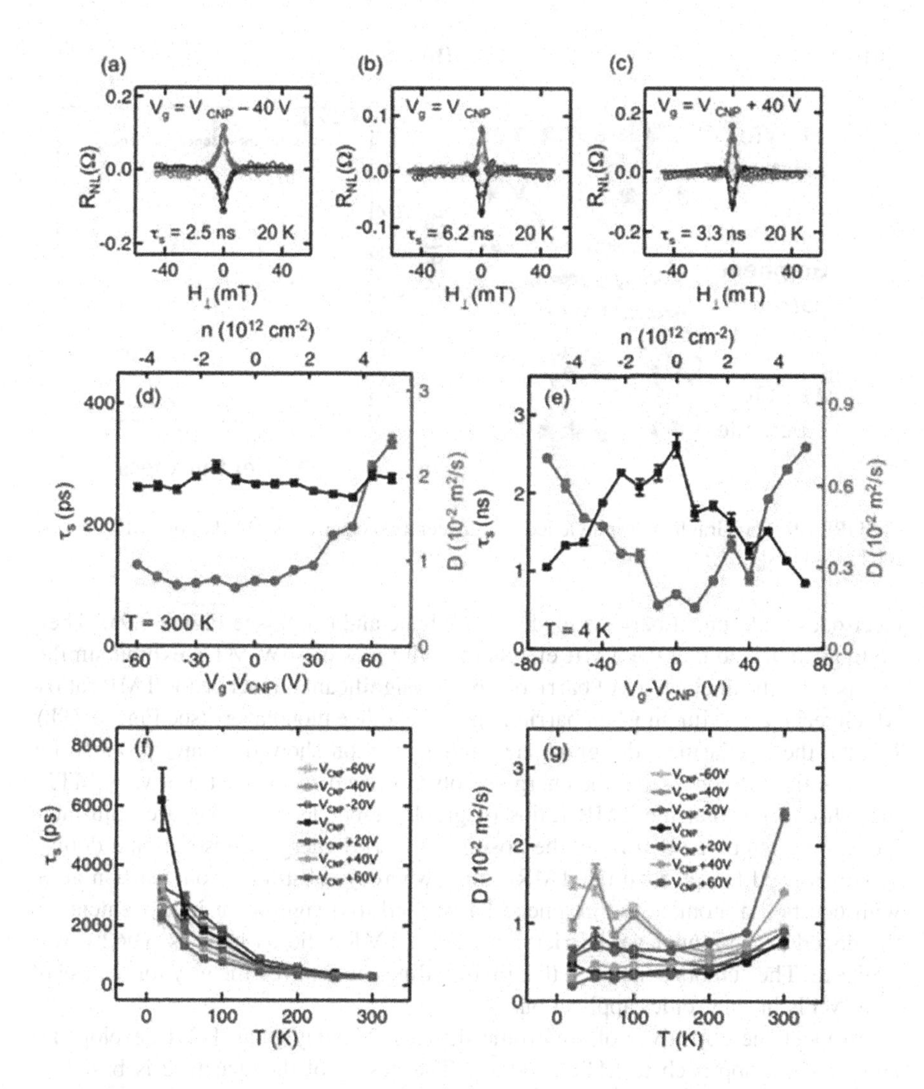

FIGURE 7.8 Spin relaxation in BLG (a–c) Hanle measurement, (d and e) Spin lifetime and diffusion coefficient against gate voltage, (f) Spin lifetime against temperature, (g) Spin diffusion coefficient against temperature [107].

SLG to Dyakanov-Perel dominated BLG could be the result of a significant decrease in the Elliot-Yafet contribution due to increased screening of the impurity potential in thicker graphene [130,137], and the smaller surface-to-volume ratio [107].

7.5.2.3 TMR in Graphene

Through the use of two-dimensional (2D) van der Waals (vdW) heterostructures by Zhou et al. [138], spintronic applications with novel properties were made possible. In their study, they employed density functional theory (DFT) calculations to investigate the spin-dependent transport in vdW MTJs composed of 1T-$CrTe_2$ FM

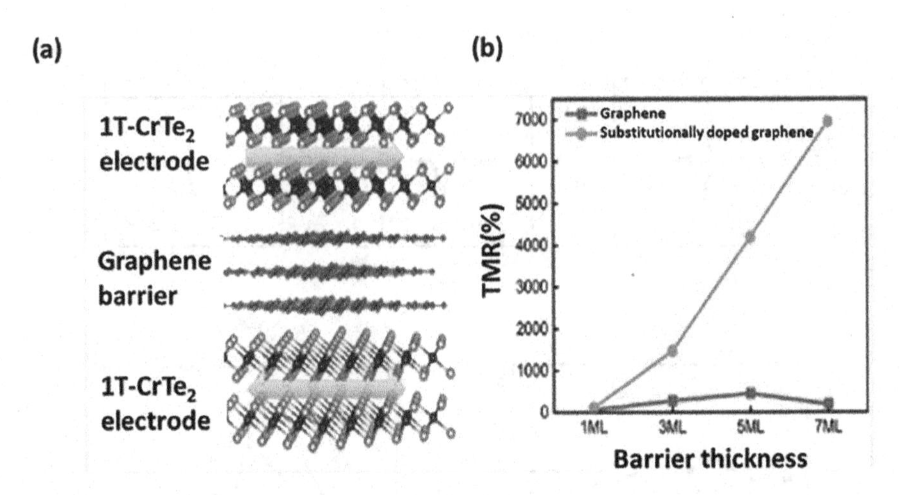

FIGURE 7.9 (a) Graphene sandwiched in between two electrodes. (b) Plot of TMR against thickness [138].

electrodes with tunnel barriers made of graphene and h-BN (see Figure 7.9a). Their findings indicated that the TMR effects of both types of vdW MTJs exhibit similar trends, with the thicknesses of barriers having a significant impact on the TMR ratios, which reach a maximum when barriers increase to five monolayers (see Figure 7.9b). Despite the similarities, the graphene-barrier junction showed greater promise for optimization. By observing the energy-resolved transmission spectra of vdW MTJs, they discovered that the TMR ratios of graphene-barrier junctions are adjustable and can be improved by tuning the position of Fermi energy. Substitutional doping was then used to optimize the TMR ratios, whereby substituting one carbon atom with one boron atom in the graphene barrier led to a significant improvement. In the doped seven-monolayer-barrier junction, a TMR ratio as high as 6,962% was achieved. The authors concluded that their findings have paved the way for the use of vdW MTJs in spintronics applications.

To meet the objectives of spintronic devices, Makdey et al. [139] developed a new efficient approach to MTJ structure. The design of the structure is based on molybdenum disulfide (MoS_2)/graphene quantum dots (GQDs)/MoS_2, which offers better efficiency and lower power consumption. The two-dimensional MoS_2 acts as FM electrodes in the MTJ due to its magnetic properties, while GQDs serve as the barrier. The current-voltage (*I-V*) characteristics estimated for both parallel (P) and antiparallel (AP) MTJ junctions resulted in a MTJ TMR design of 1,450% at zero bias voltage. The performance of the system was evaluated by measuring the rate of temperature at $1/T_1$ and $1/T_2$. It was observed that the two temperature levels exhibit different relaxation times in terms of $1/T_1$, where the relaxation time is reduced, and $1/T_2$, where the relaxation time remains constant. This inverse relaxation time method led to an increase in signal power, indicating that the spintronics device does not consume any power during relaxation time. As a result, the signal power, lifetime, and magnetization were all enhanced. The results of the simulation demonstrate the effectiveness of the proposed MTJ structure for spintronics.

7.5.2.4 MR in Graphene

The phenomenon of MR, or changes in electrical resistance under an external magnetic field, is a topic of significant interest in both fundamental and technological research. Two-dimensional layered materials, such as graphene, offer a unique opportunity to investigate MR during the early stages of structural formation. The study of MR has practical applications, as MR sensors are widely used in everyday devices, where the MR value is an important measure of performance. In the case of graphene, there have been various attempts to comprehend MR, with a particular focus on the large MR values that can be obtained in weak magnetic fields at RT. Extensive research has been carried out on MR in single-layer and multilayer graphene over a temperature range of 1.9–400 K and in magnetic fields up to 14 T by researchers [140–142].

Studies have shown that MR values for HOPG with varying thickness can range from 20% to 40% at 300 K and a 0.5 T magnetic field, depending on the sample's cross-size [140]. In contrast, the MR value for multilayer epitaxial graphene grown on a SiC substrate is less than 5% [142]. Research by Gopinadhan et al. [143] has revealed that the MR of SLG does not reach saturation and equals 275% at 300 K in a 9 T magnetic field, while the MR value at 0.5 T does not exceed 10% and is less than 1% when the current is perpendicular to the graphene layer plane [141]. However, in few-layer graphene/BN heterostructures, an exceptionally high MR value of 880% at 9 T and 400 K was recently measured by Gopinadhan et al. [144], but at 300 K and a 0.5 T magnetic field, the MR value was only around 25%. In epitaxial multilayer graphene, a linear MR of 80–250% was reported at 2 K in a normal magnetic field of 12 T, whereas in chemical vapor deposition-grown few-layer graphene with current perpendicular to the film plane, a linear and quadratic MR of 60% was observed at 300 K and a 14 T magnetic field [141]. Despite the explanation based on the quantum theory, the observed MR at RT cannot be accounted for by quantum effects, as they are not expected to occur at that temperature [145].

Researchers are exploring new materials for MR sensors that are highly sensitive to low magnetic fields, energy-efficient, cost-effective, and widely available [146,147]. Graphene, an atomically thin material composed of single-layer carbon atoms arranged in a hexagonal lattice with weak interlayer interaction, has the potential to exhibit high MR values. It is a simple system for understanding the origin of MR, thanks to the discovery of semimetallic graphene and other low-dimensional conducting structures [100,135,146]. However, defects, impurities, and grain boundaries can limit the mobility of the carriers, reducing MR. To increase mobility, graphene peeled from kish graphite can be used [148]. Although highly ordered pyrolytic graphite samples with sizes of 100 mm exhibit a very small MR [140], recent research has reported a significant MR at RT from single-layer graphene due to enhanced scattering from charged impurities [143]. A negative MR has been projected [149] and observed in graphene nanoribbons [150], but the MR value is limited. Recently, finite MR has been reported on graphene/BN vertical heterostructures [151,152]. In magnetic storage applications, MR read sensors are used to retrieve data from magnetic hard disks that are highly sensitive to stray magnetic fields.

7.5.2.5 Magnetization in Graphene

Graphene's long spin diffusion length makes it an attractive material for new spintronic devices, which has led to a search for ways to combine its charge and spin properties. While ideal graphene is nonmagnetic due to its delocalized π bonding network, there is an urgent need to synthesize FM graphene or its derivatives with high magnetization for both fundamental and technological reasons. However, despite significant progress in this area, the magnetization of graphene is often low because defect-induced magnetization depends mainly on the creation of magnetic moments at the edges of the graphene sheet. The clustering of ad-atoms and vacancy reconstruction at these edge sites, as well as thermal stability and structural integrity issues, make it difficult to continuously increase the underlying spin density [153]. On the contrary, the defect-rich structure of GO, which is partially retained even after being reduced to rGO, provides an attractive option for increasing the density of magnetic moments in graphene-related materials. Studies have reported a significant boost in magnetization ranging from 0.5 to 1.66 emu g^{-1} for N-doped rGO [154] and GO [155]. In addition, by creating sp-type defects on the basal plane of the graphene sheet, the magnetization of GO-based materials has been further increased to 2.4 emu g^{-1}, overcoming the limitations of edge-type defect magnetism [156]. Specifically, high magnetization has been achieved in rGO by inducing hydroxyl groups to create magnetic moments in GO, whose density can be controlled by high-temperature annealing. The isothermal magnetization data analysis of ultrasonically exfoliated graphene laminates [157], pristine GO [156], and other similar cases revealed the presence of high spin states ($S=2$ and 5/2), indicating a significant departure from the spin-half paramagnetism caused by point defects in graphene materials [153]. The theoretical predictions of large magnetic moments in graphene oxide (GO) and reduced graphene oxide (rGO) were attributed to the stabilization of seven OH-group clusters on the basal graphene plane with a high spin $S=5/2$ ground state [158]. However, the specific relationship between magnetic moments and oxygen-containing and hydroxyl groups in these materials remained unclear. Recent studies have indicated that magnetic behavior in rGO upon reduction from GO is primarily associated with $C\ 2p(\sigma^*)$-derived states involving edge defects or vacancies rather than $C\ 2p(\pi^*)$ states related to oxygen functional groups [159–161]. The variability in oxidation conditions and defective structures of GO account for the inevitable scattering of magnetic response data. Experimental techniques such as ESR that are sensitive to spin could provide valuable evidence for the presence of high-spin magnetic clusters in GO-based magnetic materials, aiding in the understanding and customization of their magnetic properties [162]. In a study by Diamantopoulou et al. [163], the evolution of magnetism in GO and chemically rGO was examined using static magnetization and ESR spectroscopy. The researchers discovered that pristine GO displayed strong paramagnetism with a saturation magnetization of approximately 1.2 emu g^{-1} and weak anti-FM interactions. The results of ESR spectroscopy suggest that in addition to spin-half defect centers, high spin states are also excited, which is consistent with the presence of high spin ($S=2$) magnetic moments observed in the magnetization analysis. This supports the creation of spatially "isolated" magnetic clusters in GO. Upon chemical reduction of GO by sodium borohydride, a significant decrease in magnetization (~0.17 emu g^{-1}) and a considerable increase in diamagnetism (-2.43×10^6 emu g^{-1} Oe) were observed. This suggests the removal of paramagnetic defects and the growth of sp^2 domains in rGO.

7.6 CONCLUSION, CHALLENGES, AND PERSPECTIVES

This chapter discusses the advances in PCM memories, CNTs, and grapheme-based spintronics. It is noted that PCM faces challenges in terms of scalability, endurance, thermal cross-talk, reset speed, and cost, but it has promising future perspectives in terms of high-density storage, NVM, neuromorphic computing, and hybrid memory systems. Continued research, material development, and optimization are essential to overcome these challenges and unlock the full potential of PCM for memory devices.

Spin currents have been successfully injected into SWCNT and MWCNTs using various FM electrodes. These nanotubes exhibit large MR and have a long spin diffusion length, making them promise for spintronic devices. Graphene has received significant attention and development due to its unique structure and properties. Also, spin injection has been achieved in graphene, with spin valves exhibiting large MR at RT. CNTs and graphene-based MTJs have potential in spintronics, which utilizes electron spin for information processing. Challenges with CNTs include synthesizing high-quality CNTs with controlled chirality, integrating them into spintronic devices, creating low-resistance electrical contacts, addressing spin relaxation and coherence issues, and achieving reproducibility and scalability. Advancement in this field involving CNTs requires the development of chirality engineering techniques, exploring hybrid structures, finding ways to manipulate and control spins in CNTs, and integrating CNT-based devices with conventional electronics.

Regarding grapheme-based MTJs, achieving high spin polarization and efficient spin injection/detection, maintaining the magnetic stability of FM electrodes, and optimizing tunnel barrier properties are very crucial. Future perspectives involve material engineering to enhance spin polarization, optimizing device design and fabrication techniques, exploring multifunctional devices with additional capabilities, and addressing scalability and integration issues. Zigzag-edged graphene nanoribbon heterojunctions are predicted to be useful in future spintronics [164–166], but further experimental investigation is needed.

However, spin relaxation mechanisms in these materials (CNTs and graphene) are complicated and require further investigation to enhance spin polarization and lifetime.

REFERENCES

[1] S.M. Yakout, Spintronics: future technology for new data storage and communication devices, *Journal of Superconductivity and Novel Magnetism*. 33 (9) (2020) 2557–2580.
[2] R. Sbiaa, S.N. Piramanayagam, Recent developments in spin transfer torque MRAM, *Physica Status Solidi (RRL)-Rapid Research Letters*. 11 (12) (2017) 1700163.
[3] J.S. Meena, S.M. Sze, U. Chand, T.Y. Tseng, Overview of emerging nonvolatile memory technologies, *Nanoscale Research Letters*. 9 (1) (2014) 1–33.

[4] G. Min, Q. Zhang, S. Lamon, Nanomaterials for optical data storage, *Nature Reviews Materials*. 1 (12) (2016) 1–14.
[5] A. Hirohata et al., Review on spintronics: principles and device applications, *Journal of Magnetism and Magnetic Materials*. 509 (2020) 166711.
[6] T. Alexoudi, G.T. Kanellos, N. Pleros, Optical RAM and integrated optical memories: a survey, *Light: Science & Applications*. 9 (1) (2020) 91.
[7] H. Yang et al., Two-dimensional materials prospects for non-volatile spintronic memories, *Nature*.606 (7915) (2022) 663–673.
[8] M. Si, H.-Y. Cheng, T. Ando, G. Hu, P.D. Ye, Overview and outlook of emerging non-volatile memories, *MRS Bulletin*. 46 (10) (2021) 946–958.
[9] K. Singh et al., A review on GeTe thin film-based phase-change materials, *Applied Nanoscience*. 13 (1) (2023) 95–110.
[10] M.H.R. Lankhorst, Modelling glass transition temperatures of chalcogenide glasses. Applied to phase-change optical recording materials, *Journal of Non-Crystalline Solids*. 297 (2-3) (2002) 210–219.
[11] M. Chen, K.A. Rubin, R.W. Barton, Compound materials for reversible, phase-change optical data storage, *Applied Physics Letters*. 49 (9) (1986) 502–504.
[12] M. Wuttig, Towards a universal memory?, *Nature Materials* 4 (4) (2005) 265–266.
[13] S. Raoux, W. Wełnic, D. Ielmini, Phase change materials and their application to non-volatile memories, *Chemical Reviews* 110 (1) (2010) 240–267.
[14] M. Wuttig, N. Yamada, Phase-change materials for rewriteable data storage, *Nature Materials*. 6 (11) (2007) 824–832.
[15] M. Wuttig, S. Raoux, The science and technology of phase change materials, *Zeitschrift für anorganische und allgemeine Chemie*. 638 (15) (2012) 2455–2465.
[16] Z. Zhang et al., Memory materials and devices: from concept to application, *InfoMat*. 2 (2) (2020) 261–290.
[17] N. Raeis-Hosseini, J. Rho, Metasurfaces based on phase-change material as a reconfigurable platform for multifunctional devices, *Materials*. 10 (9) (2017) 1046.
[18] Y. Zhai et al., Toward non-volatile photonic memory: concept, material and design, *Materials Horizons*. 5 (4) (2018) 641–654.
[19] P. Fantini, Phase change memory applications: the history, the present and the future, *Journal of Physics D: Applied Physics*. 53 (28) (2020) 283002.
[20] N. Kraft, G. Wang, H. Bryja, A. Prager, J. Griebel, A. Lotnyk, Phase and grain size engineering in Ge-Sb-Te-O by alloying with La-Sr-Mn-O towards improved material properties, *Materials & Design*. 199 (2021) 109392.
[21] X. Zhou et al., Carbon-doped Ge2Sb2Te5 phase change material: a candidate for high-density phase change memory application, *Applied Physics Letters*. 101 (14) (2012) 142104.
[22] Y. Lu et al., Performance improvement of Ge-Sb-Te material by GaSb doping for phase change memory, *Applied Physics Letters*. 102 (24) (2013) 241907.
[23] K. Peng et al., Tailorable fragile-to-strong kinetics features of metal oxides nanocomposite phase-change antimony films, *Acta Materialia*. 234 (2022) 118013.
[24] R. Sinha-Roy, A. Louiset, M. Benoit, L. Calmels, Electronic structure and conductivity of off-stoichiometric and Si-doped Ge 2 Sb 2 Te 5 crystals from multiple-scattering theory, *Physical Review B*. 99 (24) (2019) 245124.
[25] W. Leng et al., Crystallization kinetics of monatomic antimony, *Applied Physics Letters*. 119 (17) (2021) 171908.
[26] S. Xu et al., Performance improvement of Sb phase change thin film by Y doping, *ECS Journal of Solid State Science and Technology*. 10 (9) (2021), 93002.
[27] S. Gao, Y. Hu, Simultaneously higher thermal stability and lower resistance drifting for Sb/In 48.9 Sb 15.5 Te 35.6 nanocomposite multilayer films, *CrystEngComm*. 24 (8) (2022) 1638–1644.

[28] Y. Xu et al., Study on the performance of superlattice-like thin film V2O5/Sb in phase change memory, *ECS Journal of Solid State Science and Technology*. 9 (3) (2020) 33003.

[29] Y. Huang, W. Wu, S. Xu, X. Zhu, S. Song, Z. Song, Thickness effect on the crystallization characteristic of RF sputtered Sb thin films, *Journal of Materials Science: Materials in Electronics*. 32 (19) (2021) 24240–24247.

[30] D.T. Yimam, B.J. Kooi, Thickness-dependent crystallization of ultrathin antimony thin films for monatomic multilevel reflectance and phase change memory designs, *ACS Applied Materials & Interfaces*. 14 (11) (2022) 13593–13600.

[31] Y. Jiao, G. Wang, A. Lotnyk, T. Wu, J. Zhu, A. He, Designing Sb phase change materials by alloying with Ga2S3 towards high thermal stability and low resistance drift by bond reconfigurations, *Journal of Alloys and Compounds*. 953 (2023) 169970.

[32] U. Russo, D. Ielmini, A.L. Lacaita, Analytical modeling of chalcogenide crystallization for PCM data-retention extrapolation, *IEEE Transactions on Electron Devices*. 54 (10) (2007) 2769–2777.

[33] H.-Y. Cheng, S. Raoux, Y.-C. Chen, The impact of film thickness and melt-quenched phase on the phase transition characteristics of Ge 2 Sb 2 Te 5, *Journal of Applied Physics*. 107 (7) (2010) 74308.

[34] P.D. Szkutnik et al., Impact of In doping on GeTe phase-change materials thin films obtained by means of an innovative plasma enhanced metalorganic chemical vapor deposition process, *Journal of Applied Physics*. 121 (10) (2017) 105301.

[35] J. Shen, W. Song, K. Ren, Z. Song, P. Zhou, M. Zhu, Toward the speed limit of phase-change memory, *Advanced Materials*. 35 (11) (2023) 2208065.

[36] S. Raoux, F. Xiong, M. Wuttig, E. Pop, Phase change materials and phase change memory, *MRS Bulletin* 39 (8) (2014) 703–710.

[37] T. Ravsher et al., Self-rectifying memory cell based on SiGeAsSe ovonic threshold switch, *IEEE Transactions on Electron Devices*. 70 (5) (2023) 2276–2281.

[38] T. Ravsher et al., Polarity-induced threshold voltage shift in ovonic threshold switching chalcogenides and the impact of material composition, *Physica Status Solidi (RRL)-Rapid Research Letters*. 17 (8) (2023) 2200417.

[39] W. Choi et al., Parallel integration of nanoscale atomic layer deposited Ge2Sb2Te5 phase-change memory with an indium gallium zinc oxide thin-film transistor, *ACS Applied Electronic Materials*. 5 (3) (2023) 1721–1729.

[40] T.-Y. Yang, I.-M. Park, B.-J. Kim, Y.-C. Joo, Atomic migration in molten and crystalline Ge 2 Sb 2 Te 5 under high electric field, *Applied Physics Letters*. 95 (3) (2009) 32104.

[41] T.-Y. Yang, J.-Y. Cho, Y.-J. Park, Y.-C. Joo, Influence of dopants on atomic migration and void formation in molten Ge2Sb2Te5 under high-amplitude electrical pulse, *Acta Materialia*. 60 (5) (2012) 2021–2030.

[42] W. Kim et al., ALD-based confined PCM with a metallic liner toward unlimited endurance, in: *2016 IEEE International Electron Devices Meeting (IEDM)*, 2016: pp. 2–4.

[43] S.C. Ray, *Magnetism and Spintronics in Carbon and Carbon Nanostructured Materials*, Elsevier, 2020.

[44] R.C. Sousa, I.L. Prejbeanu, Non-volatile magnetic random access memories (MRAM), *Comptes Rendus Physique*. 6 (9) (2005) 1013–1021.

[45] B. Tudu, A. Tiwari, Recent developments in perpendicular magnetic anisotropy thin films for data storage applications, *Vacuum*. 146 (2017) 329–341.

[46] S.P. Mathew, P.C. Mondal, H. Moshe, Y. Mastai, R. Naaman, Non-magnetic organic/inorganic spin injector at room temperature, *Applied Physics Letters*. 105 (24) (2014) 242408.

[47] P.M. Levy, S. Zhang, Spin dependent tunneling, *Current Opinion in Solid State and Materials Science*. 4 (2) (1999) 223–229.

[48] A.M. Sahadevan, Study of spin-dependent transport phenomena in magnetic tunneling systems?, Doctoral Dissertation, National University of Singapore, 2012.
[49] M. Julliere, Tunneling between ferromagnetic films, *Physics Letters A*. 54 (3) (1975) 225–226.
[50] L. Berger, Low-field magnetoresistance and domain drag in ferromagnets, *Journal of Applied Physics*. 49 (3) (1978) 2156–2161.
[51] D. Huertas-Hernando, F. Guinea, A. Brataas, Spin-orbit coupling in curved graphene, fullerenes, nanotubes, and nanotube caps, *Physical Review B*. 74 (15) (2006) 155426.
[52] F. Kuemmeth, H.O.H. Churchill, P.K. Herring, C.M. Marcus, Carbon nanotubes for coherent spintronics, *Materials Today*. 13 (3) (2010) 18–26.
[53] P. Chen, G. Zhang, Carbon-based spintronics, *Science China Physics, Mechanics and Astronomy*. 56 (2013) 207–221.
[54] K. Tsukagoshi, B.W. Alphenaar, H. Ago, Coherent transport of electron spin in a ferromagnetically contacted carbon nanotube, *Nature*. 401 (6753) (1999) 572–574.
[55] J.-R. Kim, H.M. So, J.-J. Kim, J. Kim, Spin-dependent transport properties in a single-walled carbon nanotube with mesoscopic Co contacts, *Physical Review B*. 66 (23) (2002) 233401.
[56] N. Tombros, S.J. Van Der Molen, B.J. Van Wees, Separating spin and charge transport in single-wall carbon nanotubes, *Physical Review B*. 73 (23) (2006) 233403.
[57] H. Yang et al., Nonlocal spin transport in single-walled carbon nanotube networks, *Physical Review B*. 85 (5) (2012) 52401.
[58] A. Jensen, J.R. Hauptmann, J. Nygård, P.E. Lindelof, Magnetoresistance in ferromagnetically contacted single-wall carbon nanotubes, *Physical Review B - Condensed Matter and Materials Physics*. 72 (3) (2005) 1–5.
[59] L.E. Hueso et al., Transformation of spin information into large electrical signals using carbon nanotubes, *Nature*. 445 (7126) (2007) 410–413.
[60] B. Nagabhirava, T. Bansal, G.U. Sumanasekera, B.W. Alphenaar, L. Liu, Gated spin transport through an individual single wall carbon nanotube, *Applied Physics Letters*. 88 (2) (2006) 23503.
[61] K.M. Borysenko, Y.G. Semenov, K.W. Kim, J.M. Zavada, Electron spin relaxation via flexural phonon modes in semiconducting carbon nanotubes, *Physical Review B*. 77 (20) (2008) 205402.
[62] Y.G. Semenov, K.W. Kim, G.J. Iafrate, Electron spin relaxation in semiconducting carbon nanotubes: the role of hyperfine interaction, *Physical Review B*. 75 (4) (2007) 45429.
[63] Y.G. Semenov, J.M. Zavada, K.W. Kim, Electron spin relaxation in carbon nanotubes, *Physical Review B*. 82 (15) (2010) 155449.
[64] H.T. Man, I.J.W. Wever, A.F. Morpurgo, Spin-dependent quantum interference in single-wall carbon nanotubes with ferromagnetic contacts, *Physical Review B*. 73 (24) (2006) 241401.
[65] S. Sahoo et al., Electric field control of spin transport, *Nature Physics*. 1 (2) (2005) 99–102.
[66] G. Gunnarsson, J. Trbovic, C. Schoenenberger, Large oscillating nonlocal voltage in multiterminal single-wall carbon nanotube devices, *Physical Review B*. 77 (20) (2008) 201405.
[67] A. Makarovski, A. Zhukov, J. Liu, G. Finkelstein, Four-probe measurements of carbon nanotubes with narrow metal contacts, *Physical Review B*. 76 (16) (2007) 161405.
[68] J.S. Moodera, L.R. Kinder, T.M. Wong, R. Meservey, Large magnetoresistance at room temperature in ferromagnetic thin film tunnel junctions, *Physical Review Letters*. 74 (16) (1995) 3273.

[69] S.S.P. Parkin et al., Exchange-biased magnetic tunnel junctions and application to non-volatile magnetic random access memory, *Journal of Applied Physics.* 85 (8) (1999) 5828–5833.
[70] Z.H. Xiong, D. Wu, Z. Valy Vardeny, J. Shi, Giant magnetoresistance in organic spin-valves, *Nature.* 427 (6977) (2004) 821–824.
[71] H. Shimada, K. Ono, Y. Ootuka, Magneto-Coulomb oscillation in ferromagnetic single electron transistors, *Journal of the Physical Society of Japan.* 67 (4) (1998) 1359–1370.
[72] P.L. McEuen, M.S. Fuhrer, H. Park, Single-walled carbon nanotube electronics, *IEEE Transactions on Nanotechnology.* 1 (1) (2002) 78–85.
[73] J.M. De Teresa, A. Barthelemy, A. Fert, J.P. Contour, F. Montaigne, P. Seneor, Role of metal-oxide interface in determining the spin polarization of magnetic tunnel junctions, *Science.* 286 (5439) (1999) 507–509.
[74] S. Yuasa, T. Nagahama, A. Fukushima, Y. Suzuki, K. Ando, Giant room-temperature magnetoresistance in single-crystal Fe/MgO/Fe magnetic tunnel junctions, *Nature Materials.* 3 (12) (2004) 868–871.
[75] I. Bergenti et al., Small angle neutron investigation of Au-Fe alloys with GMR behaviour, *Journal of Magnetism and Magnetic Materials.* 272 (2004) 1554–1556.
[76] B.-C. Min, K. Motohashi, C. Lodder, R. Jansen, Tunable spin-tunnel contacts to silicon using low-work-function ferromagnets, *Nature Materials.* 5 (10) (2006) 817–822.
[77] D.J. Monsma, S.S.P. Parkin, Spin polarization of tunneling current from ferromagnet/A12O3 interfaces using copper-doped aluminum superconducting films, *Applied Physics Letters.* 77 (5) (2000) 720–722.
[78] W.A. Hofer, A.S. Foster, A.L. Shluger, Theories of scanning probe microscopes at the atomic scale, *Reviews of Modern Physics.* 75 (4) (2003) 1287.
[79] M.R. Buitelaar, A. Bachtold, T. Nussbaumer, M. Iqbal, C. Schönenberger, Multiwall carbon nanotubes as quantum dots, *Physical Review Letters.* 88 (15) (2002) 156801.
[80] S. Sahoo, T. Kontos, C. Schönenberger, C. Sürgers, Electrical spin injection in multiwall carbon nanotubes with transparent ferromagnetic contacts, *Applied Physics Letters.* 86 (11) (2005) 112109.
[81] T. Schäpers, J. Nitta, H.B. Heersche, H. Takayanagi, Interference ferromagnet/semiconductor/ferromagnet spin field-effect transistor, *Physical Review B.* 64 (12) (2001) 125314.
[82] J. Barnaś, J. Martinek, G. Michałek, B.R. Bułka, A. Fert, Spin effects in ferromagnetic single-electron transistors, *Physical Review B.* 62 (18) (2000) 12363.
[83] M.A. Mohamed, N. Inami, E. Shikoh, Y. Yamamoto, H. Hori, A. Fujiwara, Fabrication of spintronics device by direct synthesis of single-walled carbon nanotubes from ferromagnetic electrodes, *Science and Technology of Advanced Materials.* 9 (2) (2008) 025019.
[84] N. Inami, M.A. Mohamed, E. Shikoh, A. Fujiwara, Synthesis-condition dependence of carbon nanotube growth by alcohol catalytic chemical vapor deposition method, *Science and Technology of Advanced Materials.* 8 (4) (2007) 292.
[85] M. Sagnes et al., Probing the electronic properties of individual carbon nanotube in 35 T pulsed magnetic field, *Chemical Physics Letters.* 372 (5-6) (2003) 733–738.
[86] S. Kenane, J. Voiron, N. Benbrahim, E. Chaînet, F. Robaut, Magnetic properties and giant magnetoresistance in electrodeposited Co-Ag granular films, *Journal of Magnetism and Magnetic Materials*, 297 (2) (2006) 99–106.
[87] Y. Li, T. Kaneko, T. Ogawa, M. Takahashi, R. Hatakeyama, Magnetic characterization of Fe-nanoparticles encapsulated single-walled carbonnanotubes, *Chemical Communications.* (3) (2007) 254–256.
[88] F.C. Fonseca et al., Superparamagnetism and magnetic properties of Ni nanoparticles embedded in SiO 2, *Physical Review B.* 66 (10) (2002) 104406.

[89] S. Linderoth, L. Balcells, A. Labarta, J. Tejada, P.V Hendriksen, S.A. Sethi, Magnetization and Mössbauer studies of ultrafine Fe-C particles, *Journal of Magnetism and Magnetic Materials.* 124 (3) (1993) 269–276.
[90] J.A. Oke et al., Electronic, electrical, and magnetic behavioral change of SiO 2 - NP-decorated MWCNTs, *ACS Omega.* 4 (2019) 14589–14598.
[91] S.C. Ray, D.K. Mishra, A.M. Strydom, P. Papakonstantinou, Magnetic behavioural change of silane exposed graphene nanoflakes, *Journal of Applied Physics.* 118 (11) (2015) 115302.
[92] M. Mostafa, S. Banerjee, Effect of functional group topology of carbon nanotubes on electrophoretic alignment and properties of deposited layer, *The Journal of Physical Chemistry C.* 118 (21) (2014) 11417–11425.
[93] S.S. Rao, A. Stesmans, J.V Noyen, P. Jacobs, B. Sels, ESR evidence for disordered magnetic phase from ultra-small carbon nanotubes embedded in zeolite nanochannels, *Europhysics Letters.* 90 (5) (2010) 57003.
[94] E.G. Gerstner, P.B. Lukins, D.R. McKenzie, D.G. McCulloch, Substrate bias effects on the structural and electronic properties of tetrahedral amorphous carbon, *Physical Review B.* 54 (20) (1996) 14504.
[95] S.C. Ray, S. Sitha, P. Papakonstantinou, Change of magnetic behaviour of nitrogenated carbon nanotubes on chlorination/oxidation, *International Journal of Nanotechnology.* 14 (1-6) (2017) 356–366.
[96] S.C. Ray, D.K. Mishra, A. Tagliaferro, Hall effect studies and magnetic behaviour in Fe-nanoparticle embedded multi wall CNTs, *Journal of Nanoscience and Nanotechnology.* 17 (12) (2017) 9167–9171.
[97] K. Lipert, F. Kretzschmar, M. Ritschel, A. Leonhardt, R. Klingeler, B. Büchner, Nonmagnetic carbon nanotubes, *Journal of Applied Physics.* 105 (6) (2009) 63906.
[98] L. Del Bianco et al., Evidence of spin disorder at the surface-core interface of oxygen passivated Fe nanoparticles, *Journal of Applied Physics.* 84 (4) (1998) 2189–2192.
[99] H. Zhang, N. Du, P. Wu, B. Chen, D. Yang, Functionalization of carbon nanotubes with magnetic nanoparticles: general nonaqueous synthesis and magnetic properties, *Nanotechnology.* 19 (31) (2008) 315604.
[100] Y. Zhang, Y.-W. Tan, H. L. Stormer, P. Kim, Experimental observation of the quantum Hall effect and Berry's phase in graphene, *Nature.* 438 (7065) (2005) 201–204.
[101] A.K. Geim, K.S. Novoselov, The rise of graphene, *Nature Materials.* 6 (3) (2007) 183–191.
[102] E.W. Hill, A.K. Geim, K. Novoselov, F. Schedin, P. Blake, Graphene spin valve devices, *IEEE Transactions on Magnetics.* 42 (10) (2006) 2694–2696.
[103] N. Tombros, C. Jozsa, M. Popinciuc, H.T. Jonkman, B.J. Van Wees, Electronic spin transport and spin precession in single graphene layers at room temperature, *Nature.* 448 (7153) (2007) 571–574.
[104] M. Ohishi, M. Shiraishi, R. Nouchi, T. Nozaki, T. Shinjo, Y. Suzuki, Spin injection into a graphene thin film at room temperature, *Japanese Journal of Applied Physics.* 46 (7L) (2007) L605.
[105] S. Cho, Y.-F. Chen, M.S. Fuhrer, Gate-tunable graphene spin valve, *Applied Physics Letters.* 91 (12) (2007) 123105.
[106] M. Shiraishi et al., Robustness of spin polarization in graphene-based spin valves, *Advanced Functional Materials.* 19 (23) (2009) 3711–3716.
[107] W. Han et al., Spin transport and relaxation in graphene, *Journal of Magnetism and Magnetic Materials.* 324 (4) (2012) 369–381.
[108] G. Schmidt, D. Ferrand, L.W. Molenkamp, A.T. Filip, B.J. Van Wees, Fundamental obstacle for electrical spin injection from a ferromagnetic metal into a diffusive semiconductor, *Physical Review B.* 62 (8) (2000) R4790.

[109] X. Lou et al., Electrical detection of spin transport in lateral ferromagnet-semiconductor devices, *Nature Physics*. 3 (3) (2007) 197–202.
[110] B.T. Jonker, G. Kioseoglou, A.T. Hanbicki, C.H. Li, P.E. Thompson, Electrical spin-injection into silicon from a ferromagnetic metal/tunnel barrier contact, *Nature Physics*. 3 (8) (2007) 542–546.
[111] T. Sasaki, T. Oikawa, T. Suzuki, M. Shiraishi, Y. Suzuki, K. Tagami, Electrical spin injection into silicon using MgO tunnel barrier, *Applied Physics Express*. 2 (5) (2009) 53003.
[112] M. Tran et al., Enhancement of the spin accumulation at the interface between a spin-polarized tunnel junction and a semiconductor, *Physical Review Letters*. 102 (3) (2009) 36601.
[113] K. Zou, X. Hong, D. Keefer, J. Zhu, Deposition of high-quality HfO 2 on graphene and the effect of remote oxide phonon scattering, *Physical Review Letters*. 105 (12) (2010) 126601.
[114] X. Wang, S.M. Tabakman, H. Dai, Atomic layer deposition of metal oxides on pristine and functionalized graphene, *Journal of the American Chemical Society*. 130 (26) (2008) 8152–8153.
[115] B. Lee et al., Conformal Al 2 O 3 dielectric layer deposited by atomic layer deposition for graphene-based nanoelectronics, *Applied Physics Letters*. 92 (20) (2008) 203102.
[116] E.I. Rashba, Theory of electrical spin injection: tunnel contacts as a solution of the conductivity mismatch problem, *Physical Review B*. 62 (24) (2000) R16267.
[117] A. Fert, H. Jaffres, Conditions for efficient spin injection from a ferromagnetic metal into a semiconductor, *Physical Review B*. 64 (18) (2001) 184420.
[118] C. Kaiser, S. van Dijken, S.-H. Yang, H. Yang, S.S.P. Parkin, Role of tunneling matrix elements in determining the magnitude of the tunneling spin polarization of 3 d transition metal ferromagnetic alloys, *Physical Review Letters*. 94 (24) (2005) 247203.
[119] R. Meservey, P.M. Tedrow, Spin-polarized electron tunneling, *Physics Reports*. 238 (4) (1994) 173–243.
[120] C.H. Kant, J.T. Kohlhepp, H.J.M. Swagten, W.J.M. de Jonge, Intrinsic thermal robustness of tunneling spin polarization in Al/Al 2 O 3/Co junctions, *Applied Physics Letters*. 84 (7) (2004) 1141–1143.
[121] F.J. Jedema, M.S. Nijboer, A.T. Filip, B.J. Van Wees, Spin injection and spin accumulation in all-metal mesoscopic spin valves, *Physical Review B*. 67 (8) (2003) 85319.
[122] A. Fert, S.-F. Lee, Theory of the bipolar spin switch, *Physical Review B*. 53 (10) (1996) 6554.
[123] Y. Yao, F. Ye, X.-L. Qi, S.-C. Zhang, Z. Fang, Spin-orbit gap of graphene: first-principles calculations, *Physical Review B*. 75 (4) (2007) 41401.
[124] D. Huertas-Hernando, F. Guinea, A. Brataas, Spin-orbit-mediated spin relaxation in graphene, *Physical Review Letters*. 103 (14) (2009) 146801.
[125] H. Min, G. Borghi, M. Polini, A.H. MacDonald, Pseudospin magnetism in graphene, *Physical Review B*. 77 (4) (2008) 41407.
[126] C. Józsa et al., Linear scaling between momentum and spin scattering in graphene, *Physical Review B*. 80 (24) (2009) 241403.
[127] K. Pi, W. Han, K.M. McCreary, A.G. Swartz, Y. Li, R.K. Kawakami, Manipulation of spin transport in graphene by surface chemical doping, *Physical Review Letters*. 104 (18) (2010) 187201.
[128] W. Han et al., Electrical detection of spin precession in single layer graphene spin valves with transparent contacts, *Applied Physics Letters*. 94 (22) (2009) 222109.
[129] J. Fabian, A. Matos-Abiague, C. Ertler, P. Stano, I. Zutic, Semiconductor spintronics, *arXiv preprint arXiv:07111461*, 2007.

[130] T. Maassen, F.K. Dejene, M.H.D. Guimarães, C. Józsa, B.J. Van Wees, Comparison between charge and spin transport in few-layer graphene, *Physical Review B.* 83 (11) (2011) 115410.
[131] D.A. Abanin, K.S. Novoselov, U. Zeitler, P.A. Lee, A.K. Geim, L.S. Levitov, Dissipative quantum Hall effect in graphene near the Dirac point, *Physical Review Letters.* 98 (19) (2007) 196806.
[132] F. Guinea, Spin-orbit coupling in a graphene bilayer and in graphite, *New Journal of Physics.* 12 (8) (2010) 83063.
[133] M.I. Dyakonov, V.I. Perel, Spin relaxation of conduction electrons in noncentrosymmetric semiconductors, *Soviet Physics Solid State, USSR.* 13 (12) (1972) 3023–3026.
[134] T.-Y. Yang et al., Observation of long spin-relaxation times in bilayer graphene at room temperature, *Physical Review Letters.* 107 (4) (2011) 47206.
[135] A.H.C. Neto, F. Guinea, Impurity-induced spin-orbit coupling in graphene, *Physical Review Letters.* 103 (2) (2009) 26804.
[136] V.K. Dugaev, E.Y. Sherman, J. Barnaś, Spin dephasing and pumping in graphene due to random spin-orbit interaction, *Physical Review B.* 83 (8) (2011) 85306.
[137] F. Guinea, Charge distribution and screening in layered graphene systems, *Physical Review B.* 75 (23) (2007) 235433.
[138] H. Zhou, Y. Zhang, W. Zhao, Tunable tunneling magnetoresistance in van der Waals magnetic tunnel junctions with 1 T-CrTe 2 electrodes, *ACS Applied Materials & Interfaces.* 13 (1) (2020) 1214–1221.
[139] S. Makdey, R. Patrikar, M.F. Hashmi, Modeling and design of magnetic tunneling junction using MoS 2/graphene quantum dots/MoS 2 approach, *Journal of Nanoparticle Research.* 22 (2020) 1–13.
[140] J.C. González et al., Sample-size effects in the magnetoresistance of graphite, *Physical Review Letters.* 99 (21) (2007) 216601.
[141] Z. Liao et al., Large magnetoresistance in few layer graphene stacks with current perpendicular to plane geometry, *Advanced Materials.* 24 (14) (2012) 1862–1866.
[142] A.L. Friedman et al., Quantum linear magnetoresistance in multilayer epitaxial graphene, *Nano Letters.* 10 (10) (2010) 3962–3965.
[143] K. Gopinadhan, Y.J. Shin, I. Yudhistira, J. Niu, H. Yang, Giant magnetoresistance in single-layer graphene flakes with a gate-voltage-tunable weak antilocalization, *Physical Review B.* 88 (19) (2013) 195429.
[144] K. Gopinadhan et al., Extremely large magnetoresistance in few-layer graphene/boron-nitride heterostructures, *Nature Communications.* 6 (1) (2015) 8337.
[145] A.A. Abrikosov, Quantum linear magnetoresistance, *Europhysics Letters.* 49 (6) (2000) 789.
[146] K.S. Novoselov et al., Two-dimensional gas of massless Dirac fermions in graphene, *Nature.* 438 (7065) (2005) 197–200.
[147] S.S.P. Parkin et al., Giant tunnelling magnetoresistance at room temperature with MgO (100) tunnel barriers, *Nature Materials.* 3 (12) (2004) 862–867.
[148] S.V Morozov et al., Giant intrinsic carrier mobilities in graphene and its bilayer, *Physical Review Letters.* 100 (1) (2008) 16602.
[149] W.Y. Kim, K.S. Kim, Prediction of very large values of magnetoresistance in a graphene nanoribbon device, *Nature Nanotechnology.* 3 (7) (2008) 408–412.
[150] J. Bai et al., Very large magnetoresistance in graphene nanoribbons, *Nature Nanotechnology.* 5 (9) (2010) 655–659.
[151] M.V. Kamalakar, A. Dankert, J. Bergsten, T. Ive, S.P. Dash, Enhanced tunnel spin injection into graphene using chemical vapor deposited hexagonal boron nitride, *Scientific Reports.* 4 (1) (2014) 1–8.
[152] J.-J. Chen et al., Layer-by-layer assembly of vertically conducting graphene devices, *Nature Communications.* 4 (1) (2013) 1921.

[153] R.R. Nair et al., Spin-half paramagnetism in graphene induced by point defects, *Nature Physics*. 8 (3) (2012) 199–202.
[154] Y. Liu et al., Increased magnetization of reduced graphene oxide by nitrogen-doping, *Carbon*. 60 (2013) 549–551.
[155] Y. Liu et al., Realization of ferromagnetic graphene oxide with high magnetization by doping graphene oxide with nitrogen, *Scientific Reports*. 3 (1) (2013) 2566.
[156] T. Tang et al., Identifying the magnetic properties of graphene oxide, *Applied Physics Letters*. 104 (12) (2014) 123104.
[157] M. Sepioni et al., Limits on intrinsic magnetism in graphene, *Physical Review Letters*. 105 (20) (2010) 207205.
[158] D.W. Boukhvalov, M.I. Katsnelson, sp-Electron magnetic clusters with a large spin in graphene, *ACS Nano*. 5 (4) (2011) 2440–2446.
[159] K. Bagani, M.K. Ray, B. Satpati, N.R. Ray, M. Sardar, S. Banerjee, Contrasting magnetic properties of thermally and chemically reduced graphene oxide, *The Journal of Physical Chemistry C*. 118 (24) (2014) 13254–13259.
[160] K. Bagani et al., Anomalous behaviour of magnetic coercivity in graphene oxide and reduced graphene oxide, *Journal of Applied Physics*. 115 (2) (2014) 23902.
[161] Y.F. Wang et al., Visualizing chemical states and defects induced magnetism of graphene oxide by spatially-resolved-X-ray microscopy and spectroscopy, *Scientific Reports*. 5 (1) (2015) 15439.
[162] S.R. Singamaneni, A. Stesmans, J. van Tol, D.V Kosynkin, J.M. Tour, Magnetic defects in chemically converted graphene nanoribbons: electron spin resonance investigation, *AIP Advances*. 4 (4) (2014) 47104.
[163] A. Diamantopoulou, S. Glenis, G. Zolnierkiwicz, N. Guskos, V. Likodimos, Magnetism in pristine and chemically reduced graphene oxide, *Journal of Applied Physics*. 121 (4) (2017) 43906.
[164] L. Cao, X. Li, C. Jia, G. Liu, Z. Liu, G. Zhou, Spin-charge transport properties for graphene/graphyne zigzag-edged nanoribbon heterojunctions: a first-principles study, *Carbon*. 127 (2018) 519–526.
[165] X.Q. Deng, Z.H. Zhang, G.P. Tang, Z.Q. Fan, C.H. Yang, Spin filter effects in zigzag-edge graphene nanoribbons with symmetric and asymmetric edge hydrogenations, *Carbon*. 66 (2014) 646–653.
[166] X. Li, L. Cao, H.-L. Li, H. Wan, G. Zhou, Spin-resolved transport properties of a pyridine-linked single molecule embedded between zigzag-edged graphene nanoribbon electrodes, *The Journal of Physical Chemistry C*. 120 (5) (2016) 3010–3018.

8 Integration of Optoelectronics and Spintronics in Smart Thin Films for Advance Technology

8.1 INTEGRATION OF OPTOELECTRONICS AND SPINTRONICS AS AN EMERGING TECHNOLOGY

Optoelectronics and spintronics are two closely related fields of study that involve the manipulation of light and spin, respectively, to create new materials and devices with unique properties. Both fields have the potential to revolutionize the way we think about and use technology, with applications in fields such as computing, telecommunications, energy, and medicine.

Optoelectronics is the study of electronic devices that convert electrical energy into light or vice versa. Examples of optoelectronic devices include light-emitting diodes (LEDs), solar cells, laser diodes, and optical fibers. On the other hand, spintronics, short for spin electronics, is a branch of electronics that deals with the study of the spin of electrons and their interactions with magnetic fields to store and process information. It has been proposed as a possible solution to the limitations of traditional electronics, such as the power consumption and heat dissipation of transistors. Examples of spintronics devices include magnetoresistive random access memory (MRAM) and spin valves.

Smart thin films are thin layers of material that could sense, process, and respond to external stimuli and can have specific properties that can be used in various applications. They have many potential applications in areas such as optoelectronics to create thin films for solar cells, LEDs, and laser diodes, and spintronics to create thin films for magnetic sensors, MRAM, and spin valves.

One of the key areas of research in optoelectronics and spintronics is the development of smart, thin films. Smart thin films are thin layers of material that have unique optical and magnetic properties. These properties can be used to create new materials and devices that are lightweight, flexible, and energy-efficient. The main benefit of smart thin films is their high efficiency. Smart thin films can absorb and utilize a larger amount of light and spin, which can lead to higher efficiency in the devices they are used in. They are also lightweight and flexible, which makes them ideal for use in portable and wearable devices. Another key area of research in optoelectronics

DOI: 10.1201/9781003331940-8

and spintronics is the integration of these materials and devices into existing technology. This includes the development of new, more efficient solar cells, LED lights, and spintronic data storage devices. It also includes the integration of optoelectronic and spintronic devices into existing technology, such as smartphones and laptops.

The integration of optoelectronics and spintronics (opto-spintronics) in smart thin films is an emerging technology that has the potential to revolutionize many areas of science and technology. The combination of these two fields exploits the unique properties of light-matter, spin-orbit, and magnetic interactions for the development of new types of devices that have both optical and spintronic devices with improved performance and functionality. An example of this is the use of smart thin films in the development of spintronic LEDs. These devices would use the spin of electrons to control the emission of light, rather than the traditional method of controlling the flow of electrical current. This interdisciplinary field is at the forefront of technology as it also has numerous potential applications in solar cells, memory storage, and sensors.

Memory Storage

Opto-spintronics has the potential to revolutionize data storage. Traditional data storage methods such as magnetic hard disk drives are facing limitations in terms of capacity, speed, and stability. Opto-spintronic memory storage is based on the interaction between light and spins in magnetic materials, which can store and retrieve data much faster and with a higher density than conventional methods.

Solar Cėlls

Opto-spintronics can also be used to improve the performance of solar cells. The integration of spins and photonics in a single material system can result in increased light absorption, thus improving the efficiency of solar cells. Opto-spintronics can also improve the stability of solar cells, which is crucial for their long-term use.

Research in this field is still in its early stages, and many challenges still need to be overcome. However, the potential benefits of this technology make it a very promising area of research. Some key points to consider when studying optoelectronics and spintronics in smart thin films include:

- The properties of the thin-film material and how they affect device performance.
- Techniques for depositing and patterning thin films.
- Characterization methods for measuring the optoelectronic and spintronic properties of thin films.
- The potential applications of optoelectronics and spintronics in smart thin films, such as energy-efficient lighting and data storage.
- The challenges and limitations of using thin films in optoelectronics and spintronics, such as scaling up to large-area devices.

8.1.1 Emerging Technology in Smart Thin Films for Solar Cells

Opto-spintronics is an emerging technology that combines optics and spintronics to enhance the efficiency and performance of solar cells. By leveraging the interaction between light and electron spin, opto-spintronics offers new possibilities for improving solar cell functionality. While there were no specific search results directly addressing opto-spintronics in the context of smart thin films for solar cells, we can draw upon related information to discuss the potential applications and benefits.

8.1.1.1 Principle of Opto-Spintronics for Solar Cells

Opto-spintronics focuses on the manipulation and control of both the charge and spin of electrons. While traditional solar cells convert light energy into electrical energy using the photoelectric effect, opto-spintronics goes beyond this by utilizing the spin of electrons. By harnessing the interaction between light and spin, opto-spintronics offers new possibilities for improving the efficiency and functionality of solar cells.

8.1.1.2 Thin-Film Technology

Thin-film technology is known to reduce the cost of solar cells by conserving materials and allowing for deposition on various substrates, including flexible ones [1]. This flexibility opens new applications and possibilities for solar cells, including the integration of opto-spintronic functionalities. Opto-spintronics can enhance carrier generation, transport, and collection, leading to higher conversion efficiencies in solar cells [2].

8.1.1.3 Applications of Opto-Spintronics in Solar Cells

Smart thin films can benefit from the integration of opto-spintronics by enabling spin-selective contacts that reduce recombination losses and improve charge extraction efficiency [3]. Additionally, opto-spintronics techniques can be utilized to control the carrier lifetime within thin films, thus extending the time available for carrier extraction and enhancing overall solar cell efficiency [4]. Integrating organic and inorganic materials in hybrid thin films can further optimize solar cell performance, and opto-spintronics can facilitate spin manipulation and control within these films [4].

8.1.2 Emerging Technology in Smart Thin Films for Memory Storage

Opto-Spintronics, the fusion of optics and spintronics, is an emerging technology that holds great promise for advancing memory storage capabilities in smart thin films. By harnessing the synergy between light and electron spin, opto-spintronic devices have the potential to revolutionize information processing and storage.

A recent discovery made by scientists at Purdue University could bring about a significant breakthrough in the development of new photonic and spintronic devices by linking the spin and momentum of light waves. The discovery involves a property of light waves known as "spin momentum locking," where the direction of the rotating electric field accompanying light aligns with the photons' momentum. Specifically, light waves that rotate counter-clockwise only move forward, while those that rotate clockwise only move backward. This finding is noteworthy because it implies that light, which was previously only utilized for communication purposes in technology, can now be utilized for memory and logic operations in computers by means of photonic spin [5,6]. This means that information stored and processed by magnetic spin can be converted into photonic spin and processed by light signals at room temperature, resulting in faster processing due to light's high velocity. Ultimately, this breakthrough could enable the seamless integration of electronic, optoelectronic, and magnetoelectronic multifunctionality on a single device, surpassing the capabilities of current microelectronic devices and realizing the dream of spintronics. In this chapter, we also explore the key principles of opto-spintronics and discuss its applications in the development of smart thin films for memory storage.

8.1.2.1 Principle of Opto-Spintronics for Memory Storage

Opto-spintronics is based on the manipulation and control of both the charge and spin of electrons. In conventional spintronics, the spin of electrons is utilized to encode and process information. However, opto-spintronics expands upon this by incorporating the interaction of light with electron spins. This interaction can enable efficient spin manipulation, detection, and transfer, leading to enhanced memory storage capabilities.

8.1.2.2 Smart Thin Films for Memory Storage

Smart thin films refer to thin layers of materials that exhibit unique optical, electrical, or magnetic properties, making them ideal for various applications. When it comes to memory storage, smart thin films enable high-density data storage, low-power operation, and fast read and write speeds. Opto-spintronics provides an avenue for further enhancing these attributes.

8.1.2.3 Applications of Opto-Spintronics in Memory Storage

8.1.2.3.1 Spintronic Memory Devices

Opto-spintronics can enhance the performance of spintronic memory devices, such as MRAM. By leveraging the interaction between light and spin, opto-spintronic MRAM can achieve faster switching speeds and lower power consumption, leading to more efficient memory storage [7].

8.1.2.3.2 Optical Data Storage

Opto-spintronics can also revolutionize optical data storage systems. By integrating spintronic elements into the storage medium, it becomes possible to control and manipulate the spin of electrons using light. This opens new possibilities for high-capacity and high-speed data storage systems with improved energy efficiency [8].

8.1.2.3.3 Photonic Spin-Transfer Torque Memory

Photonic spin-transfer torque memory (PSTTM) is an emerging memory concept that combines spintronics and photonics. It utilizes circularly polarized light to excite spin currents, enabling efficient spin manipulation. Opto-spintronics plays a vital role in PSTTM by facilitating the generation and detection of spin currents using light, leading to novel memory storage solutions [9].

8.1.2.3.4 Hybrid Organic-Inorganic Thin Films

Opto-spintronics can be integrated into hybrid organic-inorganic thin films, combining the advantages of organic and inorganic materials. Such films can exhibit enhanced optical properties and efficient spin manipulation, making them suitable for applications in memory storage [10].

8.2 BENEFITS AND PERSPECTIVES

The integration of opto-spintronics into smart thin films for solar cells can offer several benefits. It can improve the efficiency and performance of solar cells by enhancing carrier generation, transport, and collection. Opto-spintronics can also enable the development of novel functionalities, such as spin-selective contacts and spin-dependent carrier lifetime control. Furthermore, the integration of organic and inorganic materials in hybrid thin films can be optimized using opto-spintronics techniques. On the other hand, opto-spintronics into smart thin films for memory storage represents a promising direction for advancing data storage technologies. However, specific research on opto-spintronics in this context is limited, and the principles and applications discussed in the broader field highlight the potential benefits and possibilities. While opto-spintronics in solar cells is still in its early stages, continued research and development efforts are crucial to fully exploring and harnessing the capabilities of opto-spintronics in smart thin films for efficient and advanced photovoltaic devices and memory storage.

8.3 CONCLUSION

Opto-spintronics is a rapidly growing field with numerous potential applications in solar cells, memory storage, and sensors. The integration of spins and photonics in a single thin-film material system can offer new opportunities for improved performance, efficiency, and stability in these technologies and allow the production of lightweight and flexible devices. With continued research and development, opto-spintronics is poised to play a crucial role in shaping our technological future.

REFERENCES

[1] A.A.F. Husain, W.Z.W. Hasan, S. Shafie, M.N. Hamidon, S.S. Pandey, A review of transparent solar photovoltaic technologies, *Renewable and Sustainable Energy Reviews*. 94 (2018) 779–791.

[2] E.T. Efaz, M.M. Rhaman, S.A. Imam, K.L. Bashar, F. Kabir, M.D. Ehasan Mourtaza, S.N. Sakib, F.A. Mozahid, A review of primary technologies of thin-film solar cells, *Engineering Research Express*. 3 (3) (2021) 032001.

[3] Z. Chen, R. Xu, S. Ma, Y. Ma, Y. Hu, L. Zhang, Y. Guo et al., Searching for circular photo galvanic effect in oxyhalide perovskite Bi4NbO8Cl, *Advanced Functional Materials*. 32 (47) (2022) 2206343.

[4] D. Giovanni, *Optical-Spin Dynamics in Organic-Inorganic Hybrid Lead Halide Perovskites*, Nanyang Technological University, 2017.

[5] New Discovery May Allow Us to Harness the Power of a Photon's Spin, https://futurism.com/new-discovery-may-allow-us-harness-power-photons-spin, (accessed 20 May 2023).

[6] Agarwal Group Puts Spin on Photons, https://www.photonics.com/Articles/Agarwal_Group_Puts_Spin_on_Photons/a63962, (accessed 20 May 2023).

[7] P.J. Rajput, S.U. Bhandari, G. Wadhwa, A review on-spintronics an emerging technology, *Silicon*. 14 (15) (2022) 9195–9210.

[8] Y. Huang, V. Polojärvi, S. Hiura, P. Höjer, A. Aho, R. Isoaho, T. Hakkarainen et al., Room-temperature electron spin polarization exceeding 90% in an opto-spintronic semiconductor nanostructure via remote spin filtering, *Nature Photonics*. 15 (6) (2021) 475–482.

[9] P. Dey, J.N. Roy, P. Dey, J.N. Roy, Basic elements of spintronics, in: P. Dey, J. N. Roy, eds., *Spintronics*, Springer, 2021: pp. 23–73.

[10] J.F. Sierra, J. Fabian, R.K. Kawakami, S. Roche, S.O. Valenzuela, Van der Waals heterostructures for spintronics and opto-spintronics, *Nature Nanotechnology*. 16 (8) (2021) 856–868.

Index

Note: *Italic* page numbers refer to figures.

Taylor & Francis eBooks

www.taylorfrancis.com

A single destination for eBooks from Taylor & Francis with increased functionality and an improved user experience to meet the needs of our customers.

90,000+ eBooks of award-winning academic content in Humanities, Social Science, Science, Technology, Engineering, and Medical written by a global network of editors and authors.

TAYLOR & FRANCIS EBOOKS OFFERS:

A streamlined experience for our library customers

A single point of discovery for all of our eBook content

Improved search and discovery of content at both book and chapter level

REQUEST A FREE TRIAL

support@taylorfrancis.com

Printed in the United States
by Baker & Taylor Publisher Services

Printed in the United States
by Baker & Taylor Publisher Services